Environmental Biotechnology and Cleaner Bioprocesses

Environmental Biotechnology and Cleaner Bioprocesses

Edited by

EUGENIA J. OLGUÍN, GLORIA SÁNCHEZ and ELIZABETH HERNÁNDEZ

Instituto de Ecologia, A.C., Xapala, Veracruz, Mexico

CRC Press
Taylor & Francis Group
Boca Raton London New York

CRC Press is an imprint of the
Taylor & Francis Group, an **informa** business
A TAYLOR & FRANCIS BOOK

First published 2000 by Taylor & Francis

Published 2019 by CRC Press
Taylor & Francis Group
6000 Broken Sound Parkway NW, Suite 300
Boca Raton, FL 33487-2742

© 2000 by Taylor & Francis Group, LLC
CRC Press is an imprint of Taylor & Francis Group, an Informa business

First issued in paperback 2019

No claim to original U.S. Government works

ISBN 13: 978-0-367-45555-2 (pbk)
ISBN 13: 978-0-7484-0729-3 (hbk)

Visit the Taylor & Francis Web site at
http://www.taylorandfrancis.com

and the CRC Press Web site at
http://www.crcpress.com

Typeset in Times by Graphicraft Limited, Hong Kong

British Library Cataloguing in Publication Data
A catalogue record for this book is available from the British Library

Library of Congress Cataloging in Publication Data
Environmental biotechnology and cleaner bioprocesses / edited by
   Eugenia J. Olguin, Gloria Sanchez, Elizabeth Hernandez.
       p.   cm.
   Includes bibliographical references and index.
     1. Bioremediation. 2. Biotechnology. 3. Production management—Environmental
aspects. 4. Sustainable development. I. Olguin, Eugenia. II. Sanchez, Gloria.
III. Hernandez, Elizabeth.
TD192.E584 1999
628.5—dc21
                                                                              99–31648
                                                                                  CIP

Concern for man himself and his fate must always form the chief interest of all technical endeavor. Never forget this in the midst of your diagrams and equations.

Albert Einstein

Contents

Contributors   *page* xiii
Foreword   xvii
Preface   xix

PART ONE General Aspects and Case Studies

1   Cleaner Bioprocesses and Sustainable Development   3
Eugenia J. Olguín

1.1   Introduction   3
1.2   Cleaner Bioprocesses   4
1.3   The Five R Policies   7
1.4   The Growth in Demand for Environmental Biotechnologies and
      Cleaner Processes   9
1.5   International Standards and Competitiveness as Promoters of
      Cleaner Processes   11
1.6   Research and Technological Organizations as Promoters of
      Clean Production Schemes   15
      References   16

2   Environmental Policy for Small and Medium Enterprises   19
Víctor L. Urquidi

2.1   The Current Situation on Sustainable Development   19
2.2   The Situation in Mexico   20
2.3   Industrial Pollution   22
2.4   A Policy of Economic Incentives   23
2.5   Economic Incentives in Mexico   24
2.6   Some Conclusions   26
      References   27

3   **The Vital Issues Process: Managing Critical Infrastructures in the**
    Global Arena   29
    Dennis Engi

    3.1   Introduction   29
    3.2   Approach   30
    3.3   Stakeholder Panellists   32
    3.4   Panel Session Format   34
    3.5   Pairwise Comparison Results/Data Presentation   37
    3.6   The Puerto Rico Water Resources Management Initiative   39
    3.7   Concluding Remarks   43
          References   44

4   **Environmental Impact of Nitrogen Fertilizers in the 'Bajío'**
    Region of Guanajuato State, Mexico   45
    Oscar A. Grageda, Fernando Esparza-García and Juan J. Peña-Cabriales

    4.1   Introduction   45
    4.2   Materials and Methods   47
    4.3   Results and Discussion   48
    4.4   Conclusion   53
          References   53

5   **Impermeable Barrier Liners in Containment Type Landfills**   55
    Emir J. Macari and Carlos H. Ortiz-Gómez

    5.1   Introduction   55
    5.2   Compacted Clay Liners   56
    5.3   Amended Soil Liners   57
    5.4   Synthetic or Composite Liners   57
    5.5   Factors Relevant to all Types of Liners   59
    5.6   Permeability Testing   59
    5.7   Conclusion   60
          References   61

6   **Control of Submicron Air Toxin Particles after Coal Combustion**
    Utilizing Calcium Magnesium Acetate   63
    Jianxi Zhao, Donald L. Wise, Edgar B. Gutoff, Joseph D. Gresser and
    Yiannis A. Levendis

    6.1   Introduction   63
    6.2   Background   64
    6.3   Materials and Methods   69
    6.4   Results and Discussion   71
    6.5   Conclusions and Recommendations   75
          References   79

PART TWO **Recycling and Treatment of Organic Wastes**

7 **Duckweed-Based Wastewater Treatment for Rational Resource
Recovery and Reuse** 83
Huub J. Gijzen and Siemen Veenstra

7.1 Introduction 83
7.2 Characteristics of Duckweeds 85
7.3 Duckweed and Domestic Wastewater Treatment 86
7.4 Integrated Concepts 91
7.5 Conclusions 96
References 97

8 **Anaerobic Treatment of Tequila Vinasse** 101
Kuppusamy Ilangovan, Josefina Linerio, Roberto Briones and
Adalberto Noyola

8.1 Introduction 101
8.2 Materials and Methods 101
8.3 Results and Discussion 103
Acknowledgements 106
References 106

9 **Immobilization of Living Microalgae and their Use for Inorganic
Nitrogen and Phosphorus Removal from Water** 107
Carlos Garbisu, Itziar Alkorta, María J. Llama and Juan L. Serra

9.1 Introduction 107
9.2 Microalgae and Cyanobacteria 108
9.3 Biological Wastewater Treatment 108
9.4 Utilization of Inorganic Nitrogen and Phosphorus by Cyanobacteria 109
9.5 Immobilization Techniques 110
9.6 Concluding Remarks 117
Acknowledgements 118
References 118

10 **Engineered Reed Bed Systems for the Treatment of Dirty Waters** 123
A.J. Biddlestone and K.R. Gray

10.1 Introduction 123
10.2 Basis of Treatment 124
10.3 Horizontal Flow Beds 126
10.4 Downflow Beds 128
10.5 Overland Flow Beds 129
10.6 Sludge Treatment Beds 130
10.7 Application of Reed Bed Systems 130
10.8 Conclusions 131
References 131

PART THREE **Removal of Recalcitrant Compounds**

**11 Immobilization of Non-viable Cyanobacteria and their use for
    Heavy Metal Adsorption from Water**   135
    Alicia Blanco, Miguel Angel Sampedro, Begoña Sanz, María J. Llama
    and Juan L. Serra

    11.1 Introduction   135
    11.2 Microbial Mechanisms for Removal of Metal Ions   136
    11.3 Biomass Immobilization   142
    11.4 Reactors for the Treatment of Metal-containing Effluents   145
    11.5 Metal Biosorption by Immobilized Biomass   145
    11.6 Conclusion   151
          Acknowledgements   151
          References   151

12 Bioremediation: Clean-up Biotechnologies for Soils and Aquifers   155
    Susana Saval

    12.1   Introduction   155
    12.2   The Soil: Where Contaminants and Microorganisms Meet   155
    12.3   Microorganism Survival in Adverse Conditions   156
    12.4   Advantages of Bioremediation   159
    12.5   Knowing the Contaminated Site   159
    12.6   Suitability of the Site for Biotreatability Tests   160
    12.7   From Laboratory to Field   161
    12.8   Bioremediation Monitoring in the Field   164
    12.9   Bioremediation as a Clean Technology   164
    12.10 Management Technology Needs   165
           References   166

**13 Increasing Bioavailability of Recalcitrant Molecules in
    Contaminated Soils**   167
    Mariano Gutiérrez-Rojas

    13.1 Introduction   167
    13.2 Soil Bioremediation: An Emerging Technology   168
    13.3 Bioavailability Constraints   169
    13.4 General Strategies to Increase Bioavailability   171
    13.5 Addition of Synthetic Surfactants   172
    13.6 Increasing Bioavailability: General Recommendations   174
    13.7 Conclusions   175
          Acknowledgements   175
          References   175

14 Bioremediation of Contaminated Soils   179
    María Trejo and Rodolfo Quintero

    14.1 Introduction   179
    14.2 Current Market for Bioremediation   180

14.3 Bioremediation Systems 180
14.4 Concluding Remarks 188
References 188

15 Environmental Oil Biocatalysis 191
Rafael Vazquez-Duhalt

15.1 Introduction 191
15.2 Pathways in Hydrocarbon Degradation 192
15.3 Genetics of Aromatic Hydrocarbon Biodegradation 197
15.4 Mechanisms of Genetic Adaptation 200
15.5 Final Remarks 202
References 203

PART FOUR **Cleaner Bioprocesses**

16 Clean Biological Bleaching Processes in the Pulp and Paper Industry 211
Juan M. Lema, M. Teresa Moreira, Carolyn Palma and
Gumersindo Feijoo

16.1 Introduction 211
16.2 New Bleaching Processes 214
16.3 Enzymatic Bleaching 216
16.4 Production of Manganese-Dependent Peroxidase 220
16.5 Future Perspectives 221
Acknowledgements 222
References 222

17 The Cleaner Production Strategy Applied to Animal Production 227
Eugenia J. Olguín

17.1 Introduction 227
17.2 Cleaner Pig Production Units 231
17.3 Integrated System for Recycling Pig Wastewater, and
Recovering Biogas, *Spirulina* and *Lemna* and Biomass
(Biospirulinema System) 233
17.4 Final Remarks 239
References 241

**18 Clean Technologies through Microbial Processes for Economic
Benefits and Sustainability** 245
Horst W. Doelle, Aran Hanpongkittikun and Poonsuk Prasertsan

18.1 Introduction 245
18.2 The Sugarcane and Sugar Processing Industry 245
18.3 The Palm Oil Industry 249
18.4 The Seafood Processing Industry 254
18.5 Concluding Remarks 259
References 259

19 Cleaner Biotechnologies and the Oil Agroindustry   265
     Agustín López-Munguía

     19.1   Technology and Raw Materials   265
     19.2   The Market   266
     19.3   Structure and Application of Fats and Oils   267
     19.4   Fat and Oil Biotechnology   267
     19.5   The Coconut Industry: A Case Study   270
     19.6   Fat Substitutes   271
     19.7   Conclusions   271
            References   273

20 In Search of Novel and Better Bioinsecticides   275
     Argelia Lorence and Rodolfo Quintero

     20.1   Introduction   275
     20.2   Bioinsecticides based on *Bt*   277
     20.3   Mode of Action of *Bt* δ-Endotoxins   278
     20.4   Structure and Function of δ-Endotoxins   279
     20.5   Transgenic Plants Resistant to Insects   281
     20.6   Novel Systems using *Bt*   282
     20.7   Concluding Remarks   283
            References   283

**21 *Bacillus thuringiensis*: Relationship Between *cry* Gene Expression
     and Process Conditions**   285
     Reynold R. Farrera and Mayra de la Torre

     21.1   Introduction   285
     21.2   Mode of Action and Specificity of *Bt* δ-Endotoxins   285
     21.3   Molecular Biology of *Bt*   286
     21.4   Production of *Bt*   288
     21.5   Conclusions   295
            References   295

22 Cleaner Production Activities in Zimbabwe   299
     Richard T. Tawamba, R. Gurajena and Christopher J. Chetsanga

     22.1   Background   299
     22.2   Project Inputs   300
     22.3   Institutional Arrangement   300
     22.4   Demonstration Projects   301
     22.5   Information Dissemination   303
     22.6   Cleaner Production Manual   304
     22.7   Barriers Encountered During Demonstration Projects   304
     22.8   Conclusions   305
            References   305

     Appendix A Sample Calculation   307
     Index   309

Contributors

ITZIAR ALKORTA, Departamento de Bioquímica y Biología Molecular, Universidad del País Vasco, Bilbao, Spain.

A. JOE BIDDLESTONE, School of Chemical Engineering, University of Birmingham, UK.

ALICIA BLANCO, Departamento de Bioquímica y Biología Molecular, Universidad del País Vasco, Bilbao, Spain.

ROBERTO BRIONES, Coordinación de Bioprocesos Ambientales Instituto de Ingeniería, UNAM, México, D.F.

CHRISTOPHER J. CHETSANGA, Environment and Remote Sensing Institute, Scientific and Industrial Research and Development Centre, Harare, Zimbabwe.

MAYRA DE LA TORRE, Department of Biotechnology and Bioengineering CINVESTAV-I.P.N., México, D.F.

HORST W. DOELLE, MIRCEN-Biotechnology Brisbane and Pacific Regional Network, University of Queensland, St. Lucia, Australia.

DENNIS ENGI, Strategic Initiatives Department, Sandia National Laboratories, Albuquerque, NM, USA.

FERNANDO ESPARZA-GARCÍA, Department of Biotechnology and Bioengineering, CIVESTAV-I.P.N., México, D.F.

REYNOLD R. FARRERA, Department of Biochemical Engineering, Escuela Nacional de Ciencias Biológicas, Instituto Politécnico Nacional, México, D.F. Becario de COFFA.

GUMERSINDO FEIJOO, Department of Chemical Engineering, University of Santiago de Compostela, Spain.

CARLOS GARBISU, Departamento de Bioquímica y Biología Molecular, Universidad del País Vasco, Bilbao, Spain.

HUUB J. GIJZEN, International Institute for Infrastructural Hydraulic and Environmental Engineering, Delft, The Netherlands.

OSCAR A. GRAGEDA, Department of Biotechnology and Bioengineering, CINVESTAV-I.P.N., México, D.F.

K.R. GRAY, School of Chemical Engineering, University of Birmingham, UK.

JOSEPH D. GRESSER, Cambridge Scientific Inc., Velmont, MA, USA.

R. GURAJENA, Environment and Remote Sensing Institute, Scientific and Industrial Research and Development Centre, Harare, Zimbabwe.

MARIANO GUTIÉRREZ-ROJAS, Departamento de Biotecnología, UAM- Iztapalapa, México, D.F.

EDGAR B. GUTOFF, Chemical Engineering Department, Northeastern University, Boston, MA, USA.

ARAN HANPONGKITTIKUN, Department of Industrial Biotechnology, Faculty of Agro-Industry, Prince of Songkla University, Hat Ya, Songkla, Thailand.

KUPPUSAMY ILANGOVAN, Coordinación de Bioprocesos Ambientales, Instituto de Ingeniería UNAM, México, D.F.

JUAN M. LEMA, Department of Chemical Engineering, University of Santiago de Compostela, Spain.

YIANNIS A. LEVENDIS, Mechanical Engineering Department, Northeastern University, Boston, MA, USA.

JOSEFINA LINERIO, CIATEJ, A.C., Guadalajara, Jal., México.

ARGELIA LORENCE, Instituto de Biotecnología, UNAM, Cuernavaca, Mor., México.

AGUSTÍN LÓPEZ-MUNGUÍA, Instituto de Biotecnología, UNAM, Cuernavaca, Mor., México.

MARÍA J. LLAMA, Departamento de Bioquímica y Biología Molecular, Universidad del País Vasco, Bilbao, Spain.

EMIR J. MACARI, School of Civil and Environmental Engineering, Georgia Institute of Technology, Atlanta, GA, USA.

M. TERESA MOREIRA, Department of Chemical Engineering, University of Santiago de Compostela, Spain.

SITOO MUKERJI, Office of Science and Technology, Organization of American States, Washington, DC, USA.

ADALBERTO NOYOLA, Coordinación de Bioprocesos Ambientales, Instituto de Ingeniería, UNAM, México, D.F.

EUGENIA J. OLGUÍN, Department of Environmental Biotechnology, Institute of Ecology, Xalapa, Ver., México.

CARLOS H. ORTÍZ-GÓMEZ, School of Civil and Environmental Engineering, Georgia Institute of Technology, Atlanta, GA, USA.

CAROLYN PALMA, Department of Chemical Engineering, University of Santiago de Compostela, Spain.

JUAN J. PEÑA-CABRIALES, Laboratorio de Ecología Microbiana CINVESTAV-I.P.N., Unidad Irapuato, Irapuato, Guanajuato, México.

POONSUK PRASERTSAN, Department of Industrial Biotechnology, Faculty of Agro-Industry, Prince of Songkla University, Hat Ya, Songkla, Thailand.

RODOLFO QUINTERO, Instituto de Biotecnología, UNAM, Cuernavaca, Mor., México.

MIGUEL ANGEL SAMPEDRO, Departamento de Bioquímica y Biología Molecular, Universidad del País Vasco, Bilbao, Spain.

BEGOÑA SANZ, Departamento de Bioquímica y Biología Molecular, Universidad del País Vasco, Bilbao, Spain.

SUSANA SAVAL, Coordinacion de Bioprocesos Ambientales, Instituto de Ingeniería, UNAM, México, D.F.

JUAN L. SERRA, Departamento de Bioquímica y Biología Molecular, Universidad del País Vasco, Bilbao, Spain.

RICHARD T. TAWAMBA, Environment and Remote Sensing Institute, Scientific and Industrial Research and Development Centre, Harare, Zimbabwe.

MARÍA TREJO, Centro de Investigación en Biotecnología, UAEM, Cuernavaca, Mor., México.

VÍCTOR L. URQUIDI, Colegio de México, México, D.F.

RAFAEL VAZQUEZ-DUHALT, Instituto de Biotecnología, UNAM, Cuernavaca, Mor., México.

SIEMEN VEENSTRA, International Institute for Infrastructural, Hydraulic and Environmental Engineering, Delft, The Netherlands.

DONALD L. WISE, Chemical Engineering Department, Northeastern University, Boston, MA, USA.

JIANXI ZHAO, Chemical Engineering Department, Northeastern University, Boston, MA, USA.

Foreword

There is little disagreement among nations about the necessity of following a development strategy that would be sustainable over a long period. Northern industrialized countries view the solving of the environmental problems which will reestablish and maintain the harmony of the natural order of the planet as a condition for sustainable development. Countries of the South on the other hand, faced with the debilitating poverty of the majority of their population, believe economic and social development as necessary conditions that will lead to sustainability.

Both the countries of the North and the South, however, realize that reduction in pollution and elimination of the practice of indiscriminate exploitation of natural resources are necessary steps in order to avoid committing irrevocable environmental damage to this planet. The outcome of the United Nations Conference on the Environment and Development (UNCED) held in Rio de Janeiro in 1992, Agenda 21, shows a clear recognition of the problems which we face, including the existing global disparities. On one hand, there is concern about the devastating social and economic situation in many countries of the South; on the other hand, there is the worsening perspective of the global environment. Integration of environment and development concerns, and greater attention to these, are required to meet basic needs in the South, to improve living standards, to achieve better protection and management of ecosystems and a safer, prosperous future. These tasks reflect the complexity of addressing both environmental and developmental issues as a single overall objective, but this is necessary as we prepare for the challenges of the next century. Nations need to acknowledge the relative scarcity and total value of resources and introduce practices which will help to prevent environmental degradation. This requires researching technologies for prevention, and investigating different alternatives for different needs and conditions. Clearly the earlier attempts of 'end of pipe' solutions have not proven satisfactory. To respond to the problems of adverse climate change or ozone depletion requires introduction of technologies which will use resources efficiently, reduce or eliminate potential pollution at the source, and contribute towards economic development.

Developing technologies which would consume fewer resources, incorporate recycling and reuse of components and reduce production of wastes, while improving production efficiency, is challenging. However, it is a challenge that cannot be ignored. The introduction of cleaner technologies is not only socially responsible but has also been shown to lead to increased productivity, competitiveness and profitability. During production processes a variety of conditions could lead to potential environmental contamination, waste of raw materials and other inputs through their inefficient use. Some of these conditions could be eliminated without substantial capital investment. A simple production audit could identify areas where environmental efficiency could be introduced in use of raw materials and resources such as water and energy through education and active participation of the work force and through introduction of best management practices. Such audit could also identify alternative, less harmful raw materials for the production process.

Introduction of best-management practices and production audit, however, are only partial measures. Design, production and introduction of cleaner technological alternatives may very well require application of most up-to-date scientific and technological knowledge. Many firms are taking advantage of advances in biotechnology to modify their industrial processes. Some have substituted synthetic inputs for biological inputs, while others are replacing existing chemical processes with enzyme technologies which work at normal temperature and pressure, thus eliminating many hazardous chemical reactions.

Biotechnology can also offer a 'natural' way of addressing environmental problems ranging from the identification of biohazards to bioremediation techniques for industrial, agricultural and municipal effluents and residues. Thus environmental use of biotechnology includes the development, biosecurity use and regulation of biological systems for remediation of contaminated environments (land, air, water), as well as for use as environmentally sound processes leading to cleaner manufacturing technologies and sustainable development.

This publication attempts to provide information about the potential role that biological processes could play in achievement of sustainable development. It looks at the conditions such as infrastructure and favourable policy requirements necessary to encourage adoption of biotechnology. It looks especially at the role bioprocessing could play in recycling and treatment of waste, bioremediation and gives examples of cleaner biological processes which could replace problematic existing technologies. This publication is not an exhaustive survey of potential contributions of bioprocessing in the development process but rather is an attempt to provide the readers with selective examples of research findings which have, or can make, a difference. It is my belief that this publication will prove to be an important reference to both research community as well as practitioners.

Sitoo Mukerji, PhD
Director, Office of Science and Technology
Organization of American States
Washington, DC

Preface

Biotechnology is a multidisciplinary field which has been considered in several National Development Programmes as one of the strategic areas, involving edge technologies and promoting new bio-industries as source of a considerable amount of new products with high impact in agriculture, the food and chemical industry and also in the health sector. Among the various subfields within biotechnology, the one dealing with environmental issues is currently one of the major demand and development.

Environmental biotechnologies are competing with great success against traditional technologies and are providing solutions to acute problems through the so-called 'end of pipe' treatment technologies and bioremediation.

On the other hand, a young and yet very relevant field dealing with the prevention of pollution as a better and more effective answer to the current environmental problems, is the strategy of Cleaner Production. Some countries have already adopted this new style of production and have launched national programmes aimed at its development and promotion. However, there is still a lack of knowledge concerning this emergent field among most people in the industry, government and even the academic world in most countries in which severe damage to the environment is currently occurring.

This volume is aimed at providing information at various levels, from introductory to advanced aspects, including general concepts of environmental policy, applications and specific case studies related to both environmental biotechnology and cleaner bioprocesses. It is also aimed at benefiting readers of different backgrounds, including engineers, consultants, scientists and decision makers working for industry, government, academia and social or private organizations.

Part I covers general aspects and case studies, starting with an overview of the concepts and facts behind the development of cleaner bioprocesses within the framework of sustainable development. Some aspects of environmental policy oriented towards the needs of small and medium enterprises are discussed from the perspective of an expert in economy. Various case studies in the USA and in Mexico are also presented in this section.

Part II of this volume comprises chapters dealing with the recycling and treatment of organic wastes. The use of aquatic plants and microalgae for wastewater treatment and recovery of nutrients is extensively covered by experts from the Netherlands, the United

Kingdom and Spain. A specific case of an anaerobic treatment for the wastewater (stillage) from the Tequila industry is illustrated by researchers from Mexico.

In Part III, relevant and updated information about bioremediation technologies can be founded. Emphasis is given to the removal of recalcitrant compounds from contaminated water and soils. Four different research groups from Mexico and one from Spain present in-depth discussions of this currently flourishing topic.

Finally, in Part IV, seven chapters present wide ranging and updated information related to the emergent field of cleaner bioprocesses. Applications of this novel preventive approach to various industries are discussed by researchers from four continents, including countries such as Spain, Australia, Thailand, Zimbabwe and Mexico. Current information about environmentally sound products such as the bioinsecticides is also presented in this section.

The editors of this volume acknowledge the effort made by all contributors for preparing excellent manuscripts and wish that this valuable material reaches many potential users and practitioners who might be laying bricks around the world to build the foundations of Industrial Sustainable Development.

Eugenia J. Olguín
Gloria Sánchez
Elizabeth Hernández
Xalapa, Veracruz, México

PART ONE

General Aspects and Case Studies

1

Cleaner Bioprocesses and Sustainable Development

EUGENIA J. OLGUÍN

1.1 Introduction

The prevailing industrial development style from the last century up to the beginning of the 1970s was characterized by production of goods and services aimed at satisfying ever-growing consumer demands of certain social groups with high purchasing power, within monopolic and closed economies, regardless of the consequent negative impacts on the environment. Furthermore, wealth and health was enjoyed only by a minority of the world population, while a growing percentage of it, especially in the less developed countries, was maintained in poverty.

International concern for such a trend led to the organization of the Conference on Environment at Stockholm by the United Nations in 1972. After recognizing that conservation of environment was crucial for human permanence on this planet, several actions were undertaken and stringent environmental legislation and regulations were enacted soon afterwards, mainly in developed countries. Such regulations, based on the 'command-and-control' approach, generally prescribe very rigid standards and define compliance in terms of 'end of pipe' requirements (Graedel and Allenby, 1995). Similar legislation was adopted by most developing countries at the end of the 1980s, or soon after the international agreements were made at the Rio Summit in 1992.

Thus, within the 20 years between the two crucial international conferences, the results in terms of technology development were the commercialization of 'end of pipe' technologies for the treatment of municipal and industrial wastewater and the development of remediation technologies for cleaning up contaminated soils, mainly in developed countries. Some of these technologies were also commercialized in developing countries and countries in transition, although to a lesser extent.

The World Commission on Environment and Development launched the concept of sustainable development in 1987: 'Development that meets the needs of the present without compromising the ability of future generations to meet their own needs' (WCED, 1987). Although many countries have adopted such a concept as the basis of their development programmes, studies are still in progress and have not been fully analysed. However, several positive international initiatives have been associated with sustainable development, such as the United Nations Programme for Cleaner Production, launched in 1990.

Figure 1.1   The challenge for the twenty-first century

Thus, the challenge for the next century is to achieve a high degree of sustainability in such a way that industrial and rural development can grow steadily while preserving the environment and social development can counteract poverty by offering a better quality of life for all (Figure 1.1). Industrial development should be sustainable and environment-ally sound through cleaner production.

1.2   Cleaner Bioprocesses

Two of the premises of sustainable development are that economic growth has to be in harmony with the environment and that a rational and sustainable use of natural resources has to be implemented. In congruence with such premises, industrial development has to change from the degradative to the sustainable style. To meet such a purpose, adoption of cleaner production systems is essential.

The United Nations Environment Programme (UNEP) defines the cleaner production concept as 'the continuous application of an integrated preventive environmental strategy to processes, products and services to increase eco-efficiency and reduce risks to humans and the environment' (UNEP, 1996). One of its distinctive features is that reduction of the quantity and toxicity of all emissions and wastes is made before they leave the process stream. Also, the entire life cycle of the product, from raw material extraction to the ultimate disposal of the product, is evaluated to reduce negative impacts. In the case of services, environmental concerns should be incorporated into design and delivery.

Eco-efficiency is a concept promoted by the Business Council for Sustainable Devel-opment. It involves 'the delivery of competitively-priced goods and services that satisfy human needs and bring quality of life while progressively reducing ecological impacts and resource intensity, throughout the life cycle, to a level at least in line with the Earth's estimated capacity' (UNEP, 1994).

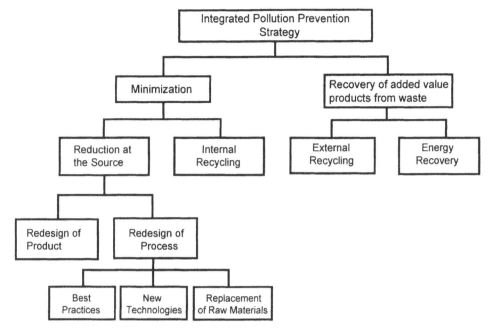

Figure 1.2 Different actions for the implementation of the cleaner production strategy (adapted from Centro de Iniciativas para la Producción Limpia, 1998)

The cleaner production strategy may be implemented through a sequential path (Figure 1.2) according to the specific needs and availability of funds, always within an integrated approach.

It is clear that adoption of clean production systems by industries calls for fundamental changes, not only at the technological level but also at the legislative level. Innovation and adoption of clean technologies should be a paramount target of research and development groups world-wide. Accordingly, new incentives and policies should be promoted in the near future (Olguín *et al.*, 1995).

Although new regulation systems will be required to accelerate the adoption of cleaner technologies, the striking feature of this movement is that industries are expected to adopt them voluntarily, moved by the motivation of considerable savings in water, energy and other inputs, as well as in savings in pollution control costs. Such is the case of some processes like the 'wet process' for coffee depulping and demucilagination, in which the introduction of new equipment that does not require water gives considerable savings in water cost and wastewater treatment.

Cleaner bioprocesses are under intensive research and development following the general guidelines for cleaner production. The most common approach has been to use enzymes as substitutes for chemical catalysts, and enzyme applications have been introduced in the textile industry (for bio-stonewashing of jeans), in the leather and tanning industry (for degreasing hides), in the food industry (for production of instant coffee, among other products) and in pulp and paper bleaching. Significant reduction or complete elimination of harsh chemicals may be achieved and the cost of production reduced by saving water and energy (Table 1.1).

The use of enzymes in the animal feed industry to improve digestibility of raw material, mainly cereals, results in several relevant benefits: saving in the cost of raw material,

Table 1.1   Use of enzymes for the replacement of harsh chemicals in leather and textile processing

Type of process	Unit operation	Type of enzyme	Advantages
Leather processing	Liming/unhairing	Alkaline protease	40 % Reduction of sodium sulphide 50 % Reduction of the use of lime Reduction of BOD and COD of wastewater
	Degreasing	Lipase	Reduction of non-biodegradable tensides
Textile	Desizing	Amylase	Removal of starch size from fabric after weaving. Processing avoiding the use of agressive chemicals (acid, alkali or oxidation agents)
	Stone washing	Cellulase	Replacement of the use of pumice stone
	Bleach clean up	Catalase	Efficient removal of hydrogen peroxide without the use of thiosulphate For a fabric volume of 1500 t/year, savings are as follows: water savings: 13 500 m^3 energy savings: 2780 GJ/year reduction of CO_2 emissions: 160 t/year
Pulp and paper	Bleach boosting	Hemicellulase	1–2 g of enzyme is required to treat 1 t pulp. It substitutes for up to 10–15 kg of chlorine Significant reduction of chlorinated organic effluent

Adapted from *Novo Nordisk A/S and Products for 'Cleaner Production'* (1994)

significant reduction in waste volume and reduction in the nitrogen and phosphorus content of the waste (Table 1.2).

The use of enzymes in the food indutry (Table 1.3) allows the replacement of acid for starch hydrolysis, with concomitant reduction in energy consumption and aggressive chemicals in the wastewater. In beer production the benefits from using enzymes are manifold and lower the cost of production.

Another approach to clean production is the generation of new and cleaner products such as bio-insecticides, for example the production of *Bacillus thuringiensis* as a source of potent insecticide crystals. In this case commercialization has already begun, although much research is still in progress to optimize production, as described in other chapters of this book.

The incentives to invest human and financial resources in the research and development of cleaner bioprocesses are high, considering the benefits which might be achieved in terms of environment protection and manufacturing costs. In the near and medium term, the development of bioprocesses for waste recycling and resource recovery might be one of the most viable options, considering much research work has already been done.

Table 1.2 Use of enzymes in the feed industry

Type of industry	Unit operation	Type of enzyme	Advantages
Animal feed	Feed additives	Xylanase	Used to improved digestibility of wheat-based feeds
		Beta-glucanase	For barley-based feeds
		Alpha-Amylase	For improved digestion
		Protease	For improved digestibility of protein
		Carbohydrase	Three main groups of cereals may be identified based on their non starch polysaccharides (NSP) content. Group I comprising oats and barley may show a digestibility improvement of 10–15 % from enzyme addition. Group II, including wheat, triticale and rye, may show an improvement of 5–7 % and group III, comprising maize and sorghum, in the range of 3–5 %
			The use of enzymes allows a wider choice of cereals, by-products and protein sources, reducing the dependence on more costly (imported) raw materials
			Improved uptake of nutrients results in the reduction in the amount of manure produced and in a lower excretion of nitrogen and phosphorus to the environment
			For poultry and pigs savings in manure volume can be up to 30 % and dry matter excretion reduced by 10–15 %

Adapted from *Novo Nordisk A/S and Products for 'Cleaner Production'* (1994)

In the long term, the application of enzymes and the intensive use of bio-insecticides might be economically feasible.

1.3 The Five R Policies

The waste minimization strategy should include the five R policies (Figure 1.3). Thus, reduction at source (i.e. during the process) implies at least four options:

1 Process modifications aimed at the *reduction* of waste

2 Feedstock substitutions seeking the *replacement* of toxic or hazardous raw materials for more environment-friendly inputs

3 Efficient use of water, energy or inputs by means of *reuse* and/or *recovery* practices

4 Good housekeeping and management practices, applying the *reuse* and/or *recovery* practices

In the case of bioprocesses in which waste is generated even after applying the options discussed above, or if several constraints are present (lack of capital, of incentives, etc.),

Table 1.3 Use of enzymes for cleaner production in the food industry

Type of industry	Unit operation	Type of Enzyme	Advantages
Starch processing	Liquefaction Saccharification	Amylase Glucoamylase Amylase Pullulanase Maltogenic amylase	Replacement of acid for starch hydrolysis, with concomitant reduction in energy consumption, salt in wastewater, sulphur dioxide and elimination of aggresive acids
	Isomerization	Glucose isomerase	causing corrosion problems
Beer production	Liquefaction Barley brewing	Amylase Alpha-amylase Protease Beta-glucanase	Partial or complete replacement of malt by barley to reduce energy consumption and wastewater. The liquefaction power of alpha-amylase is 100 times that of malt. Minimization of the size of the adjunct mash for cooking and reduction of energy consumption Reduction of energy during cooling by reducing the processing time in fermenting and storing Reduction of energy consumption and wastewater due to reduction in beta-glucan and increase of filterability

Adapted from *Novo Nordisk A/S and Products for 'Cleaner Production'* (1994)

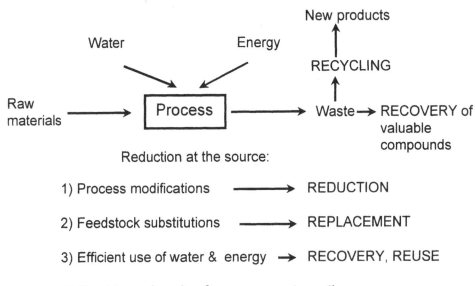

Figure 1.3 The five R policies

recycling and *recovery* of waste should be the practice to follow. Furthermore, a practical and realistic strategy should consider waste recycling and recovery as the core of short- and medium-term policies, envisaging reduction at the source as the long-term policy.

Clean bioprocesses may be developed applying one or more of the R policies. Currently, as discussed earlier and as may be appreciated from the information presented in several chapters of this book, biotechnological processes are based either on the *replacement* policy, such as in the case of the use of enzymes or bio-insecticides, or on the *recycling* and *recovery* of the organic non-toxic fractions of wastes.

A systematic working method at the plant level for cleaner production or assessment comprises various stages according to van Berkel (1995). The first consists of a 'source identification', in which a process flow diagram allows identification of all sources of waste and emission generation. This stage is followed by a 'cause evaluation', in which an investigation into the factors that influence the volume and composition of the waste and emissions is made. The next step is to develop alternative methods or 'cleaner production options' for eliminating or controlling the causes of waste generation. It is within this stage that the five R policies are effectively applied. The assessment is completed after selecting the most suitable options. Feasibility studies allow final decisions to be made, proceeding to the final stage: implementation and continuation.

1.4 The Growth in Demand for Environmental Biotechnologies and Cleaner Processes

During the last few years major global strategies have either directly or indirectly promoted the growth in the demand of environmental biotechnologies and cleaner bioprocesses. Economic globalization, follow-up programmes from Agenda 21, environmental legislation within an international context comprising international standards and linked to competitiveness are some of these global strategies imposing a positive effect.

1.4.1 *Globalization and the US–Mexico Border*

Globalization as a modern economic strategy has influenced several other aspects of development, such as the environmental policies. Regional trade agreements such as NAFTA (North America Free Trade Agreement) have promoted new trends in the environmental framework, demanding new infrastructures and more strict environmental controls. The result of these new trends is a growth in the demand for environmental biotechnologies and cleaner processes.

The Maquiladora Programme, an export processing zone for foreign-owned assembly operations located mainly at the US–Mexico border, has been in operation for over 25 years. Rapid urban growth to house approximately 12 million people in 14 pairs of cities with a poor investment in infrastructure has led to a critical situation (Crane, 1997). Large numbers of people have inadequate drinking water or municipal solid-waste-disposal facilities. Projected needs for sewage collection and wastewater treatment are perhaps the most acute of the region's environmental problems. In the Nuevo Laredo/Laredo area alone, 27 million gallons per day of untreated wastewater are being discharged directly into the Rio Grande (World Bank, 1994).

As a result of NAFTA and the pressure of the current situation, the World Bank approved, in 1995, US$918 million in loans to Mexico to build and improve the infrastructure along the Mexican side of the border (Crane, 1997). It is expected that the North American Development Bank (NADBank) and the Border Environmental Cooperation Commission (BECC) will also approve large loans for the same purpose. In addition, the Mexican government has decided to increase the current 1 % of Mexico's gross domestic product dedicated to environmental infrastructure projects, at a rate of 10–15 % per year (Deju, 1997).

It is clear that current economic policies and strategies are also creating a demand for environmental technologies. Adequate biotechnologies and cleaner bioprocesses are a clear answer to such demands. All cities along the border have a common water supply problem because the region has an arid or semi-arid climate. Water reuse and saving should become an essential feature of appropriate technologies for the region and municipal solid-waste recycling should be a priority.

1.4.2   *Follow-up Programmes from Agenda 21*

'Partnership for Pollution Prevention' is the 23rd initiative in the plan of action developed at the Summit of the Americas. It calls for a partnership in building cooperative efforts to develop and improve frameworks for environment protection and mechanisms for implementing and enforcing environmental regulations (Pumarada-O'Neill, 1997). The Intergovernmental Meeting of Technical Experts, held in Puerto Rico in November 1995, was convened as one of the actions within the Partnership for Pollution Prevention Initiative. The working group on Water Quality proposed the following projects:

> Cleaner production, using public/private cooperation to develop training, economic incentives and internal audits. Study the different sectors in terms of water usage, pollution prevention and recycling.

> Bringing ownership of water supply and treatment systems to the smaller communities and the communities that are more remote, e.g. through low cost, low maintenance options for portable water treatment and wastewater treatment in the Latin American and Caribbean (LAC) region, such as augmented or constructed wetlands, overland treatments, anaerobic treatments.

Thus, international collaborative projects are promoting the demand for environmental biotechnologies and clean bioprocesses. It is essential, then, that innovative and non-conventional technologies reach their maturity level and become available in the near future.

1.4.3   *Environmental Legislation within an International Context*

Developed countries have established strong mechanisms and organizations (such as the Environmental Protection Agency – EPA – in the USA) to ensure environmental legislation compliance ever since such laws were launched at the beginning of the 1970s. However, most developing countries enacted environmental laws only at the beginning of the 1990s, mainly due to the international pressure and the agreements derived from the Earth Summit Conference in 1992. Thus, enforcement of environmental laws is still in progress and pollution control technologies are in higher demand in these countries.

As a result of such trends, the environmental market growth rates in the Latin American region, Eastern Europe and Asia are between 12 and 17 %, in contrast with growth

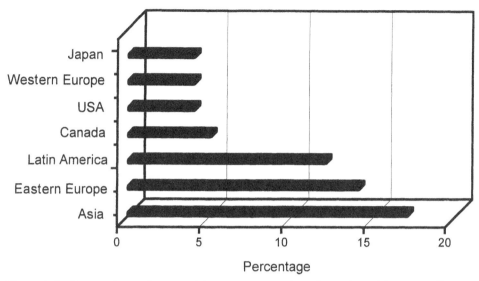

Figure 1.4 Expected growth rates of the environmental market (adapted from Portelli, 1997)

rates of 4 and 5 % for countries such as Japan, the USA, Canada and others from Western Europe (Figure 1.4).

Among the environmental technologies available, biotechnological processes are quite competitive in terms of cost and energy requirements. Data from the current situation in the Netherlands (Hesselink and Enzing, 1994), show that biological treatment of wastewater (Table 1.4), bioremediation of contaminated soils (Table 1.5) and biofiltration of polluted air (Table 1.6), are the technological options showing lower costs and lower energy requirements.

1.5 International Standards and Competitiveness as Promoters of Cleaner Processes

While legislation has been imposed to industries by governments generally within a rigid command-and-control approach, new factors such as globalization and competitiveness are moving industries – voluntarily – into certification programmes such as the Environmental Management Systems regulated by the set of international standards known as ISO 14000.

At a general glance (Figure 1.5), ISO 14000 involves standards either for organization and processes or for products. Within the first group, there are standards for management systems, for environmental auditing and for environmental performance evaluation. Product-oriented standards involve lifecycle assessment and environmental labelling among others. This wide spectrum of standards forces industries to seek new processes and products even at the cost of new investment. Thus, clean production schemes are key to achieving higher competitiveness and more exclusive markets within the current globalization context.

In addition, the relationship between companies has become increasingly self-regulating since suppliers are requested to implement quality and environmental management systems (Foley, 1996). Such trends will eventually drive demand for cleaner technologies.

Table 1.4 Indicative cost, energy consumption and materials balance of some water treatment techniques for the Netherlands

Technique	Raw materials	Waste materials and emissions	Energy requirements (kWh m^{-3})	Cost (Dff m^{-3})
Biological water treatment	Oxygen	Biomass, biogas, carbon dioxide and organic degradation products	0.1–0.3	0.10–1
Sedimentation	Chemicals	Precipitated sludge (polluted)	0.1–0.3	0.05–30
Flotation	Chemicals and/or air	Flotation sludge, odours	0.1–0.4	0.10–2
Adsorption	Adsorbent (solvent for regeneration)	Polluted adsorbent (polluted regeneration solvent)	0.1 (100–200 kWh t^{-1} VOC* for regeneration)	1–10
Chemical oxidation	Hydrogen peroxide, Ozone, ultraviolet light	Carbon dioxide, nitrogen degeneration products	2.25 kWh kg^{-1} (Ozone, water)	0.50 to >5
Micro/ultrafiltration		Concentrated wastewater	0.5–20	<1 to >20

* VOC: Volatile organic compounds
Hesselink and Enzing, 1994

Table 1.5 Indicative cost, energy consumption and materials balance of some soil-cleaning methods for the Netherlands

Technique	Raw materials	Waste materials and emissions	Energy requirements (kWh t^{-1})	Cost (Dfl m^{-3})
In situ				
Biorestoration	Water, oxygen, compost	Carbon dioxide, biomass and organic degradation products, water	20	70–150
Chemical extraction	Adsorbent, water (regeneration solvent)	Polluted water, polluted adsorbent (and generation solvent)	20	125–150
Electro-reclamation	Rinsing solution	Polluted rinsing solution concentrated contaminants	100–500	150–300
Steam stripping	Water	Polluted water and air	85–200	250–300
On/off site				
Land-farming	Fertilizer	Carbon dioxide, water	10–30	50–140
Chemical extraction	Adsorbent regeneration	Polluted adsorbent and regeneration solvent, sludge	20–30	120–240
Thermal treatment		Exhaust gases, 'dead soil'	250–700	100–300

Hesselink and Enzing, 1994

Table 1.6 Indicative cost, energy consumption and materials balance of some air treatment techniques for the Netherlands

Technique	Raw materials	Waste materials and emissions	Energy requirements (kWh m^{-3})	Cost (Dfl m^{-3})
Biofiltration	Compost, water	Compost, carbon dioxide, water and organic degradation products	≤ 1	0.50–5.00
Bioscrubbing	Water	Biomass, carbon dioxide and water	1.6–3.0	3.00–6.00
Scrubbing	Water or solvent and chemicals (for regeneration)	Polluted water or solvent, concentrated contaminants	5–10 (for regeneration)	1.00 to > 20.00
Adsorption (by active carbon)	Active carbon (regeneration solvent)	Active carbon with contaminants, polluted regeneration solvent	100–200 kWh t^{-1} VOC 2–4 t steam per tonne VOC* for regeneration	1.00–10.00
Incineration	Fuel	Exhaust gases	1–3 $(+0$–$2.10 \times 10^6)$ kJ/1000 m^3	2.50–25.00
Catalytic incineration	Catalyst (fuel)	Exhaust gases and spilt/poisoned catalyst	1.2 $(+0.10^6)$ kJ/1000 m^3	

* VOC: Volatile organic compounds
Hesselink and Enzing, 1994

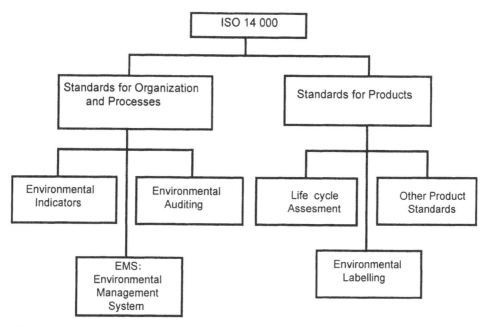

Figure 1.5 ISO 14 000 structure

ISO 14000 has been adopted by large and transnational industries at a fast rate, as the following figures show: from September 1996 to March 1997, 257 enterprises around the world had been certificated under these regulatory standards (Shrestha, 1997). Unfortunately, because only those companies which have been certified for quality assurance under ISO 9000 and have enough infrastructure and resources are able to go into more sophisticated certification programmes such as ISO 14000, a large number of small and medium-sized enterprises are still out of this certification scheme. In countries such as Mexico, only 6 % of a total of 21 800 enterprises (excluding 244 200 microenterprises), have been envisaged as possible candidates for certification with this set of regulations (Moncada, 1997).

1.6 Research and Technological Organizations as Promoters of Clean Production Schemes

Legislation, international agreements in favour of environment, trade agreements, globalization and competitiveness were discussed as major promoters of clean production schemes earlier in this chapter. However, the research and technical organizations (RTOs) also play a major role in favour of such schemes. Actually, RTOs are the source of knowledge and the developers of clean technologies and their participation is essential for the required transformations in industry.

The most successful RTOs are those which have implemented research and technological programmes oriented towards the satisfaction of concrete demands from industries. Thus, only RTOs closely linked to industries are able to grow and approach self-sufficiency. Within this context, countries which have invested in the development of this kind of RTO have been rewarded accordingly. There are several examples of developed countries which have strong and large RTOs playing a key role in the promotion of edge technologies. Also, in developing countries the number of research groups engaged in research and development programmes in the field of environmental biotechnology is growing, such as in the case of Mexico (Olguín, 1996a).

Technology transfer is one of the most important activities in which the support from RTOs is required. They can provide experts or 'brokers' who can identify and match clean technologies to users' requirements (Grovlen and Aarvak, 1997). Such technologies might be transferred between countries or between companies in the same country. Furthermore, technology transfer and human resource formation might be successfully achieved within South–South cooperation programmes (Olguín, 1996b) and not only within the traditional North–South programmes.

The World Association of Industrial and Technological Research Organizations (WAITRO), with headquarters in Denmark, is one of the various international associations which may be approached for information and future contact with RTOs dealing with environment and clean production. Currently, WAITRO is helping in the capacity building of the International Centre for Cleaner Technologies and Sustainable Development – CITELDES – recently created in Mexico as a non-governmental organization. The mission of this new centre is dissemination of information related to cleaner production and sustainable development as well as transfer of innovative technologies.

References

BERKEL, R. VAN, 1995, Introduction to cleaner production assessments with applications in the food processing industry, *UNEP Industry and Environment*, January–March, 8–15.

CENTRO DE INICIATIVAS PARA LA PRODUCCIÓN LIMPIA, 1998, http://junres.es/cipn/castella/epneta_i.htm

CRANE, R., 1997, Water and Waste at the U.S./Mexico Border: Post-NAFTA Issues, in Macari, J. and Saunders, M. (eds) *Environmental Quality, Innovative Technologies and Sustainable Economic Development*, New York: American Society of Civil Engineers, pp.49–60.

DEJU, R.A., 1997, Sustainable development and Environmental Conservation in the Americas, in Macari, J. and Saunders, M. (eds) *Environmental Quality, Innovative Technologies and Sustainable Economic Development*, New York: American Society of Civil Engineers, pp.61–66.

FOLEY, C., 1996, 'Common Ground in North America: Issues facing companies implementing ISO 14000', presentation at Workshop on Environmental quality, innovative technologies and sustainable economic development: a NAFTA perspective, Mexico City, Mexico, February.

GRAEDEL, T.E. and ALLENBY, B.R., 1995, *Industrial Ecology*, New Jersey: Prentice Hall.

GROVLEN, M. and AARVAK, K., 1997, Technology transfer from R & D institutions to SMEs, The Norwegian TEFT Programme, *Industry & Higher Ed.* February: 50–53.

HESSELINK, P.G.M. and ENZING, C.M., 1994, Biotechnology for a clean environment. OECD Report.

MONCADA, G., 1997, Lo negro de la certificación verde, *Manufactura*, **4**, 57–61.

NOVO NORDISK A/S, 1994, *Novo Nordisk and Products for 'Cleaner Production'*.

OLGUÍN, E.J., 1996a, Current status and potential of Environmental Biotechnology in Mexico, in Moo-Young et al. (eds) *Environmental Biotechnology: Principles and Applications*, Dordrecht/Boston/London: Kluwer Academic Publishers, pp.723–734.

OLGUÍN, E.J., 1996b, Training young researchers through co-operative R&D programmes, in Qureshi, M.A. (ed.) *Human Resource Needs for Change in R&D Institutes*, New Delhi, India: WAITRO & NISTADS, pp.83–87.

OLGUÍN, E.J., MERCADO, G., and SÁNCHEZ, G., 1995, Incentives and policies for the promotion of clean technologies for agroindustries in Mexico, in Maltezou, S.P., Kogerler, R. and Osterauer, M.

(eds) *Incentives and Policies for Clean Technology, Legislative and Educational Frameworks*, Vienna, Austria: Federal Ministry for Economic Affairs. International Association for Clean Technology, pp.139–143.

PORTELLI, R.V., 1997, Canadian Environment Industry Overview, Proceed. Round Table on Agri-Food Sector. Inter-American Program for Environment Technology Cooperation in Key Industry Sectors. San Jose, Costa Rica, August 6–8.

PUMARADA-O'NEILL, L.F., 1997, The intergovernmental meeting of hemispheric environmental technical experts 11/1995: Projects and Priority Issues, in Macari, J. and Saunders, M. (eds) *Environmental Quality, Innovative Technologies and Sustainable Economic Development*, New York: American Society of Civil Engineers, pp.91–96.

SHRESTHA, L.K., 1997, ¿Matrimonio de Conveniencia?, *Soluciones Ambientales*, 1(6), 19–22.

UNEP, 1994, The eco-efficiency challenge facing industry, *Industry & Environment*, 17(4), 12–13.

UNEP, 1996, Sustainable production and consumption, *Industry & Environment*, 19(3), 4–5.

WORLD BANK, 1994, Mexico: Northern Border Environmental Project, Staff Appraisal Report No. 12603-ME. Washington, D.C.

WORLD COMMISSION ON ENVIRONMENT AND DEVELOPMENT (WCED), 1987, *Our Common Future*. Oxford: Oxford University Press.

2

Environmental Policy for Small and Medium Enterprises

VÍCTOR L. URQUIDI

2.1 The Current Situation on Sustainable Development

Environmental policy for the disposal of industrial waste is generally based on the command-and-control (c-and-c) strategy, by which the public authorities may impose fines and, eventually, close down non-complying enterprises. However, in recent years at the OECD and various countries, the use of supplementary economic instruments has been considered to induce environmentally favourable behaviour by enterprises to reduce or even eliminate hazardous and toxic emissions (see Section 2.4).

Despite the conventions signed at the 1992 Rio de Janeiro summit, it is still not possible to identify a single country that has effectively interrelated its environmental policy and its development and economic growth policies in the new concept of sustainable development, not even partially or approximately. There has been no lack of declarations, speeches and documents referring to the issue. Good intentions abound; they appear expressed, for example, in the preamble of the North American Free Trade Agreement, which lists as one of the Agreement's purposes 'to promote sustainable development', and in the tripartite Agreement for Environmental Cooperation, which states that 'cooperation . . . is an essential element for attaining sustainable development, for the benefit of present and future generations.'

In some countries, such as Japan, Sweden, Germany and the Netherlands, relatively effective environmental programmes have been developed, which could be integrated into a sustainable development policy. It is still not possible to include Canada and the USA in that list, much less the great majority of developing countries, including Mexico. Nor are countries complying effectively with international commitments except, to some extent, the Montreal Protocol on the reduction and elimination of chlorofluorocarbons that harm and partially destroy the thin ozone layer which protects the Earth's atmosphere and animals from ultraviolet rays.

The World Bank's Global Environment Fund has already been endowed with additional resources. The European Union directs investments increasingly to enhance environmental policies of its member countries. The Inter-American Development Bank has started a support programme for environmental policy. The North American Development Bank, based in San Antonio, Texas, whose objective is to finance environmental projects in an area that covers up to 100 km on each side of the Mexico–US border, has commenced operations. Even so, all of these actions and programmes, taken together, signify not even the start of a movement toward sustainable development. The conventions on climate change and on biodiversity (signed in Rio) are advancing but slowly in their application, and commitments to cutting back deforestation and arresting soil erosion are still lacking.

Basically, national and regional policies conducive to sustainable development – apart from their cost and the related national organizational problems – can emerge only from a broad public knowledge of the determining factors of environmental degradation, of the development trends, and of the environmental situation prevailing at the beginning.

The heart of the matter, according to the Brundtland Report, lies in the use of fossil fuels – coal, oil and its byproducts, natural gas. As long as those energy sources are not substituted in a large measure – especially the first two, whose combustion generates the main gases contributing to global atmospheric deterioration through the 'greenhouse effect' and its consequences – the first important step toward the creation of conditions permitting sustainable development will not have been taken. Ability to attain sustainable development has to be evaluated to a large extent in a historical context: the fact is that modern development of industry and transport, and the growth of urban concentrations, have been based precisely on the utilization of those energy sources. Accordingly, the issue has been raised in terms of bringing about a fundamental change in the relationship of available energy sources to the major needs in almost all economies and to economic and social activity as a whole. Such sources generate local production for local use, or, through international commerce, supply the economies that lack them.

2.2   The Situation in Mexico

Once these problems are transferred to the sphere of a country such as Mexico, we can clearly see that it will not be easy, in environmental and economic development aspects alone, to move in the direction of sustainable development. Added to this is the social component, which among other aspects is characterized by an acute inequality. Thus the objective of equity remains even more distant, however much the concept may be present in official, private-sector and academic rhetoric.

In Mexico, important measures for an environmental policy were enacted as far back as 1972, and awareness was promoted at the political administrative and general levels of the importance of improving such a policy. Nevertheless, without detracting from the efforts made over a period of more than 20 years, in particular the new orientation that the Ministry of Environment, Natural Resources and Fisheries (SEMARNAP) and the National Institute of Ecology seem to have adopted for the period of 1996 onward, it must be recognized that an integrated policy to counter the country's general environmental degradation has only recently begun to be conceived.

In government circles, the seriousness of the situation was clearly recognized in the late 1980s, and in the 1990–1994 National Programme for Environmental Protection under the then Ministry of Urban Development, Housing and Ecology (SEDUE). This programme was designed to implement provisions of Mexico's 1988 General Law on Ecological Equilibrium and Environmental Protection (SEDUE, 1990). This document stated that every ecosystem in Mexico was already jeopardized by agricultural, industrial or urban development, as well as by the impact of still very rapid population growth. The main areas that at that time were affected by environmental pollution included the large cities, the industrial ports, a large number of river basins and certain specific ecosystems. The Programme had very little to say, incidentally, about land used for agricultural and livestock production.

The most recent biennial report, issued by the Ministry for Social Development (before the creation of SEMARNAP), contains much disturbing information as to the current state of the environment in Mexico. A short summary follows (SEDESOL, 1994).

The report lists 25 critical areas, including certain agricultural zones (see Chapter 3). (The whole of the northern border area, not explicitly included in the list, should obviously be added.) In addition, 15 areas are identified as risk areas for human health, including Mexico City and its Metropolitan Area, and five are shown to be high-risk areas for the local population due to the presence of cancer-causing substances (see Chapter 3). Moreover, the quality of urban water supplies in general has declined, deforestation continues at an annual rate in excess of 1 %, and energy intensity in production has increased instead of falling. It is estimated that the polluting intensity of manufacturing industry multiplied by a factor of 20 between 1950 and 1989, without any indication that its growth rate may have slowed down (see Chapter 11). In 1993, according to the report, municipal solid waste emissions (excluding those arising from industrial processes) averaged an estimated 839 kg per inhabitant for Mexico as a whole (in the Federal District, 1259 kg). Most of the 28 million tonnes of municipal waste generated in 1993 originated in the central areas of Mexico and the Federal District (60 %), in the north (21 %) and along the northern border area (6 %). Over one-half of all municipal waste consisted of organic waste, nearly 20 % was glass, paper and plastics, and the remainder metals and rubber. It has been reckoned that 82 % of such waste ended up in open-air dumps, in the main unregulated. In 1994, moreover, the aggregate volume of hazardous industrial waste was estimated at 7.7 million tonnes, of which 38 % was solvents, 43 % oils, paint, solder waste, resins, acids and petroleum derivatives, and the remaining 19 % a variety of other waste products (see Chapter 18).

Programmes have been put in place to build waste confinements, including special projects for toxic waste, water treatment and recycling facilities. A special regime is in operation along the northern border area applicable to *maquila* (subcontracting) manufacturing establishments, and there is growing concern in that area about groundwater pollution and other acute environmental issues. Broadly speaking, there is still much to be done in Mexico, for example, in respect of drawing up an updated national inventory of hazardous waste. Moreover, the authorities have been unable to prevent a large percentage of hazardous waste from finding its way into municipal sewage systems, rivers and other waterways, coastal lagoons and estuaries, bays, and other unregulated dumping areas, including soils from which liquid and particulate waste filters down into the subsoil and the groundwater sources.

A full diagnosis of the environmental situation in Mexico, especially in a possible context of sustainable development as a medium- and long-term objective, has not yet been made. However, much new knowledge has been developed as to the ravaging of nature, the condition of protected areas, forestry decline, river basin pollution, and so on. The question of industrial and household waste, however, lacks systematic assessment, and seems to be getting worse by the year. A national policy for municipal waste disposal and treatment has yet to be adopted. Much of the official emphasis on environmental damage seems to be concerned especially with natural resources; on the other hand, very little is said about the problem of industrial and municipal waste disposal, even though the system of c-and-c regulations is designed primarily to contain or reduce pollution from hazardous and toxic waste emissions arising from agricultural and industrial processes

– and even, to a certain extent, from business enterprises in the service sector. Unques-
tionably, natural resources should be protected and rationally utilized, including the wildlife – tortoises, whales, butterflies and many others. However, the problem of municipal
and industrial waste is equally important because it directly and indirectly affects human
health.

It can be safely assumed that the environmental situation in Mexico has continued to
deteriorate through 1995 and that the same diagnosis is likely for 1999. Moreover, even
if resources allocated to current programmes were substantially increased, it would take
many years before positive results could be seen. It is essential that the environment is
given top priority at the highest level of government, in order to successfully coordinate
programmes and actions within the public sector and between the public and business
sectors. Isolated measures under the responsibility of different government departments,
with no integrated policy framework, are not likely to yield measurable results. The
environmental authorities must cooperate and work more closely with the private productive sector as a whole who, through their operations, largely determine the degree of
pollution and environmental degradation. It is not possible to avoid including in an
integrated environmental policy framework an adequate interaction with the vast, wide
world of households that directly generate waste contributing to soil and water pollution,
or that intervene in the intensification and the dissemination of atmospheric pollution
through urban and interurban transport or in other ways (SEMARNAP, 1996).

2.3 Industrial Pollution

The polluting effect of industrial processes arises from solid and liquid waste and gas
emissions, characterized as hazardous – and sometimes toxic – which are mainly dis-
posed of in sewage systems and in the so-called unregulated, open-air dumps, as well as
into rivers and streams and other waterways, coastal lagoons and estuaries, ravines, and
so on. Gas emissions are released into an atmosphere already charged with motor vehicle
fumes.

Among industrial and service establishments, particularly small and medium enterprises and microenterprises, there is a lack of sufficient awareness and knowledge of the
environmental problems. This was borne out by the results of a survey at El Colegio de
México (Mercado *et al.*, 1995; Urquidi, 1997). This survey was derived in part from the
assumption that in Mexico, given the ineffectiveness of the regulatory measures in their
early stages, the adoption of economic incentives as supplementary to c-and-c measures
could be justified as part of an environmental policy. In official government circles, in
1992, for the first time the use of possible economic instruments was mentioned in
documents issued by the Ministry for Social Development (SEDESOL-INE, 1992), at the
technical level and, in a more general way, in the 1993–1994 biennial report on the
environmental situation in Mexico (SEDESOL-INE, 1994). However, hardly any empirical research was available on the possible application of such economic instruments, for
example, to the problem of industrial waste in a developing country. In the absence of
studies on the use of economic instruments for environmental protection in Latin America,
and after a fruitless search for literature on the subject in developing countries of other
regions, El Colegio de México towards the end of 1992 decided to undertake the necessary research on business behaviour towards the environment, with a view to exploring
the use of economic incentives to reinforce regulatory measures.

With financial support from Mexico's National Science and Technology Council and from the Canadian International Development Research Center – the survey was conducted in 1994–1995 by means of questionnaires distributed among 116 manufacturing and service enterprises in the Mexico City Metropolitan Area, of which 90 were in nine manufacturing branches showing evidence of a high volume of hazardous and toxic emissions and 26 were in three service activity branches (on the latter there had been an almost total absence of information). Among the industrial enterprises, the following branches were covered: chemical products, pulp and paper, alcoholic and non-alcoholic beverages, tanneries, printing, paints and varnishes, metallurgy and metalworking, pharmaceuticals, and electronics. The service establishments were chosen from three branches: hotels, private hospitals, and medical laboratories. The selection of the enterprises surveyed included large, small and medium enterprises, both Mexican- and foreign-owned or joint ventures. In addition, ten case studies were carried out to obtain a deeper understanding of policies of large and medium-sized corporations in other regions of Mexico: Tijuana, Guadalajara, San Juan del Río, Monterrey, Naucalpan, Cuernavaca, and Mexico City itself.

Mexican environmental policy on industrial waste disposal is based, as is that of all member countries of the OECD, on the adoption of standards and regulatory measures (the c-and-c policy). Even though the administration of this policy has improved, in Mexico only 80 environmental standards have been adopted, the application of which is far from being strict and effective. As one could have assumed, the survey showed that large enterprises, whether national, foreign or joint ventures, tend to comply with the emission standards but frequently barely enough to be within the limits and certainly not enough to contribute noticeably to environmental improvement. For example, compliance may be on the order of 90 over a maximum permitted of 100, but it does not decrease to 15 or 25; in other words, large enterprises do not always move from the 'end of the pipe' to the 'front of the pipe', which would imply changes in technological processes sometimes requiring heavy investment. Enterprises are not yet moving sufficiently from environmentally dirty technologies to cleaner, or less 'dirty', ones.

Hazardous emissions generated by small and medium-sized enterprises, as the Colegio de México survey clearly shows, frequently exceed the applicable standards. This may be due to poor information, technical impossibility, lack of adequate financing, or low quality of environmental management policies.

2.4 A Policy of Economic Incentives

One of the main conclusions of the survey is that the environmental policy concerning waste disposal of industrial and service establishments could be improved by introducing a set of temporary incentives as a supplement to the c-and-c régime, especially for small and medium-sized enterprises.

The possible use of economic instruments, specifically fiscal, financial and other incentives, as a necessary supplement to the regulatory measures would induce enterprises to assume pro-environmental business behaviour, which in turn may prove profitable and competitive. These new policies, which have already begun to arouse the interest and support of the OECD, are recommended in Chapter 8 of Agenda 21, which was adopted at the 1992 Rio de Janeiro Conference. Furthermore, a fair number of academic institutions have initiated research studies on this particular subject matter. The purpose would be to induce businesses, especially the medium and small enterprises which tend to be at a disadvantage from many points of view, to make the necessary investments in equipment, technology and new processes to economize in the use of water and fuels, to recycle waste, to reduce or avoid hazardous waste emissions and to adopt clean technologies – thus improving their efficiency and competitiveness. Mexico has yet to embark on this new stage, the study of which has barely begun.

Given the industrial crisis which the Mexican economy is presently experiencing it is even more necessary to concentrate, from the environmental point of view, on medium and small enterprises because they have scant access to bank loans, are often in a technologically inferior position, and in general face very limited and frankly declining domestic market prospects. They require not only financing and access to information but also assistance to train their technical and managerial staff.

2.5   Economic Incentives in Mexico

Paradoxically, Mexico has had a long experience in the use of fiscal and financial incentives to stimulate industrial investment, although the authorities have also much abused such instruments in the past. One would expect to find, as in other countries (including Southeast Asian countries), that the appropriate instruments are in place, in the context of an environmental policy applied to industrial waste disposal.

Mexico has implemented three economic instruments of environmental persuasion, unrelated to each other, which perhaps may be described as economic incentives to induce improved business behavior towards the environment. These are summarized below.

2.5.1   *Fees*

A schedule of fees has been introduced for the use of water and for discharging liquid waste into federal hydraulic sources, under the control of the federal water authorities. The rate has been determined by the Ministry of Finance, apparently with no regard to any policy guidance from the environmental viewpoint, and is administered by the federal water authorities, now a part of the new Ministry for Environment. It is doubtful that such fees constitute a true 'incentive' to improve business behaviour towards the environment. The fee would have to be very high and applied effectively.

2.5.2   *Tax Allowances*

Accelerated depreciation tax allowances, which in general have never been an instrument among the Finance Ministry's preferred policy options, were announced in 1993 as an 'environmental instrument' to give a name to an incentive that the business sector demanded in view of what they then considered, in general, a heavy fiscal burden in the presence of stagnation of the GDP. This accelerated depreciation, conceived for application to purchases of pollution control devices and equipment, to be fully applied almost instantly, is granted exclusively to high-bracket taxpayers (in 1994 business enterprises whose gross revenues amounted to over 1.8 million Mexican new pesos); that is, enterprises which are required to submit full financial statements, including depreciation accounts, to the tax authorities for determining the taxable income and applying the respective tax rate. The Finance Ministry does not have the means to verify if the accelerated depreciation truly refers to pollution control devices and equipment – and, furthermore, tax audit tests are carried out on only 5 % of business enterprises, so that a company's accountants are quite capable of passing off as intended for pollution control the purchase of any equipment in order to obtain the accelerated depreciation tax allowance. Low-bracket taxpayers – with respect to the indicated limit, which presumably is periodically updated according to the rate of inflation – are not required to submit returns to the tax authorities other than a ledger (instead of full financial statements); consequently, they do not submit their depreciation accounts even if they carry them. The low-bracket taxpayer pays income tax as a tax on personal income, on the difference between outlay and income when the resulting balance turns out to be positive.

The El Colegio de México survey found few enterprises that had applied for the accelerated depreciation tax allowances granted by the Finance Ministry or that were aware, even among higher-bracket taxpaying enterprises, of the existence of such an incentive. It is clear that the measures adopted by the Finance Ministry, which lacked coordination with the environmental authorities, cannot be considered to be efficient economic instruments, at least not in the form in which they are currently applied or administered. This situation requires remedy.

2.5.3 Long-term Credit

In 1991, the national development bank, Nacional Financiera (NAFIN), established a special line of long-term credit at preferential interest rates for long-term environmental loans to be granted through the commercial banking system. Although in 1992 this preferential-rate loan programme was publicized by the Metropolitan Commission for Prevention and Control of Environmental Pollution in the Valley of Mexico with regard to atmospheric pollution alone, it was applicable in any state of Mexico and for any type of pollution. The loans could be granted for periods of up to 20 years and to cover up to 100 % of the proposed investment, to enterprises that applied for such loans for 'any environmental purpose' – ranging from the introduction of some pollution control equipment to a change of technology and processes, and even for a plant's relocation. The loans could be granted at rates below the average cost of funds (CCP) – although it should not be forgotten that banks usually charge commissions, require reciprocal deposits, deduct interest in advance when the loan is approved, and apply other charges that add to the rate the borrower pays.

To stimulate commercial bank institutions to grant such type of loans, NAFIN was empowered to make the funds available at three points below the CPP to microenterprises, two points below to small enterprises, one point below to medium-sized enterprises, and at the CPP rate to large enterprises. It was assumed that NAFIN could guarantee the lending bank up to 80 % of the loan granted to the enterprise. A substantial part of the loans was channelled to public transport motor vehicles in the Federal District to induce them to change over to engines that use better-quality fuel generating less polluting gaseous emissions. (It should be noted that the Finance Ministry considers all public transport enterprises to be 'low-bracket taxpayers', irrespective of the amount of gross revenues reported by any such enterprise.)

According to data obtained from NAFIN, the cumulative outstanding loans through March 1994 were 493 million Mexican new pesos as from 1991, in support of 1244 enterprises, plus US$26 million of dollar-denominated loans. It can thus be concluded that the system of preferential loans for the environment, especially in industrial branches strictly speaking, has not operated significantly. One can put forward, moreover, that in general commercial banking institutions show little interest in granting such loans and that in fact medium and small enterprises have had little access to bank borrowing in general, even ordinary bank loans, quite apart from their high cost. The survey also found that many of the industrial and service enterprises in the Metropolitan Area were unaware of the existence of this environmental loan facility or had not used it.

2.6 Some Conclusions

Industrial pollution is already serious, not just in Mexico but in all industrializing countries, as is evidenced in reports published by the United Nations and by numerous international agencies.

Where modern industrial structures prevail, in economic environments of outward-looking domestic competitive markets, it has been recognized that enterprises no longer have any alternative but to change their processes and technology to reduce their hazardous and toxic waste emissions to below the applicable national standards, or even below the international regulatory standards (Schmidheiny, 1992). To express it differently: it is necessary, and should be profitable, to move from the 'end of the pipe' to the front of it – that is, to adopt for environmental purposes 'cleaner' ('less dirty') technologies, as has been recommended by the International Association for Clean Technologies based in Vienna.

From the environmental point of view, the c-and-c system of regulations is not sufficient for medium and small enterprises. Fiscal and financial incentives, together with environmental training and information programmes in which both the public and the private sector should participate are also needed. The incentives should be supplementary to the application of the existing standards and should be put in place on a limited-time basis. The purpose is to achieve environmental efficiency and competitive capability at the same time. Environmentally friendly investments should be cost-effective.

In numerous Pacific Rim countries, even the USA, similar problems are to be found. The subject has hardly been researched empirically at the business sector level in Latin America or other semi-industrialized regions. An industrial environmental policy for medium and small enterprises, in Mexico as in other countries, will require not only the adoption of supplementary economic incentives but also the creation of technical and technological support mechanisms and access to training programmes, as well as promotion of awareness on the need for the private and public sectors to share responsibility.

To the extent that territorially large countries organized into a federal political system – such as Mexico – expect to implement a decentralized system of environmental administration and functions, it will be appropriate that regions and subregions, zones and subzones be defined, from the environmental viewpoint, beyond territorial limits and not perforce entirely within the domestic territory or that of the neighbouring state or municipality. Environmental policy, including its decentralization aspects, needs to be conceived within the regional and subregional frameworks of industrial and agricultural development programmes, taking into account transportation systems, supply routes, sources of available energy, and many other aspects of regional integration. Not all of the environmental problems in the different regions and subregions will be common to all municipalities, but no doubt some issues of importance will go beyond the state and municipal borders or affect zones beyond those borders and consequently will require regional administrative linkages and not just occasional joint coordination of efforts. It is obvious that the environmental policy of a particular state or municipality can be nullified by the absence of an environmental policy, its insufficiency or lack of harmonization, in another state or adjacent municipality.

Regionalization in Mexico should be the basis of coordination between state entities and municipalities for the definition and implementation of the respective environmental policies, within the national framework. In addition, Mexican environmental policy cannot and should not disassociate itself from global environmental policy and that of the countries with which it has its largest trade and economic relations (the USA and Canada, and, secondarily, Western Europe and Japan). Accordingly, regionalization of the environmental policy should especially cover the Mexico–US border on both sides and the areas along the southern borders with Guatemala and Belize should be considered as a whole. Furthermore, any area in which foreign trade plays an important role in the use of natural resources should incorporate the environmental impact into the regionalization criteria.

This approach will greatly improve upon the current attitude to research and issues of regional development, general and other environmental policies – and may lead, in time, to sustainable development.

References

MERCADO, A., DOMINGUEZ, L. and FERNÁNDEZ, O., 1995, Contaminación Industrial en la Zona Metropolitana de la Ciudad de México, *Comercio Exterior* **45**, 10 766–774.

SECRETARÍA DE DESARROLLO SOCIAL, INSTITUTO NACIONAL DE ECOLOGÍA, MÉXICO (SEDESOL-INE), 1994, Informe de la situación General en Materia de Equilibrio Ecológico y protección del Ambiente, 1993–1994, México.

SECRETARÍA DE DESARROLLO URBANO Y ECOLOGÍA (SEDUE), 1990, Programa Nacional para la protección Ambiental 1990–1994, México.

SEDESOL-INE, 1992, *Los instrumentos Económicos Aplicados al Medio Ambiente*, Series Monográficas No. 2.

SEMARNAP, 1996, *Programa Nacional de Medio Ambiente 1995–2000*, México. Chapter III.

SCHMIDHEINY, S., 1992, *Changing Course: A Global Business Perspective on Development and the Environment*, Cambridge, MA: MIT Press.

URQUIDI, V., 1997, Economic and regulatory policy instruments in developing countries, Environment and Development in the Pacific: Problems and Policy Options, South Melbourne: Addison Wesley/Pacific Trade and Development Secretariat, The Australian National University, pp.154–165.

3

The Vital Issues Process: Managing Critical Infrastructures in the Global Arena

DENNIS ENGI

3.1 Introduction

The Vital Issues process (VIp) is a strategic planning tool that is being used to develop decision support systems (DSS), policy portfolios (PP), and investment portfolios (IP), for managing critical infrastructures in the global arena, including those relating to water resources, health care, and the environment. The VIp provides an explicit and account- able method for identifying and prioritizing relevant issues, programmatic areas, or responses to a specified problem. It provides a format for identifying the information needed to properly address issues considered vital to managing critical infrastructures. It can also be used to develop portfolios of appropriate policy options and to allocate critical resources.

The VIp employs day-long panel meetings to elicit a broad range of perspectives on a particular topic in a non-confrontational manner and to facilitate the interaction and synthesis of diverse viewpoints on a specific topic. It affords a high level of key stakeholder involvement and provides an explicit and accountable method for incorporating stakeholder input into management decision making, thus predisposing the stakeholder community to acceptance of the DSS, PP, or IP.

The VIp was developed and applied with a view toward bridging the gap between qualitative values and quantitative decisions. It incorporates two primary approaches: a *qualitative* or *transactional* segment, which entails the *synthesis* of the alternatives through negotiations or discussion, and a *quantitative* or *net benefit maximization* segment, an *analytical* approach, which involves prioritization of the alternatives using pairwise com- parisons. The VIp allows stakeholders to identify vital issues, a portfolio of strategic or tactical areas, and/or the information needed to address the issues and to select and define the criteria by which the items are identified and prioritized. It also provides stakeholders with the opportunity to modify a programmatic objective or goal statement to more accurately reflect the problem or question being addressed. This combination of facilit- ated group discussion and quantitative ranking provides input to strategic management decisions in the form of stakeholder-defined and prioritized items as well as information on potential barriers to the implementation of policies and programmes.

The VIp was initially developed in 1992 and has been further developed and refined in subsequent implementations of over 60 Vital Issues panels. The VIp has been used successfully for a range of public and semipublic organizations in support of a variety of strategic purposes. For example, the US Department of Energy's (DOE) Office of Industrial Technologies in Energy Efficiency and Renewable Energy used the VIp to craft a response to a section of the US Energy Policy Act of 1992. An office in the DOE's nuclear energy organization also convened a Vital Issues panel to obtain assistance in responding to a particular section of the Energy Policy Act. The results of that VIp iteration became the basis for a response to the US Congress regarding the nuclear energy system's activities in the area of concern; the results were also used to implement a nuclear energy programme. In another instance, Riotech, a semipublic organization in New Mexico, used the VIp to develop a mission statement and to select areas of focus for an Environmentally Conscious Manufacturing Institute that Riotech was planning to establish (Riotech, 1993).

In 1993, the New Mexico Health Care Task Force Advisory Group used the VIp to obtain input to facilitate the development and implementation of a comprehensive health care delivery system for the State of New Mexico (Engi and Icerman, 1995). In 1995, the VIp was used in the Middle Rio Grande Basin Water Resources Initiative, which resulted in a proposal for the development of a comprehensive information system designed to support decision making for the future uses of water resources in New Mexico's Middle Rio Grande Basin. More recently, the VIp was used in Puerto Rico's Water Resources Management Initiative in 1996 to help develop a state-of-the-art DSS that would be responsive to the issues considered vital to the management of Puerto Rico's water resources. The results of this VIp are summarized here.

3.2 Approach

The VIp is a multistage process involving a series of day-long panel meetings, each meeting building on the results of the previous meetings (see Engi and Glicken, 1995). An initial panel clarifies the panel topic and identifies, defines, and ranks criteria used by a subsequent panel or panels to (1) select and rank the issues considered vital to the management of the infrastructure, (2) select the items in a policy portfolio, (3) identify information needed to address the vital issues, or (4) rank the programmes in an investment portfolio, depending on the application.

Each implementation of the VIp follows a specified sequence. The first step is to identify a goal, a topical area (or areas) of concern, and the stakeholder interest groups from which the panellists are to be selected. The identified topic forms the basis for an initial *straw man* statement of a suggested goal or topic to be used by the panellists to develop a working definition of the panel topic. The meetings are administered by a facilitator, who focuses the panellists on the strategic area of concern and facilitates consensus on the selection of the vital issues, the portfolio items, or the list of information needs.

Each panel meets in a day-long session. In the morning, the panellists develop a working definition of the panel topic using straw man statements and goals developed by the sponsoring organization as guidelines and then (in the case of the initial panel) select the criteria or (in the case of subsequent panels) the vital issues, portfolio items, or information needs, depending on the application. In the afternoon, the panellists rank the criteria (initial panel) or the vital issues or portfolio items (subsequent panels) using the pairwise comparisons method.

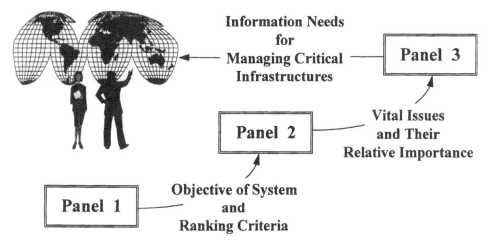

Figure 3.1   The VIp as implemented for a DSS

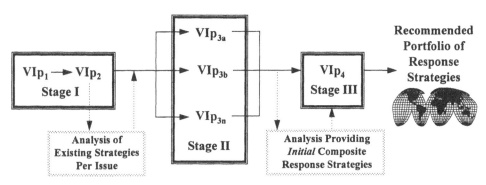

Figure 3.2   The three stages of the VIp as implemented in the development of a PP

The results of the pairwise comparisons at each meeting are used to calculate a group-averaged criteria-weighted ranking for the items (the criteria, vital issues, or portfolio items). A written report on each meeting is prepared that provides a summary of the discussions of the panellists and presents the prioritized portfolio developed by them. A measurement of the extent of (dis)agreement among the panellists as to the relative ranking of the items (the standard deviation) is also registered. After review and any necessary revision, the final report is published and distributed.

The number and type of panels convened depends on whether the objective is to develop a DSS, a PP, or a PI. The DSS application of the VIp provides input to a DSS in the form of a prioritized set of issues considered vital to the appropriate management of a critical infrastructure and preliminary lists of information considered necessary to properly address each issue. Three panel meetings are held. An initial panel develops a goal statement and criteria to be used by the second panel to identify and rank the issues considered vital to the management of the infrastructure. The second panel uses the goal statement and criteria to select and rank the vital issues. The third panel identifies a preliminary list of information needed by decision makers to address each of the vital issues (Figure 3.1).

To develop a PP, four panels or sets of panels are required (Figure 3.2). An initial panel meeting is followed by a panel that selects and ranks the vital issues (Stage I). After

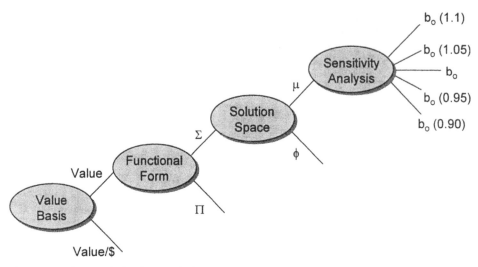

Figure 3.3 VIp optimization analysis tree

an independent analysis of the existing strategies, a set of panels, one for each vital issue, identifies response strategies or policy options for addressing each issue (Stage II). A final panel refines a composite list of strategies derived from the panels in Stage II and endorses a recommended portfolio of response strategies (Stage III).

In the IP case, the VIp is combined with classical optimization techniques to develop a value-optimized investment portfolio consisting of programmes comprising two or more elements. Two panels are utilized. After the initial panel identifies, defines, and ranks the criteria, the panellists on the second panel submit budget requests for individual programmes and programme elements, champion their programmes, and use the criteria to rank the elements. The results of the ranking are combined with classical optimization techniques to generate an investment portfolio consisting of only those programme elements that contribute to the maximum net value overall within specified budgetary constraints. Alternative portfolios may also be generated in the context of several proposed budgets and combinations of criteria.

Figure 3.3 shows the major segments of an IP optimization analysis in the form of a branched tree. A full analysis can include optimizations that maximize the sum (Σ) and the products (Π) of both the relative values and the relative values per dollar of the programme elements. These can be done for both the group-averaged criteria-weighted relative values (μ) and the criteria-weighted relative values for the individual panellists (ϕ). A sensitivity analysis of the optimization results is also included. For example, in Figure 3.3 the sensitivity to budget is indicated by the five cases at the right-hand side. More spefically, these five cases are 90, 95, 100, 105 and 110 % of the budget (bo). Relative values per dollar for the programme elements are obtained by dividing each relative value by the cost of the corresponding individual element.

3.3   Stakeholder Panellists

The VIp was designed to provide an accountable method for identifying and prioritizing relevant issues, programmatic areas, or responses to a specified problem. It was also designed to establish dialogues with key customer and stakeholder communities and, where appropriate, provide a means for incorporating stakeholder input into the decision-making process.

For purposes of the VIp,

A stakeholder is an individual or a group influenced by – and with ability to significantly impact (either directly or indirectly) – the topical area of interest.

Stakeholders can be either formally constituted organizations (such as government offices or non-profit interest groups such as the Sierra Club) or informal (i.e. not institutionalized) groups or individuals with interests and concerns in the Vital Issues topic area (such as small business owners or researchers). For a public organization, stakeholders also include individuals and groups who can directly influence the implementation of policies and activities, such as agencies that fund the implementing organization or which are capable of influencing the organization's activities in other ways.

For any public-sector organization, there are four basic groups of interests that constitute the stakeholder community and that should be represented on a Vital Issues panel. The following template of stakeholder interest groups was developed for public-sector organizations:

- Industry/private sector
- Citizen interest groups
- Government (both executive and legislative branches)
- Academe

These four groups have the ability to significantly impact, and be influenced by, public-sector activities. Their input is thus considered crucial to the successful implementation of public policies and programmatic activities.

The goal of the VIp is to solicit a broad range of perspectives and approaches to the topic. It is thus crucial that the panellists represent a diverse range of viewpoints. Ideally, there should be three individuals from each stakeholder interest group on each panel, resulting in a panel of 12 members. For panels with a public-sector programmatic or activity emphasis (such as the development of a PP), panellists should ideally be selected from both the legislative branch (such as a staffer from a key congressional committee) and the executive branch (such as someone with programmatic responsibilities). It may also be appropriate to include two individuals from the industry/private sector stakeholder interest group, one from each sector, for example, from the small and large business sectors, or from the manufacturing and service industries.

For the DSS application of the VIp, the stakeholder interest group template is expanded to include three constituencies, decision maker, modeller/analyst, and data provider, from which to select panellists for the third panel, which identifies the preliminary list of information needs for each vital issue selected and ranked by the second DSS Vital Issues panel. This panel should consist of 12 panellists, one from each constituency for each stakeholder interest group, as shown in Table 3.1. Including these three constituencies on the panel ensures that, through feedback and role exchange, the input to the DSS is not only useful to the decision maker but is also capable of being incorporated into a state-of-the-art decision support system.

It is important to recognize that, as a matter of logistics, the panellists are not delegates or representatives of the stakeholder interest groups, although they may in fact represent the interests of a particular stakeholder entity. They are, rather, members of the stakeholder communities, either as individual stakeholders themselves or as individuals affiliated with

Table 3.1   Expanded constituency/stakeholder-interest-group template

Constituency	Institutional perpectives			
	Industry	Citizens' groups	Goverment	Academe
Decision Maker	✔	✔	✔	✔
Modeller/analyst	✔	✔	✔	✔
Data provider	✔	✔	✔	✔

stakeholder organizations, who are capable of providing the perspective or vantage point of that particular stakeholder interest group.

3.4   Panel Session Format

Each panel meeting follows the same basic format, whether the panel is identifying, defining and ranking the criteria, the vital issues, or the portfolio items or identifying the information needs. Each meeting begins with discussion and clarification of the panel topic. This discussion continues until panellists reach consensus on a working definition of the topic and other key terms. The panellists then brainstorm to select and define the vital issues, portfolio items, or information needs (or the criteria).

After selecting the items, the panellists champion and rank each item using pairwise comparisons. This is done in a three-step process known as point-counterpoint-score, wherein one panellist champions the item in the context of the selected criterion or criteria, another provides counterpoint or rebuttal to the champion's presentation, and the item is then scored against all items previously presented. After all of the items are scored the panellists perform a consistency check. Figure 3.4 presents the format of a Vital Issues panel meeting session.

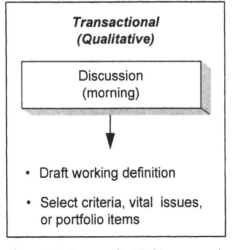
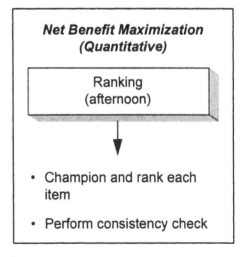

Figure 3.4   Format of a Vital Issues panel meeting

3.4.1 *Criteria Selection*

Criteria are used in the VIp to provide the panellists with a means for selecting and ranking the vital issues, portfolio items, or programme elements. The criteria are selected in a facilitated brainstorming session in which all panellists come to agreement on the final list and the scope and definition of each criterion. The criteria can be qualitative or quantitative, subjective or objective. They must, however, meet the following conditions:

- Each item on the list must be *necessary*, i.e. its elimination from the list would allow some important aspect of the topic to go unrecognized.
- Collectively, the items must be *sufficient*, i.e. all-important aspects are recognized.
- The items must be *operational*, i.e. the criteria can be effectively used to evaluate the items in the portfolio.

The criteria that are selected are used at all subsequent topical panel meetings in the process to identify and rank the vital issues, portfolio items, and/or list of information needs.

3.4.2 *Vital Issues/Portfolio Selection*

Each panel after the initial panel is charged with selecting and ranking the vital issues or programmatic areas (DSS, IP, and PP applications) or with selecting the preliminary list of information needs (third panel in the DSS application) or the response strategies (Stage II of the PP case). (Note that the information needs and the response strategies are selected but not ranked.)

In a process similar to that used by the initial panel to select the criteria, the topical panellists begin the portfolio selection portion of the meeting with a brainstorming session during which items are solicited from the group. This list is reduced in number through facilitated discussion and consensus. Items are combined if necessary and carefully defined. The facilitator ensures that all panellists are in agreement as to the scope of each of the proposed items and the definition or referent of any ambiguous terms.

The items must meet the same conditions that the criteria are required to meet: the items on the list must be *necessary*, *sufficient*, and *operational*. The items are operational if they are capable of being realistically investigated or implemented.

In the PP application, the response strategies are also selected for each vital issue in the following six categories:

- Regulations
- Fiscal incentives
- Information, education, and outreach
- Technological development and deployment
- International relations

3.4.3 *Criteria Ranking*

The criteria, as stated above, must be ranked before the items in the portfolio are selected and ranked. The criteria are ranked in the context of their own criterion, which is,

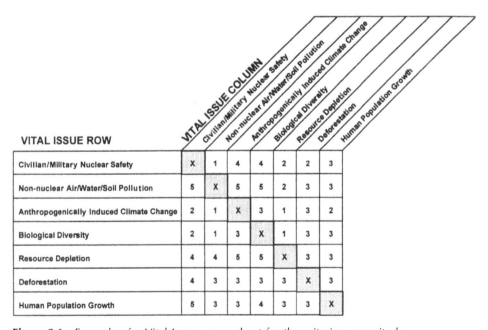

Figure 3.5 Example of criteria score sheet

Criteria	Likelihood	Magnitude	Time Frame	Robustness	Feasibility
Likelihood		2	3	4	4
Magnitude	4		4	5	5
Time Frame	3	2		5	5
Robustness	2	1	1		3
Feasibility	2	1	1	3	

VITAL ISSUE ROW	Civilian/Military Nuclear Safety	Non-nuclear Air/Water/Soil Pollution	Anthropogenically Induced Climate Change	Biological Diversity	Resource Depletion	Deforestation	Human Population Growth
Civilian/Military Nuclear Safety	X	1	4	4	2	2	3
Non-nuclear Air/Water/Soil Pollution	5	X	5	5	2	3	3
Anthropogenically Induced Climate Change	2	1	X	3	1	3	2
Biological Diversity	2	1	3	X	1	3	3
Resource Depletion	4	4	5	5	X	3	3
Deforestation	4	3	3	3	3	X	3
Human Population Growth	5	3	3	4	3	3	X

Figure 3.6 Example of a Vital Issues score sheet for the criterion *magnitude*

generally, the relative *importance* of the criteria *vis-à-vis* the goal statement. This ranking provides the relative values of the criteria and allows for the calculation of a weighted ranking of the portfolio items. The weighted criteria are used to select and rank the items at subsequent panel meetings for that particular VIp iteration. The VIp uses a separate score sheet for each criterion by which the items are ranked. An example score sheet for the five criteria *likelihood, magnitude, time frame, robustness,* and *feasibility* ranked in the context of their relative importance appears in Figure 3.5.

3.4.4 Item Ranking

The items (Vital Issues or portfolio items) are ranked in a process similar to that used by the initial panel to rank the criteria. The ranking begins after the selection of the items has been completed, and the panellists address each item in turn using the point-counterpoint-score process. The items are ranked in the context of each criterion. A sample score sheet for seven portfolio items ranked in the context of the criterion *magnitude* appears in Figure 3.6. Where the items are to be ranked using more than one criterion, a set of score sheets will be necessary.

3.5 Pairwise Comparison Results/Data Presentation

The results of the pairwise comparisons are compiled after each panel adjourns. The relative weights of the criteria are determined first in order to make them available to the topical panellists for use in the portfolio selection process. The relative weights are calculated from the relative values of the criteria, which are obtained from the results of the pairwise comparison performed on the criteria at the initial panel meeting. The relative weights for each criterion are then averaged over all the panellists to obtain group-averaged relative weighting of the criteria. The standard deviation for each criterion's weight is also calculated. These results are provided with the criteria definitions formulated by the initial panellists to the topical panellists to use in selecting and ranking the portfolio items. Figure 3.7 presents a sample score sheet for the five criteria showing row sums, relative values, and the criteria weights for one panellist.

Criteria	Likelihood	Magnitude	Time Frame	Cost/Benefit	Feasibility	Relative Value	Weight
Likelihood		2	3	4	4	13	3.25	0.21
Magnitude	4		4	5	5	18	4.50	0.30
Time Frame	3	2		5	5	15	3.75	0.25
Cost/Benefit	2	1	1		3	7	1.75	0.12
Feasibility	2	1	1	3		7	1.75	0.12
					Total:	60		1.0

Figure 3.7 Criteria scoring sheet showing row sums, relative values, and criteria relative weights

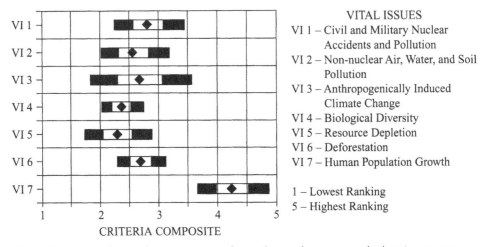

Figure 3.8 Example simple composite ranking of a Vital Issues panel, showing means and standard deviations for each vital issue

3.5.1 *Calculation of Criteria-Weighted Ranking of Portfolio Items*

To obtain the rankings of the vital issues or portfolio items, the relative values are first calculated for all items on all the score sheets for each panellist. All the relative values of the items for a given criterion are then multiplied by that criterion's corresponding weight. The resulting criteria-weighted values are then summed across each criterion for each item for each panellist. For example, the criteria-weighted value for item one for the criterion *magnitude* is summed with the weighted value for item one for the criterion *likelihood*, and with that for item one for the criterion *time frame*, and so on. The resulting values provide a criteria-weighted ranking of the items for each panellist. These weighted values are then averaged for each item over all of the panellists to obtain a composite or group-averaged criteria-weighted ranking for each portfolio item. The standard deviations for each composite value are also calculated.

3.5.2 *Data Presentation*

The ranking of the vital issues and the portfolio items can be graphically presented to show the relative positions of the items and the extent of (dis)agreement between panellists as to each ranking, as shown in Figure 3.8. Diamonds represent the composite value; the item with its diamond furthest to the right is the one ranked highest by the panel. The bars represent the panel's (dis)agreement as to the ranking of each item (equal to two standard deviations). The longest bar indicates the item for which there was the greatest disagreement as to the ranking.

The composite values and the standard deviations provide useful information. For example, when choosing between two programmatic areas or other items, the one with a lower rank but less disagreement may be preferred because it may have a greater chance of successful implementation than the one with a higher rank but greater disagreement. An item with a lower rank and less disagreement may also be preferable to a decision maker who is risk-averse.

The bar graph shown in Figure 3.8 is but one way to present the rankings. The criteria can also be combined in logical groupings or subsets and used to generate integrated composite rankings. As an example, the four criteria time frame, magnitude, likelihood of occurrence, and benefit/cost can be combined to generate an indicator of the *potential value added* of investigating an issue. Time frame, magnitude, and likelihood can also be combined to indicate the *importance* or *net present value* of an issue. Such combined criteria, however, must have been ranked as separate units in the pairwise comparison and their relative values and weights calculated in order to provide an accurate assessment of their contribution to the determination of the relative value of an item in that context. In the IP case, two or more composite criteria can be used to generate two or more sets of interprogramme relative values, one set in the context of each composite criterion. Two or more optimized portfolios can be developed, allowing a decision maker to adopt a portfolio on the basis of considerations associated with the specified criteria, as well as the net contributed value and total cost.

Integrated composite rankings can be displayed by graphically plotting the ranking of the items for each criteria subset and the levels of disagreement. For example, if criteria are combined into two groups, the panel means and standard deviations for one group will be plotted along the x-axis, and those for the other group will be plotted along the y-axis.

3.6 The Puerto Rico Water Resources Management Initiative

The VIp was used in the Puerto Rico Water Resources Management Initiative to help develop a state-of-the-art DSS that would be responsive to issues considered vital to water management and preservation in Puerto Rico. Three Vital Issues panel meetings were held for the Initiative, one each in June, July, and October 1996. The first panel developed a goal statement and identified and ranked criteria to be used by the second panel to select and rank the issues considered vital to the management of Puerto Rico's water resources. The third panel identified information needed by decision-makers to address each of the vital issues identified and ranked by the second panel.

The first Vital Issues panel developed the following goal statement:

> *To develop an implementable decision support system for the sustainable management of water resources in Puerto Rico, emphasizing the integration of various uses. This system will serve and be used by different disciplines and all sectors – including government, industry, citizens' interest groups, the public, and academe.*

The panellists selected the following five evaluation criteria:

- Magnitude or 'the extent and range of the impact of an issue'
- Likelihood or 'the probability of the occurrence of an issue impact'
- Time frame or 'the time it takes for an impact to occur, the time it takes to respond to the impact, and the time it takes for the response to take effect'
- Robustness/site independence or 'the range of application of an issue; whether it applies only to a specific site or whether it can be applied to several sites'
- Feasibility, which 'consists in the resolvability of the issue, of whether studying the issue will make a difference in its resolution. Cost-effectiveness and risk reduction constitute essential components of an issue's feasibility'

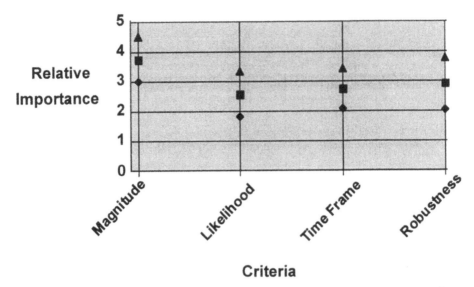

Figure 3.9 Puerto Rico Water Resources Management Initiative criteria scoring results. The triangles and diamonds represent one standard deviation above and below the means (■), respectively

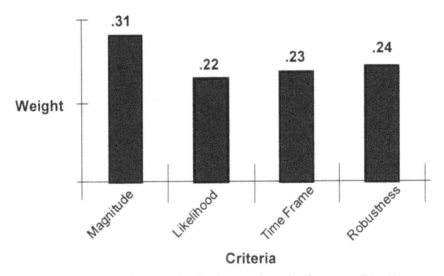

Figure 3.10 Criteria weights (normalized relative values) for the Puerto Rico Water Resources Management Initiative

The four criteria magnitude, likelihood, time frame, and robustness/site independence were scored in the context of each of the metacriteria using pairwise comparisons. The resulting scores were used to develop panel-averaged criterion weights to be used by the second panel to assess the relative importance of the vital issues in the context of the four criteria as a group, the *composite* criterion. The fifth criterion, feasibility, which was not ranked, was used by the second panel to assess the value of each vital issue in the context of that criterion alone. Figure 3.9 shows the means and standard deviations for the relative importance of the four ranked criteria. Figure 3.10 presents the criteria weights (the normalized relative values).

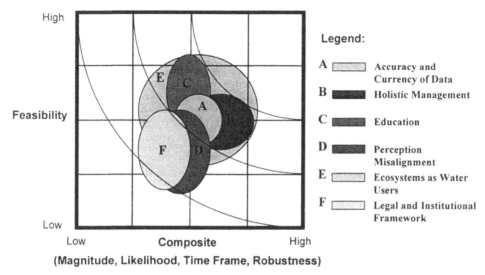

Figure 3.11   Puerto Rico Water Resources Management Initiative vital issues ranking results

The second Vital Issues panellists identified the following six issues considered vital for Puerto Rico water resources management:

- Accuracy and currency of data – 'the need to have accurate and current data on the quantity and quality of available water resources and current usage, including loss'

- Holistic management – 'the need to holistically manage water resources by: (1) a comprehensive plan, (2) a watershed approach, (3) infrastructure to tie in stakeholders, and (4) regulations required to allow the implementation of a holistically managed system'

- Education – 'the lack of a public education programme for understanding water resource management issues and the role they play'

- Perception misalignment – 'the need to align the perception of the government and public of water management'

- Ecosystems as water users – 'the need to recognize natural ecosystems as legitimate water users to maintain the regulation of water quality and quantity and their support of tourism, recreation, and wildlife'

- Legal and institutional framework – 'the need to improve the legal and institutional framework for water management'

The panellists scored the six vital issues in the context of the five criteria selected by the first panel using pairwise comparisons. The scores were combined with the criteria weights to calculate the relative importance of the vital issues in the context of the four ranked criteria as a group (the composite criterion) and in the context of the single unranked criterion (feasibility). The results were presented as an integrated composite ranking that plotted the mean relative values and standard deviations for each vital issue in the context of the composite criterion versus the assessed value of each vital issue in the context of feasibility, as shown in Figure 3.11.

The centre-points of the elliptical areas in Figure 3.11 are the plotted means of the panel's assessment of the relative importance of the issues. The ellipses represent the

Table 3.2   Preliminary list of information needed to address each of the vital issues for the Puerto Rico Water Resources Management Initiative

Vital issue	Information needed
Accuracy and currency of data	Government priorities for water resources Availability of water resources Types of data: • Historical data • Hydrological parameters of each land area • Water quality (physical, chemical, biological) • Volumes and quantity by zone • Runoff – flows, concentration of pollutants, levels • Underground water – flows, discharges, quality, injection • Production distribution and use • Economic data • Land segregation by source of water – basin, aquifers • Data available at industries – extraction, dischrages, quality, levels, accuracy Accuracy of rain gauge measurements Human impacts of environmental issues
Holistic management	Land use and public policy description River geomorphology, estuary dynamics, accurate soil data Reutilization of used water (grey and sewage, agricultural) Effects of man-made structures (e.g. dams) over influences such as coastal waters Cumulative impacts Point sources of pollution Carrying capacities Economic impacts on the industrial sector
Education	Existing laws Education needs per socioeconomic group Education needs per geographic region Census data Water demand per geographic and socioeconomic group Impact of current practices on water resources
Ecosystems as water users	An ecosystem inventory: public use, industrial use, and protected areas Priorities about the ecosystems: the public policy Expected community actions needed to protect sensitive areas Ecosystem end-use requirements (quality and quantity of water)
Legal and institutional framework	Economic cost associated with water resource managment Identification of regulations and the sector/institutions to which they apply Regulations and/or associated documents that must be addressed in environmental impact statements List of the procedures required for obtaining permits List of experts and other appropiate resources Identification of designated authorities regarding water resource management decisions Level of compilance Existing approaches to conflict resolution Case studies Interactions of government agencies with existing regulations
Perception misalignment	Degree of equality in the way water resources are managed by government, industries and residents Clarification of the accuracy of projections and descriptions of means by which the resources are managed The level of enforcement

range of the panellists' disagreement (one standard deviation above and below the mean) regarding the relative value. The vertical dimension (height) of each ellipse shows the extent of the agreement in the ranking against the feasibility criterion; the horizontal dimension (width) of each ellipse shows the range of agreement as to the ranking against the feasibility criterion. The vital issues that have the highest values along both axes are the most desirable to address. Those with the lowest values along both axes are the least desirable.

In the integrated composite ranking, the two top-ranked vital issues were education and holistic management. These two issues lie approximately equidistant from the upper right-hand corner. If the feasibility criterion and the composite criterion are considered to be of equal weight, these two issues may be considered to be of equal importance in the management of Puerto Rico's water resources. Note that these two top-ranked vital issues also have different degrees of disagreement along the two axes; for example, there was greater disagreement regarding the feasibility of addressing education than for holistic management.

The two top-ranked vital issues were followed closely in the integrated composite ranking by ecosystems as water users and accuracy and currency of data. Ecosystems as water users had a wider range of disagreement for both criteria than any other vital issue; accuracy and currency of data had the smallest range of disagreement for both criteria. Perception misalignment and legal and institutional framework were both set apart as the two least important and least feasible of the six issues for the management of Puerto Rico's water resources.

The third Vital Issues panel for the Puerto Rico Water Resources Management Initiative identified a preliminary list of information needed by decision makers to address each of the vital issues identified and ranked by the second panel. The list appears in Table 3.2.

3.7 Concluding Remarks

The three applications of the VIp – the DSS, PP, and PI – provide information that can be valuable input for the management of critical infrastructures in the global arena. The VIp provides an explicit and accountable method for identifying and prioritizing relevant issues, items in a policy portfolio, and information needed to properly address issues considered vital to managing a critical infrastructure. The VIp offers the unique combination of two problem-solving approaches: a transactional approach to problem definition, exemplified in the discussion and definition of panel topics and items for ranking, and a net benefit maximization approach, which uses pairwise comparisons to prioritize the identified items. This combination of facilitated group discussion and quantitative ranking bridges the gap between qualitative values and quantitative decision making by providing input to strategic management decisions in the form of stakeholder defined and prioritized vital issues, items in a policy portfolio, and/or information needed to address the vital issues. It also provides information on potential barriers to the implementation of programmes. Combined with classical optimization techniques, the VIp generates a value-optimized portfolio that can be used to maximize the net contributed value in anticipation of downsizing or where there are opportunities for growth, such as merger, reinvestment, or the reallocation or distribution of resources. Moreover, the VIp provides an explicit and accountable method for incorporating stakeholder input into management decision making.

References

ENGI, D. and GLICKEN, J., 1995, *The Vital Issues Process: Strategic Planning for a Changing World*, Albuquerque: Sandia National Laboratories.

ENGI, D. and ICERMAN, L., 1995, *Developing New Mexico Health Care Policy: An Application of the Vital Issues Process*, Albuquerque: Sandia National Laboratories.

RIOTECH, 1993, *Environmentally Conscious Manufacturing Institute*, videocassette, Albuquerque: Sandia National Laboratories.

4

Environmental Impact of Nitrogen Fertilizers in the 'Bajío' Region of Guanajuato State, Mexico

OSCAR A. GRAGEDA-CABRERA, FERNANDO ESPARZA-GARCIA
AND JUAN J. PEÑA-CABRIALES

4.1 Introduction

The explosive increase of world population has initiated a growing demand of food. It is estimated that in the year 2000 only 28 % of the world's nutritional needs will be satisfied through expansion of agricultural land. The remaining 72 % should be generated through the development of a more intensive agriculture, which in the short and middle term will be based upon the use of nitrogenous fertilizers (Sadik, 1988).

However, fertilizer efficiency is usually less than 50 % (Table 4.1), and the fertilizers not used by the crops may have an adverse effect on the environment. Nitrates (NO_3^-) lixiviate and pollute water, contributing to eutrophication of lakes and waterways, and making it unacceptable for human consumption. In addition, several gases (NH_3, N_2O and NO), which are produced by volatilization, nitrification and/or denitrification, have been identified as atmosphere contaminants. Some absorb thermic radiation (increasing the greenhouse effect); others partially responsible for the destruction of the stratospheric ozone layer and for acid rain (Table 4.2).

Everything indicates that as long as there exists a continuous need to employ nitrogenous fertilizers to maintain high agricultural production (due to the fact that the amounts of nutrients that the soil provides are generally small), for both economic and environmental reasons the efficiency of fertilizers should be as high as possible. It is quite

Table 4.1 Efficiency of nitrogen fertilizers

Crop	Absorption (%)
Corn	23–60
Wheat	24–60
Flooded rice	38–44
Barley	14–25
Beet	12–40

Source: Lewis, 1986

Table 4.2 The major greenhouse gases

	CO_2	CH_4	N_2O
Concentration (ppm)	350	1.7	0.30
Increase (ppb/year)	1750	19	0.75
Potential for thermic absorption	1	30	150
Biotic origin (%)	30	70	90
Persistence time (year)	100	8–12	100–200

Source: Bouwman, 1990

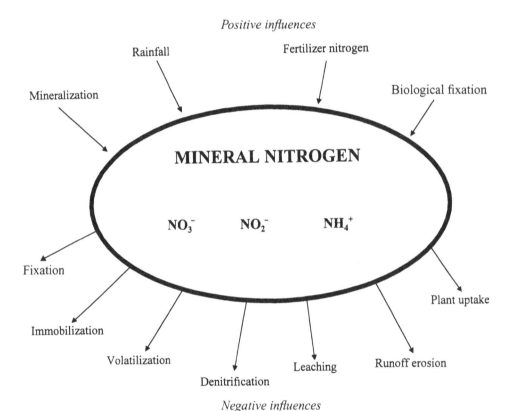

Figure 4.1 Factors positively and negatively influencing the plant-available mineral nitrogen (From Van Cleemput and Hera, 1994)

difficult to establish fertilizer efficiency, because the nitrogen cycle in agricultural systems is of a very complex nature (Figure 4.1), and depends upon many of factors, both biotic and abiotic. Therefore, to carry out a global study, the use of isotopic tracers becomes imperative. Fertilizers employed are enriched with the isotope ^{15}N, which provides exact information about the destiny of fertilizers and the transformations they are subject to (Van Cleemput and Hera, 1994).

Therefore it is very important to study the dynamics of nitrogenous fertilizers in the region, especially in view of the fact that agronomical and hydrologic local characteristics favour the processes affecting the adequate handling of nitrogen.

4.2 Materials and Methods

4.2.1 *Analysis of Available Information*

Initially, a study was carried out to collate existing information on the agricultural practices performed in the region known as 'Bajío' in the state of Guanajuato, México, over the last 30 years. The information gathered included fertilizer usage, crop rotation systems, yields, organic matter content of soils and agricultural handling practices. Information about experimental studies of the efficiency of fertilizers was also compiled.

4.2.2 *Bacterial Populations*

Denitrifying bacteria constitute one of the groups of soil microorganisms which are directly involved in low efficiency of fertilizers, as well as in the production of N_2O and NO. Thus, soils from different agricultural activity were sampled and the number of total bacteria in each sample determined using the viable count method. In order to determine the denitrifying bacteria, the technique of the most probable number was used using the Focht and Joseph (1973) method, Difco nitrate broth, and the detection of nitrate as well as nitrite (by the diazotization between sulphanilic acid and alpha-naphthylamine).

4.2.3 *Ammonia Volatilization*

In this agricultural region, high percentages of ammonium fertilizers and urea are employed; therefore the volatilization of ammonia was determined using the technique of incubation chambers (Van den Abeel *et al.*, 1990). The soil samples were collected at a depth of 0–20 cm in places with intensive-rotation sorghum–wheat–corn; broccoli–cauliflower and in a soil where no agricultural activities are carried out. Soils were moistened at 70 % of their field capacity (fc) and fertilization treatments of 0, 150 and 300 mg N kg^{-1} dry soil, in the form of ammonium sulphate, ammonium nitrate and urea, applied. The amount of liberated ammonia was quantified by the titriation of boric acid (trap to capture the NH_3) with H_2SO_4 0.005 N, using methyl red–bromocresol green as an indicator.

4.2.4 *Mineralization of Crop Residues*

In this region it is uncommon to incorporate crop residues, as these influence the activity, size and microbial diversity, causing a series of transformations of elements such as carbon and nitrogen (Clément *et al.*, 1995; Quemada and Cabrera, 1995). With the aim to explore other potential resources of nitrate, the mineralization of nitrogen was studied on soils collected from places where there is an intense cultivation of legumes–legumes and cereals–cereals. Samples were taken at a depth of 0–20 cm. The crop residues incorporated were wheat, corn, sorghum, broad bean and broccoli at doses of 0, 0.5, 1.0 and 2 %, soil moisture was adjusted to 75 % of fc and the samplings were carried out periodically over 50 days (method described by Anderson and Ingram, 1993).

4.2.5 Nitrification

Nitrification has been considered as an important source of N_2O and NO (De Groot *et al.*, 1994; Grant, 1994); for this reason, the nitrification kinetics of ammonium sulphate and urea in a dosage of 0, 180 and 240 mg N kg^{-1} dry soil were studied in material from places with different agricultural activity; the soil samples were maintained at different water content (75 and 100 % of fc), nitrate produced was determinated by the method of Cataldo *et al.* (1975), periodically over a period of 60 days.

4.3   Results and Discussion

4.3.1 Use of Fertilizers

As shown in Figure 4.2, in Mexico fertilizer consumption in the last 40 years has increased from 50 000 t to more than 5.5×10^6 t of N–P–K. The Bajío region represents 15 % of domestic consumption. In this area, intensive rotation of cereals and vegetables is employed; in addition, over the last 30 years the nitrogen dosage has increased, possibly as a consequence of drastic diminishment in the content of organic matter of the soils, given that crop rotation is predominantly cereal–cereal and vegetable–vegetable and, generally speaking, the crop residues are either removed or burnt. However, unitary yields have remained constant, which shows that the effectiveness of this agricultural system deteriorates drastically with the implementation of intensive production practices (Figure 4.3).

4.3.2 Theoretical Nitrogen Balance

Annually, in this region around 115 000 t N are applied (Table 4.3). The analysis of the amounts of nitrogen going into and coming of the main production systems (corn, wheat, sorghum and broccoli) are shown in Table 4.4. In each system, the amount of fertilizer applied and handling of crop residues are different, which affects the amount of non-accounted for nitrogen in each one of them.

   Thus, in the particular case of wheat, a system that occupies a large area and uses high doses of fertilizers (320 kg N ha^{-1} per cycle), the amount of unaccounted-for nitrogen is 24 494 t. Wheat, along with broccoli, which is cultivated in the same place as often as three times in one year using 400 kg N ha^{-1} per cycle, and where about 9000 t nitrogen are not registered each year, are agricultural production systems that can substantially contribute to contamination caused by nitrogenous products.

   In general terms, around 71 000 t of nitrogen are extracted annually, and about 45 000 t is unaccounted for. This analysis demonstrates that in many cultivation systems the efficiency of fertilizer use is less than 50 %; this corresponds with studies employing fertilizers enriched with ^{15}N in the region, where the highest efficiencies reported are less than 45 % (Vázquez-Navarro *et al.*, 1994). A large amount of this nitrogen is lost by leaching probably; in some communities of the state of Guanajuato wells have been closed, mainly because the nitrates content of the water exceeds the critical 10 ppm allowed for human consumption (SARH, 1995, personal communication).

   In other important agricultural regions of the country something similar has happened, where values of nitrates in water have been reported between 0.060 and 207.200 ppm, the

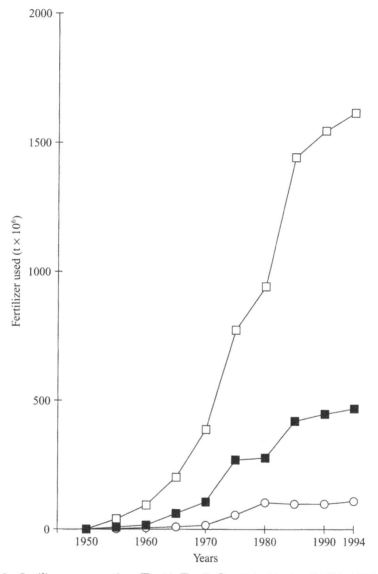

Figure 4.2 Fertilizer consumption ($\square$ = N; $\blacksquare$ = P; $\bigcirc$ = K) in Mexico (SARH, 1994)

maximum value being too high even for agricultural and livestock consumption (Castellanos and Peña-Cabriales, 1990).

The results obtained through the theoretical N balance show that the least efficient production systems are those of wheat and broccoli (Table 4.5). As an average, the loss of 100 kg N ha^{-1} per cycle cannot be justified in the area.

4.3.3 *Bacterial Populations*

Microorganisms play a very important role in the transformations that nitrogen undergoes in soil (Alexander, 1977). While comparing the microbial populations in different systems of agricultural production and non-cultivated soil, it was found that the total bacteria

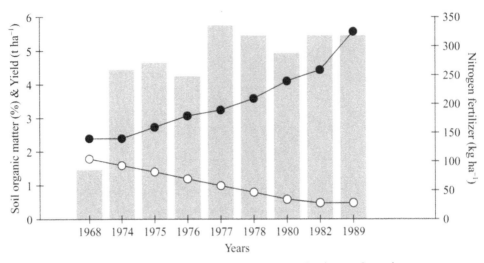

Figure 4.3 Wheat grain production (▩), use of nitrogen fertilizers (●) and percentage organic matter in soils (○) of 'Bajío', Guanajuato, México

Table 4.3 Production characteristics of the main intensive crops in the 'Bajío' region of Guanajuato, México

Crop	Area (ha)	Yield (t ha^{-1})	Production (t)	Fertilizer (t N ha^{-1})	Total fertilizer usage (t ha^{-1})
Corn	65 528	4.3	281 770	0.200	13 103
Wheat	142 095	5.5	770 522	0.320	45 470
Sorghum	162 285	6.8	1 103 538	0.240	38 948
Broccoli	15 000	10.0	150 000	0.400	18 000

Table 4.4 Theoretical nitrogen balance (t) in different agricultural production systems in the 'Bajío' region of Guanajuato, México

	Corn	Wheat	Sorghum	Broccoli
Incoming				
Mineralized residues	110	281	622	392
Fertilizer	13 105	45 470	38 948	18 000
Outgoing				
Grain	5 072	15 630	19 864	–
Straw:				
Extraction	2 725	0	0	–
Burn	–	5 627	13 242	–
Stems and leaves	–	–	–	1 750
Sprout (floret)	–	–	–	2 450
Unaccounted for N	*5 418*	*23 087*	*6 153*	*14 192*

Table 4.5 Nitrogen loss in various agricultural systems in the 'Bajío' region

System	Unaccounted-for nitrogen (kg ha^{-1} per cycle)
Corn	89
Wheat	172
Sorghum	40
Broccoli	204

population is very similar ($1.2–2.8 \times 10^7$ colony-forming units (CFU) per gram of dry soil) in the intensively cultivated and non-agricultural soils. Interestingly, denitrifying populations were found in high densities ($1.8–6.9 \times 10^5$ CFU g^{-1} dry soil) in a soil where intensive agricultural production occurred. In contrast, in non-cultivated soils the number of denitrifying bacteria was lower (1.7×10^4 CFU g^{-1} dry soil). These observations reveal that bacterial activity, both denitrification as well as nitrification, is high due to the predominantly clayish soils, excessive water application, and high doses of fertilization in this region (Tiedje, 1988).

De la Fuente and Peña-Cabriales (1988), while studying the dynamic denitrifying population of soils under wheat and sorghum cultivations, found that the denitrifying population constituted 4 % of the total heterotrophic population, higher than that reported for other ecosystems (Tiedje *et al.*, 1982).

4.3.4 Ammonia Volatilization

On a global base, agriculture is responsible for 55 % (4×10^7 t N per year) of the NH$_3$ released to the atmosphere (Duxbury, 1994). This depends on several factors, including soil pH and buffering capacity, calcium carbonate content, moisture and wind velocity (Demeyer *et al.*, 1995). The results obtained in this study under laboratory conditions indicate that the nitrogen losses, in the form of NH$_3$, of the fertilizer applied range from 19 to 83 %, increasing when urea, ammonium sulphate and ammonium nitrate are added, in that order (Figure 4.4).

4.3.5 Mineralization of Organic Residues

In intensive agricultural production systems, the soil's capacity to maintain mineral nitrogen decreases and the organic matter content falls (McKenney *et al.*, 1995). Considering that the incorporation of crop residues in this agricultural system is very uncommon, it is important to find out what happens with the nitrogen transformations when residues with different carbon/nitrogen ratios are added.

The results obtained show that the rates of ammonification-nitrification-immobilization depend to a large extent upon the chemical composition of the residues, particularly the type of carbon compound and the carbon/nitrogen ratio (Figure 4.5). The addition of organic matter stimulated nitrogen fixation, which renders the nitrogen unavailable for other processes of biological transformation or leaching.

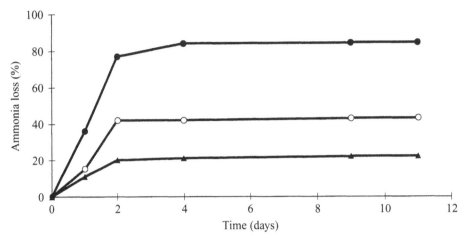

Figure 4.4   Percentage of ammonia lost upon application of urea (●), ammonium sulphate (○) and ammonium nitrate (▲) at different periods

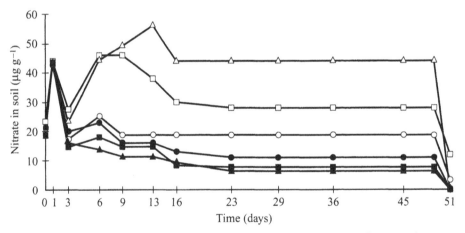

Figure 4.5   Nitrate production in soil amended with different crop residues: sorghum (■), wheat (●), faba bean (○), corn (▲), broccoli (△). □ = control

There are reports in the literature showing that emissions of N_2O and NO increase during the nitrification process of ammonium fertilizers and urea (De Groot *et al.*, 1993; Grant, 1994). In the Bajío of Guanajuato region around 84 000 t base NH_4^+ and/or urea are consumed each year. The results obtained show that the production of NO_3^- nitrogen (nitrification) was much higher in samples from sites which had been repeatedly fertilized over the past 50 years with nitrogen than in non-cultivated soils (Figure 4.6); therefore, nitrification may be very important in the production of N_2O and NO and represent an important leak of nitrogen from the agricultural system. Furthermore, environmental conditions that favour this phenomenon prevail during most of the cultivation period. This might be the cause of the high N_2O amounts detected during sorghum cultivation in this region, when the soil conditions were not most suitable for denitrification (De la Fuente-Martínez and Peña-Cabriales, 1988).

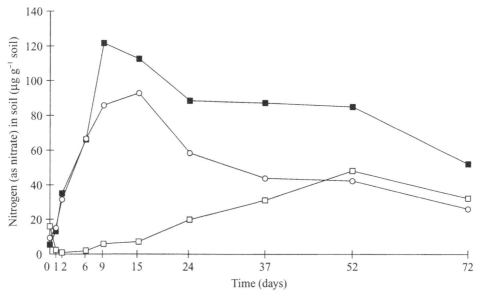

Figure 4.6 Nitrate production in soils under several crop systems, non-cultivated (□), vegetable–vegetable (■) and cereal–cereal (○), amended with urea

4.4 Conclusion

The information presented here illustrates that the management of nitrogen in this region is of serious ecological and economical relevance.

Despite all that is known about factors such as fertilizer type, application time, irrigation, crop rotation, identification of efficient genotypes in the nitrogen intake and many other variables, there are still many questions to answer about the ecological use of this important agricultural chemical.

References

ALEXANDER, M., 1977, *Introduction to Soil Microbiology*, 2nd Edn, New York: John Wiley & Sons.

ANDERSON, J.M. and INGRAM, J.S.I., 1993, *Tropical Soil Biology and Fertility. A Handbook of Methods*, 2nd Edn, Oxford CAB International.

BOUWMAN, A.F., 1990, *Soil and Greenhouse Effect*, New York: John Wiley & Sons.

CASTELLANOS, R.J.Z. and PEÑA-CABRIALES, J.J., 1990, Los nitratos provenientes de la agricultura. Una fuente de contaminación de acuíferos, *Terra*, **8**, 113–126.

CATALDO, D.A., HAROOM, M., SCHRADER, L.E. and YOUNGS, V.L., 1975, Rapid colorimetric determination in plant tissue by nitration of salicylic acid. *Communication in Soil Science and Plant Analysis*, **6**, 71–80.

CLÉMENT, A., LADHA, J.K. and CHALIFOUR, F.P., 1995, Crop residue effects on nitrogen mineralization, microbial biomass and rice yield in submerged soils, *Soil Sci. Soc. Am J.*, **59**, 1595–1603.

DE GROOT, C.J., VERMOESEN, A. and VAN CLEEMPUT, O., 1994, Laboratory study of the emission of N_2O and CH_4 out of calcareous soil, *Soil Sci.*, **158**, 335–364.

DE LA FUENTE-MARTÍNEZ, J.M. and PEÑA-CABRIALES, J.J., 1988, Dinámica poblacional desnitrificante en el Bajío Guanajuatense, *Rev. Lat-amer, Microbiol.*, **30**, 335–340.

DEMEYER, P., HOFMAN, G. and VAN CLEEMPUT, O., 1995, Fitting ammonia volatilization dynamics with a logistic equation, *Soil Sci. Soc. Am. J.*, **59**, 151–163.

DUXBURY, J.M., 1994, The significance of agricultural sources of greenhouse gases, *Fertilizer Res.*, **38**, 151–163.

FOCHT, D.D. and JOSEPH, H., 1973, An improved method for the enumeration of denitrifying bacteria, *Soil. Sci. Soc. Am. J.*, **37**, 698–699.

GRANT, B.F., 1994, Simulation of ecological controls on nitrification, *Soil Biol Biochem.*, **26**, 305–315.

LEWIS, O.A.M., 1986, *Plant and Nitrogen. Studies in Biology*, London: Edward Arnold.

MCKENNEY, D.J., WANG, S.W., DRURY, C.F. and FINDLAY, W.I., 1995, Denitrification, immobilization, and mineralization in nitrate limited and nonlimited residue-amended soil, *Soil Sci. Soc. Am. J.*, **59**, 118–124.

QUEMADA, M. and CABRERA, M.L., 1995, Ceres-N model predictions of nitrogen mineralized from cover crop residues, *Soil Sci. Soc. Am. J.*, **59**, 1059–1065.

SADIK, N., 1988, *The State of World Population 1988*, New York: United Nations Population Fund.

SARH, 1994, *Estadísticas agrícolas del Estado de Guanajuato*. Secretaría de Agricultura y Recursos Hidráulicos, México: SAHR.

TIEDJE, J.M., 1988, Ecology of denitrification and dissimilatory nitrate reduction to ammonium, in Zehnder, J.B. (ed.) *Biology of Anaerobic Microorganisms*, New York: John Wiley & Sons, pp.176–244.

TIEDJE, J.M., SEXTONE, A.J., MYROLD, D.D. and ROBINSON, J.A., 1982, Denitrification: ecological niches, competition and survival, *Antonie Van Leeuwenhoeck*, **48**, 569–583.

VAN CLEEMPUT, O. and HERA, CH., 1994, 'Fertilizer nitrogen use and efficiency in different cropping systems', presentation at the XV Congreso Internacional de la Ciencia del Suelo, Acapulco, Mexico, July.

VAN DEN ABEEL, R., CLAES, A. and VLASSAK, K., 1990, Gaseous nitrogen losses from slurry manures land, in Hansen, J. and Hendrisksen, K. (eds), London: Harcourt Brace Jovanich, pp.213–244.

VÁZQUEZ-NAVARRO, G., DE LA FUENTE-MARTÍNEZ, J.M. and PEÑA-CABRIALES, J.J., 1994, Eficiencia y pérdida vía desnitrificación de fertilizantes nitrogenados aplicados a sorgo (*Sorghum bicolor* L. MOENCH) cultivado en un vertisol, *Terra*, **12**, 345–353.

5

Impermeable Barrier Liners in Containment Type Landfills

EMIR J. MACARI AND CARLOS H. ORTIZ-GÓMEZ

5.1 Introduction

The most important component of a containment-type landfill is the barrier liner covering the base of the landfill and leachate collection system used to remove the leachate, which accumulates at the base of a landfill (shown in Figure 5.1). The amount of leachate leakage and contamination of the underlying soil layers and aquifers may be limited by constructing an 'impermeable' liner and ensuring that its integrity or structure remains relatively constant throughout the life of the landfill (controlling changes in the clay microstructure limits the change in the liner's permeability). The parameters that are critical to maintain the integrity and permeability of the liner at acceptable levels, should be tested for and controlled throughout the active life of the landfill.

The factors that should be checked and controlled to assure that the limiting permeability is maintained throughout the life of the landfill include the water content-density

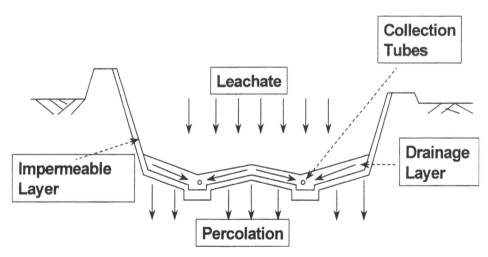

Figure 5.1 Components of a liner system in containment-type landfill

criteria, actual compaction of the liner (or placement when synthetic membrane is used), and the resulting permeability of the liner (which should be tested in the field). Other factors that should be taken into account when designing and constructing a liner system are its chemical stability, (when it comes in contact with the potential contaminants in the leachate) and the amount of degradation caused by natural processes, such as desiccation and freeze/thaw cycles, which may occur during the construction of the landfill (until the first layer of waste is placed on top of the liner).

The following sections present brief descriptions of the three types of materials commonly used for landfill liners and the factors which may affect the permeability of the liner. These materials are compacted clay, amended soils, and geosynthetics.

5.2 Compacted Clay Liners

Compacted clay has been most commonly used for barrier liners of containment type landfills. The structure of the compacted clay depends on the compaction method, compaction effort, and moisture content before compaction. The pre-compaction clod size becomes an important factor in the resulting permeability if the clay is compacted with standard Proctor compaction energy and dry-of-optimum moisture (Benson and Daniel, 1990). The importance of the compaction effort in the chemical stability, and therefore in the fabric, of barrier liners are discussed in a later section.

Clay should be compacted wet-of-optimum and using a kneading process. Kneading compaction is the type of compaction obtained using sheepsfoot-type rollers and tends to reorient the flocculated clay particles into a fabric that is more dispersed. This type of fabric is less permeable than the normal orientation of clay particles (in which the particles flocculate into large clods). An important limitation of kneading compaction is that it compacts only about 20 cm below the ground surface; hence, small compaction lifts must be used during the process.

The wet-of-optimum moisture requirement is based on the fact that the large clods are easier to break up at higher water contents (Benson and Daniel, 1990). Breaking up the large clods results in smaller interclod voids and a more discontinuous fabric. Both of these factors decrease the permeability of the clay. Benson and Daniel (1990) also show that a larger compactive effort can be used to remould the large clods (including dry, hard clods).

The density and moisture content criteria for a given liner can be obtained by performing a number of Proctor tests on the soil that will be used to determine the optimum dry density of the soil and the moisture content at which this density can be obtained. After this curve is defined, an acceptable range of moisture content and dry density values is recommended (90–95 % dry density is usually the lower boundary of this range). Even though this method is commonly used to determine the density–moisture content criteria, Daniel and Benson (1990) recommend that a modified approach be used, which accounts for different levels of compaction energy, and relates the density and moulding water contents to required hydraulic conductivity values. They propose a method by which other soil properties (e.g. shear strength) may also be taken into account.

The density and moisture content of compacted clay liners may be verified using nuclear tests (ASTM D2922-91 and D3017-88, respectively). Both methods consist of comparing a measured parameter of the liner with the calibration data of the nuclear gauge. For the density measurement, the gauge should be calibrated regularly against a direct measurement of density. Direct density measurements include the sand cone test, rubber balloon method, and drive cylinder test (Bagchi, 1994). The ASTM standard methods for these tests are D1556–90, D2167–94, and D2937–94, respectively. The water content gauge needs to be calibrated against oven-dried samples.

As would be expected, other factors – such as mineral composition, chemical composition of the pore fluid and Atterberg limits of the soil – are also important for the mechanical properties of a liner. The chemical stability issue is discussed in a later section. Regarding the mineral composition, the presence of minerals which may swell or shrink easily is not recommended as these changes will undoubtedly alter the liner's properties (e.g. permeability and shear strength). The plastic limit of the soil should be lower than the precompaction water content so that the soil can deform with a lot of cracking and degradation.

5.3   Amended Soil Liners

Soils amended with a number of admixtures and materials have also been used to construct the barrier liners of some containment landfills when a high-quality borrow source is limited or non-existent. The amending material proven to be the most useful has been bentonite, which can be used to increase the liner's water adsorption capacity. When the bentonite is wetted, it swells and creates a low-permeability layer (permeabilities in the range of 10^{-8}–10^{-9} cm s^{-1}, were reported by O'Sadnick *et al.* (1995) in their case study of a sand/bentonite liner).

As with compacted clay liners, a change in the chemical composition of the pore fluid or leachate may change the fabric of the soil and result in a decrease or increase of the liner's permeability. Therefore, the chemical stability and compatibility between the amended soil and expected leachate should also be tested before approving its use.

Bagchi (1994) recommends that central plant mixing of the admix and soil be used instead of in-place mixing with agricultural-type equipment, and provides a flow diagram describing the central plant mixing procedure which he recommends. On the other hand, O'Sadnick *et al.* (1995) reported that a 'pug mill' mixing process was used efficiently in the construction of landfill in south-western Nebraska. The pug mill process consisted of a conveyor belt, a silo used to store the bentonite, a spray bar used to add water, and computerized devices to control the amount of bentonite added.

For quality control purposes, the quality of the bentonite/soil mix should be determined. Bagchi (1994) reports that the methylene blue test (ASTM C837) and the sand equivalent test (ASTM D2419) could be used to calculate the percentage of bentonite in the mix.

5.4   Synthetic or Composite Liners

The geosynthetic materials commonly used for barrier layers are geomembranes (also known as synthetic membranes) and geosynthetic clay liners. Geomembranes are impermeable layers made from polymers; their properties will therefore be dependent on the size, shape, structure and distribution of the molecules used to create the polymer. A discussion of the effects of these parameters on the physical properties of the synthetic membrane is beyond the scope of this chapter, but one can be found in Bagchi (1994). Because of the dependence on the characteristics of the molecules used, the physical

Table 5.1 ASTM Standard Test Methods

Physical property	Method
Tensile strength	D638-94b
Tear resistance	D1004-94a
Puncture resistance	D4833-88
Low-temperature brittleness	D746 (procedure B)
Environmental stress crack resistance	D1693-94 (condition C)
Carbon black percentage	D1603-94
Carbon black dispersion	D2663-89
Accelerated heat ageing	D573, D1349
Density	D1505-90
Melt flow index	D1238

properties of each geomembrane are different and should be tested on a batch-to-batch basis.

In addition to permeability testing, which is be discussed in a separate section, some of the physical properties that may be tested when designing a geomembrane barrier are listed in Table 5.1 (adapted from Bagchi, 1994). Another relevant ASTM standard is D4354-89, which discusses the standard sampling method for geosynthetic testing.

Geosynthetic clay liners consist of a layer of low-permeability material (e.g. bentonite) 'supported by geotextiles and/or geomembranes, being mechanically held together by adhesives, needling, or stitching' (Koerner, 1995). The bentonite or low-permeability material swells when it comes in contact with water to form an impermeable layer with hydraulic conductivity ranging from 10^{-7} to 10^{-9} cm s^{-1}.

Some of the advantages that geosynthetic clay liners have over compacted clay liners are that they: cost less, can be installed in low temperatures, have higher tensile and shearing strengths and are not as affected by differential settlements and cycles of desiccation and freeze/thaw. On the other hand, their use is limited by the fact that they cannot be used in contact with most leachates.

The probability of damaging a geosynthetic liner during transportation, placement and seaming is higher than that for compacted clay liners. For this reason, geosynthetic liners should never be used alone and should be placed over a compacted clay or amended soil layer. This type of composite or double liner may be used to reduce the thickness of clay or amended soils needed to provide the same amount of containment, and usually result in lower amounts of leakage through the liner. Of course, as the complexity of the system increases, so do the concerns. Additional concerns involved with the construction and stability of double liners include the strength of the interfaces between different materials (which can be tested for with the direct shear method ASTM D5321-92) and the adverse effects of different construction or compaction techniques.

A case study of a double liner test cross-section was presented by Bergstrom *et al.* (1995). The conclusions of this study included: depth of deterioration of the liner due to freeze-thaw cycles was limited to the frost penetration depth and laboratory permeability tests on undisturbed samples of compacted clay will yield results similar to those obtained using field infiltrometers when the permeability of the liner is controlled by flow through the matrix and not through macrofabric defects (e.g. desiccation cracks). These conclusions are discussed in more detail in the following sections.

5.5 Factors Relevant to all Types of Liners

The time that elapses between the end of liner construction and the start of operation (when the first layer of waste is placed) is also critical for the efficiency of the liner. If the liner is left exposed to warm weather for too long, the shrinkage of the soil may cause desiccation cracks and increase its permeability. Desiccation cracks may be as deep as 30 cm (Bagchi, 1994) and 'can penetrate compacted clay to a depth of several inches in just a few hours' (Daniel *et al.*, 1985). Even though water can be added to close the cracks, the initial permeability will not be completely recovered. In colder regions, freeze/thaw cycles affect the integrity of the liner by degrading the soil from the surface to the depth of frost penetration. Therefore, the amount of degradation due to either of these processes has to be taken into account when choosing the liner's thickness. The liner should be thick enough so that the required impermeability is provided by the section of the liner that has not degraded. The thickness of synthetic liners is also not only dependent on the required permeability. Because of the concerns for damage of geosynthetic liners, the puncture resistance and resistance to ultraviolet rays should also be considered before choosing the thickness.

The liner's compatibility (chemical stability) with the leachate is also a critical factor, and should be extensively tested to assure the integrity of the liner. The characterization of the leachate created by a given landfill is also discussed by Bagchi (1994). To test for chemical stability, the permeability tests discussed below should be performed using the expected leachate as the permeant. Immersion tests, such as the test 9090 (USEPA), may also be used to test the stability of geosynthetic materials. The compatibility between the synthetic material and the surrounding soils should also be tested (especially if low or high pH soils are present). A short-term burial test can be used for this purpose (ASTM D3083).

The chemical stability of soil liners can be improved through processes that make the soil more dense or create cementation between soil particles. This stabilization can be done using additives or through mechanical processes (Broderick and Daniel, 1990). The additives used include lime, portland cement, and lime plus sodium silicate. When additives are used, the solubility of the cementation should be analysed to determine its stability as it comes in contact with the leachate. Broderick and Daniel (1990) also conclude that a higher compaction effort (using modified Proctor instead of standard Proctor) will improve the stability of compacted clay liners to 'chemical attack' by densifying the soil.

5.6 Permeability Testing

The presence of macrofabric defects in barrier liners can be controlled and limited with good construction techniques. During construction, the following factors should be controlled to limit the amount of defects present in the liner and to assure its integrity and 'impermeability': (a) compaction or placement of the liner, (b) desiccation or freeze/thaw degradation, (c) liner thickness, and (d) chemical composition of the liner. All of these factors, in addition to the soil properties, will control the permeability of the barrier.

Laboratory permeability tests include the flexible and rigid wall permeameters (can be falling or constant head tests), and flow pump tests. All of these tests have advantages and disadvantages that should be considered before planning the testing programme. In general, the flexible wall permeameter is more accurate than the rigid wall permeameter. The difference in results from the two tests is due to the fact that the rigid cell cannot adjust to the shape of the sample and the water or permeant will tend to flow through the sides of the sample. Therefore the measured permeability will be representative of the permeability between the sample and the wall, and not of the soil's fabric. Testing fine-grained soils with the conventional falling head permeameter may take too long; therefore flow pump tests can be used to accelerate the testing programme of fine grained soils. The use of empirical relationships between some soil properties (e.g. void ratio) and permeability are not recommended for this type of project.

The high dependency of a liner's permeability on construction technique and the fact that laboratory tests can test only small samples may lead to laboratory results which are not representative of, and are lower than, the actual field values (Daniel, 1984). For this reason, *in-situ* permeability tests are an essential component of liner permeability testing programmes. *In-situ* tests that can measure permeabilities in the range of interest for these types of projects (10^{-7}–10^{-9} cm s^{-1}) can be subdivided into four groups: borehole tests, porous probes, sealed infiltrometers and underdrains (Daniel, 1989).

Descriptions of the test methods and the respective advantages and disadvantages are also included by Daniel (1989). Of these tests, only the sealed, double-ring infiltrometer (SDRI) and pan lysimeter can test large areas of low-permeability soils. The borehole test suitable for measurement of low permeabilities analysed by Daniel was the Boutwell permeameter. The major disadvantage of this test, and of the porous probe, is that only a small volume of soil is tested. Even though the pan lysimeter tests take a very long time for low permeabilities, they can be used as part of the monitoring or quality control programmes and are recommended by Daniel for use in compacted clay liners.

Two of the major problems associated with *in-situ* testing are that the tests are not performed under fully saturated conditions and that the effect of overburden stress is not taken into account. To account for the actual degree of saturation, Daniel (1989) recommends that the relationship between degree of saturation and measured permeability be determined in the laboratory and used to extrapolate the field measurement to the fully saturated condition. The effect of overburden stress on permeability can be analysed using consolidation or flexible wall permeameters (Daniel, 1989). The permeability of a piece of geosynthetic can be tested for with ASTM E96 and using the leachate instead of water. Because geosynthetics are used as part of composite liners, test cross-sections should be built and tested to take into account the construction procedures and the actual field conditions.

5.7 Conclusion

After analysing the different construction and quality control issues involved with barrier liners of containment type landfills, some general recommendations can be made. If a good preliminary testing programme is used, high initial costs will counteract the costs and problems which may be encountered if the efficiency of the liner is not tested extensively. Even though good construction techniques and design will efficiently reduce the number of macro-defects (e.g. desiccation cracks and holes in synthetic liners), their presence should be considered inevitable. Therefore, the quality control programme should include field permeability tests, which examine large volumes of the liner. Construction of test cross-sections of the proposed liner has been proven to be a very useful and beneficial practice. Even though the preconstruction costs are greater when predesign and quality testing programmes such as those described above are used, any problems with the design criteria or construction techniques can be worked out to improve the efficiency of the liner and accelerate its construction. In the long term, the benefits of such an approach will far outweigh the initial limitations.

References

BAGCHI, A., 1994, *Design, Construction and Monitoring of Landfills*, New York: John Wiley & Sons.

BENSON, C.H. and DANIEL, D.E., 1990, Influence of clods on hydraulic conductivity of compacted clay, *Journal of Geotechnical Engineering* (ASCE), **116**(8), 1231–1248.

BERGSTROM, W.R., CREAMER, P.D., PETRUSHA, H. and BENSON, C.H., 1995, 'Field performance of a double liner test pad', in *Proceedings of Geoenvironment 2000: Characterization, Containment, Remediation, and Performance in Environmental Geotechnics*, **1**, 608–623.

BRODERICK, G.P. and DANIEL, D.E., 1990, Stabilizing compacted clay against chemical attack, *Journal of Geotechnical Engineering* (ASCE), **116**(10), 1549–1567.

DANIEL, D.E., 1984, Predicting hydraulic conductivity of clay liners, *Journal of Geotechnical Engineering* (ASCE), **110**(2), 285–300.

DANIEL, D.E., 1989, In-situ hydraulic conductivity tests for compacted clay, *Journal of Geotechnical Engineering* (ASCE), **115**(9), 1205–1226.

DANIEL, D.E. and BENSON, C.H., 1990, Water content-density criteria for compacted soil liners, *Journal of Geotechnical Engineering* (ASCE), **116**(12), 1811–1830.

DANIEL, D.E., ANDERSON, D.C. and BOYNTON, S.S., 1985, Fixed-wall versus flexible-wall permeameters, in *Hydraulic Barriers in Soil and Rock, ASTM Special Technical Publication 874*, Philadelphia: ASTM, pp.107–123.

KOERNER, R.M., 1995, 'Geosynthetics for geoenvironmental applications', in *Proceedings of Geoenvironment 2000: Characterization, Containment, Remediation, and Performance in Environmental Geotechnics*, **1**, 872–898.

O'SADNICK, D.L., SIMPSON, B.E. and KASEL, G.K., 1995, 'Evaluation and performance of a sand/bentonite layer', in *Proceedings of Geoenvironment 2000: Characterization, Containment, Remediation, and Performance in Environmental Geotechnics*, **1**, 688–701.

6

Control of Submicron Air Toxin Particles after Coal Combustion utilizing Calcium Magnesium Acetate

JIANXI ZHAO, DONALD L. WISE, EDGAR B. GUTOFF,
JOSEPH D. GRESSER AND YIANNIS A. LEVENDIS

6.1 Introduction

It is increasingly being recognized that fine particles cause severe respiratory disease and degrade the air quality. In order to decrease air pollution caused by coal combustion and increase its utilization as an alternative to oil, coal must be burned in an environmentally clean and economical manner. Much work has been done to improve coal preparation. Many coal cleaning methods reduce the formation of fly ash and the contents of trace elements. However, there are practical limitations to coal cleaning and thus flue gas cleaning becomes increasingly important (Gennrich, 1993).

Much work on the control of fine toxic particles has been done. However, there is no published work on the control of very fine toxic particles (0.01–0.1 μm). These very fine particles are produced in flue gas as it cools after leaving the combustion chamber. Previous work has shown that bag filters have the capability to remove fine particles and achieve high collection efficiency. In order to increase the collection efficiency for very fine particles, dry adsorbents were considered for injection into the combuster. In this study, the principal objective was to measure the capacities of different additives (specifically calcium magnesium acetate and calcium magnesium carbonate) to capture the very fine particles. Another objective was to find a dimensionless parameter, such as the Reynolds number or Peclet number, that correlates with collection efficiency. The relationship between the collection efficiency and such a parameter may provide a basis for the use of either calcium magnesium acetate or calcium magnesium carbonate.

Currently, the three main technologies for reducing air pollution during coal combustion involve electrostatic precipitators, scrubbers, and baghouses (Senior and Flangan, 1982). The electrostatic precipitator is designed to treat large volumetric flow rates at low pressure drop. The minimum particle size that the electrostatic precipitator can remove is less than 1 μm. Scrubbers are expensive for the removal of particulate and gaseous pollutants because of high pressure drop. The particle size that baghouses can remove is smaller than the electrostatic precipitator and scrubbers (Senior and Flangan, 1982).

However, although baghouses alone can remove a large percentage of the particulate, they cannot capture all particles or remove the acidic gases sulphur dioxide and the nitrogen oxides. The addition of calcium magnesium acetate (CMA) and calcium magnesium carbonate (CMC) to the combustion chamber were suggested both to increase the collection efficiency and to remove sulphur dioxide and nitrogen oxide in the baghouse (Shuckerow, 1995).

CMA is a chemical with many potential uses, and it can be derived from residual biomass. For example, sewage sludge can be used as a raw material to produce low-cost CMA (Shuckerow, 1995).

The dual sulphur dioxide/nitrogen oxide removal efficiency of CMA ash and CMC ash has been investigated (Raabe, 1976). The objectives of the study reported in this chapter were to measure the collection efficiency of CMA and CMC ash to fine particles and the effect of chemical engineering parameters on the collection efficiency. Ferrous sulphate particles were used to simulate very fine air toxin particles (0.01–0.1 μm). A 108 cm long pyrex glass tube with filter paper simulated the baghouse. The air velocities used were in the range of industrial values. Several trials were run varying the air flow rate and the weights of CMA ash or CMC ash on two paper filters in series. The ash from a calcium acetate gel was also briefly studied. Structural differences in ash of calcium acetate and of calcium acetate gel were examined by comparing scanning electrical microscope pictures of the two systems.

6.2 Background

The following is a brief review of air pollution caused by coal combustion, the various methods of capturing fine particles, and the physical mechanisms for particle removal. It will also summarize the previous work on which this study is based. Major air pollutants include sulphur oxides, oxides of nitrogen, carbon monoxide, ammonia, mercury, ozone and volatile organic compounds. Particulate emission is also an important air pollutant. The major sources of particulate emission include coal combustion, mineral production, paper production and chemical processing; coal combustion contributes the most to air pollution.

Coal is slow-burning fuel with high ash content. The ash consists of non-combustible inorganic constituents in the coal. In the process of combustion, coal particles break up into smaller particulates because of thermal stress (Flangan and Seinfeld, 1988). The carbon reacts with oxygen, and residual particles are left. The coal particles become more porous when the carbon is consumed. The ash particle size depends on the size of coal particles. Some ash particles are formed when ash vapourizes at high temperature and later condenses. The volatilized ash can form much smaller particles than the non-volatilized mineral inclusions in the coal. The size distribution (Flangan, 1979; Flangan and Seinfeld, 1988) of the fine particles is from 0.01 to 0.1 μm. This volatilized ash contains sodium, potassium and arsenic. The fine particles formed from volatilized ash tend to be toxic. The combustion of liquid and gaseous fuels can also form very fine particles. According to Palmer and Cullis (1965) most of the particles from gaseous combustion are in the range of 0.01–0.05 μm. Motor vehicles (Gomaa and Allawi, 1994) are also a large source of fine particle pollution, producing particles 0.01–0.03 μm in size.

The US Environmental Protection Agency has set national ambient air quality standards (Kao, 1994) for each of six pollutants: sulphur dioxide, carbon monoxide, lead, ozone, nitrogen oxides, and particulates with diameter of 10 μm or less. The first five pollutants are proven to be toxic and hazardous to human health. According to the *New York Times* (May, 1996), fine particles less than 2.5 μm can invade the human body's natural defences and cause serious ailments, such as cancer. The Natural Resources Defense Council reported that fine particles pollution causes 64 000 deaths nationwide every year (*New York Times*, May, 1996). Fine particles can accumulate in the lungs over time. In China, coal miners have to retire by the age of 36 because of lung disease (Hinds, 1982). The adverse effects of air pollutants (DeVito *et al.*, 1993) include cancer, birth defects and damage to the immune and nervous systems. Sulphur dioxide, nitrogen dioxide and mercury vapour are extremely toxic to human health, causing skin diseases and liver dysfunction (Kudlac and Farthing, 1992). Title III of the US Clean Air Act of 1990 (Akers, 1996) identifies 189 hazardous air pollutants subject to regulation. The types of compounds are produced by a wide range of emission sources and cause many detrimental health effects. There are four *other* sources (Gomaa and Allawi, 1994) for volatile organic compounds and air toxin emissions: the cleavage reaction and neutralization vessels are one; emissions also come from product transportation units such as barges, ships and freightcars. The evaluation of the health risk caused by airborne pollutants is not easy, as many airborne pollutants can undergo reactions in the atmosphere (Higley and Joffe, 1996). Based on these health issues, control of air pollutants becomes more and more important.

There are several different classes of particulate control equipment. *Gravity separation* is used in setteing chambers to let particles settle out (Flangan, 1979). In a *cyclone separator* the particles are collected by centrifugal force (Leith and Licht, 1972; Flangan, 1979). *Electrostatic precipitators* are used to charge small particles and collect them on oppositely charged plates (Tardos and Snaddon, 1984). When a layer of particles has accumulated the plates are cleaned by rapping them and letting the particles fall to the bottom of the unit for removal. *Filters* operate on the principle that the particulate-laden gas is forced through an assemblage of small openings where the larger particles cannot pass. In *wet scrubbers*, particles collide with droplets of water, and the dirty droplets are easily separated from the gas because of their large size.

Generally, settling chambers and cyclones are used to remove larger particles and are relatively inexpensive (Leith and Licht, 1972). Electrostatic precipitators, which can have a high removal efficiency, are expensive and there are some limits on the operating conditions (Raabe, 1976; Henry *et al.*, 1982; Han and Ziegler, 1984). Fabric filters have very high efficiencies and are not as expensive as electrostatic precipitators, but are generally limited to dry, low-temperature conditions (Yeh and Liu, 1974; Flangan, 1979). The great advantage of wet scrubbers is that they remove gaseous pollutants simultaneously with particles (Calvert, 1984). However, scrubbers have the same disadvantage as the electrostatic precipator in that they can be expensive. Coal cleaning can dramatically remove trace elements (Liu and Yeh, 1977; Akers and Hudyncia, 1990; DeVito *et al.*, 1993; Change and Offen, 1995). Conventional coal cleaning methods can reduce the content of lead, nickel, chromium, arsenic and cobalt, and reduces mercury and sulphur dioxide emissions. It also reduces ash and air toxins in the flue gas, by reducing the amount of ash.

Many studies have been done to control air toxic pollutants using calcium salts. Levendis and Wise (1994a,b) used the special properties of calcium salts after combustion to capture the sulphur dioxide/nitrogen oxides. These calcium salts include calcium formate, calcium acetate, calcium propionate and calcium benzoate. During combustion, calcium oxide is formed with a porous structure and large surface area, which makes it a good adsorbent for sulphur dioxide and nitrogen oxides.

Shuckerow (1995) showed that the ash formed by the addition of CMA to the combustion chamber is effective in removing gaseous mercury from an air stream. Loftus *et al.* (1992) used a confined vortex scrubber to remove fine particulates from flue gases; 99 % coefficiencies have been measured for fly ash with a mean particle size of 3 *μ*m. For 0.3 *μ*m diameter particles, 98 % efficiency has been measured. There is also some research (Muradov *et al.*, 1996) on the selective destruction of airborne hazardous organic compounds. Muradov *et al.* (1996) developed a photocatalytic method for selective oxidation of the airborne nitroglycerin.

With the development of air pollutant control technology, many collection models have been developed. This has resulted in more accurate measurements of fine particles in order to verify some assumptions that were used in these models. Hurd and Mullins (1962) used a small radial-flow parallel-plate precipitator to study particle sizes. They assumed that the fine particles carried only one electrical charge. The particle sizes were deduced from their mobilities. The direct measurement methods for wastewater and combustion processes use flux chambers or stack testing (Sinclair *et al.*, 1979). Tardos and Snaddon (1984) developed the theory of collection of monodisperse particles using a precipitator; aerosol particles can be measured by an optical particle counter using the radial-flow parallel-plate precipitator they designed in which the particle collection is totally dominated by electrostatic forces. The mean diameter of particles used in their experiments was 1.04 *μ*m. Agreement to better than 2 % was achieved between two counting systems.

With the increased requirement for control of toxic gases and particulates, relatively inexpensive advanced control technologies with higher collection efficiencies are required. High-temperature gas cleanup processes using filter bags have been able to meet the particulate control standards. Modern technologies have removed the disadvantage of the older bag filters, which were made of natural fibres and had to be operated below 250°F (120°C) (Perry *et al.*, 1984). Now fibres made from polycrystalline metal oxides can withstand 1300°F (704°C) (Steciak *et al.*, 1995). Woven alumina/boria/silica fibres can withstand temperatures of 1400°F (760°C). Porous ceramic fibre can operate at 2500°F (1370°C). Modern bag filters do not need extra dilution air, or quench-cooling of the air stream.

Currently, there are three types of fabric filters (Kudlac and Farthing, 1992; Gennrich, 1993; Singh, 1993) the shaker, the pulse jet and reverse air cleaned filters. Because shaker-cleaned filters must be shaken every 0.25–8 h, they must now be made of natural fibres, such as cotton or wool, in order to withstand the flexing and stretching. Pulse-jet bag filters trap particles on the outside of the filter, as air flows from the outside to the inside of the filter bags. In the pulse-jet filter bags dust is dislodged by short, rapid bursts of compressed air introduced at the mouth of the tubular filter bag. The pulse-jet and reverse air cleaned bag filters are usually made of glass or mineral fibres. They are designed to operate with high-temperature flue gas, and fabrics are too fragile to be cleaned by shaking. In the operation of the pulse-jet filter bags the most important factor to be controlled is the gas temperature. The collection efficiency of baghouse filters can be affected by the particle loading on the filters. Therefore time is one of the factors needing consideration when the baghouse filters are designed. In general, the collection efficiency is low at the beginning, because previously captured particles are much more effective in capturing particles than the fabric itself (Flangan, 1979). Baghouse filters are

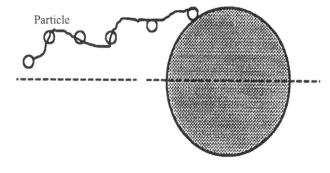

Particle

Fibre

Figure 6.1   Particle collection mechanisms of a single fibre

especially effective in capturing very fine particles and even toxic vapour (Yeh and Liu, 1974; Flangan, 1979; Bishop *et al.*, 1992). For particles less than 1 μm the collection efficiency of fabric filters can exceed 99 % but in order to satisfy the increasing require- ment on retaining toxic substances and fine particles, more efficient baghouse filters are required (Kudlac and Farthing, 1992; Gennrich, 1993; Singh, 1993). One technique is to inject dry adsorbent upstream of the bag filter. This adsorbent should capture not only toxic gas but also very fine particles in the flue gas. Adding dry adsorbent may be the most promising method of increasing the efficiency of bag filters.

Several papers have been published about the mechanism of filtration process through fibre filters. There are three stages (Gradon, 1987; Flangan and Seinfeld, 1988) in the process. The collection efficiency is low at the beginning because the new filter cloth has relatively large open spaces, and most small particles may pass through the filter bags. Once the particles accumulate the efficiency increases. Finally, the larger collected par- ticles increase the pressure drop excessively and the dust layer must be dislodged to regenerate the fabric bag. In a fabric filter, the pressure drop and collection efficiency are a function of time since the last cleaning (Flangan, 1979). Many people have developed equations for predicting outlet concentration after a fabric filter. Goren *et al.* (1968) developed a model for predicting pressure drop and filtration efficiency in fibrous media.

The most common approach in determining the collection mechanisms is focused on a single fibre (Oh *et al.*, 1981; Hinds, 1982). There are three basic mechanisms to describe the deposition of aerosol particles on a fibre: diffusion (Brownian motion), interception and inertial impaction. Figure 6.1 shows the particle collection mechanisms of a single fibre. Of course, a particle coming towards a fibre may be carried around it by the flowing air stream.

The interception mechanism does not depend on the flow velocity, u; it is only the function of the ratio, R, of D_p/D_f (D_p = diameter of particle; D_f = diameter of fibre). Inter- ception occurs when the particle hits the fibre. Usually, the interception is affected by Brownian motion and inertia. The Kuwambara hydrodynamic factor, K_u, and the dimesion- less parameter, R, are used to describe the collection efficiency because of interception, in the relation $\eta_r = [2(1 + R)\ln(1 + R) -(1 + R) + (1/(1 + R))]/2K_u$ or $\eta_r = [2(1 + R)\ln(1 + R) + (1/(1 + R))]/2k_u$ (Flangan, 1979; Hinds, 1982).

The Stokes number, St, is used to describe inertial impaction and is the ratio of the stop distance to a characteristic length scale of the flow. The velocity of a particle with an initial velocity u_0 in still air is $u_0 \exp(-t/\tau)$. When $t \to \infty$, a particle travels a distance $\chi = \tau u_0$. This is the stop distance. Then $St = \tau u_0/L = (D_p^2 \rho_p C_c u_0)/(18 D_f \mu)$, where C_c is the slip correction factor, u_0 is the fluid velocity and L is the characteristic length. With the increasing Stokes number the single fibre efficiency increases. The Stokes number can be increased by increasing D_p, ρ_p or u_0. Yeh and Liu (1974) gave the collection efficiency for the inertial impactation based on the Stokes number and Kuwahara factor.

The last mechanism is Brownian motion. The Peclet number, Pe $(= D_f u/D)$, is used here. It is a function of the fibre diameter, the flow velocity, and particle diffusion coefficient D. When the flow rate decreases, the Peclet number increases and efficiency for Brownian motion increases. Flangan and Seinfeld (1988) gave the size ranges in which the various mechanisms of collection are important. When the particle diameter, D_p, is greater than 1 μm the inertial impaction and interception mechanisms are both important; when the particle diameter is less than 0.5 μm, Brownian diffusion plays the main role. Generally, the overall collection efficiency of a single-fibre filter can be expressed as the sum of the efficiencies due to interception, impaction, and diffusion. The total efficiency of all the fibres can be calculated from the single-fibre efficiency.

Particulate air toxins from the combustion of coal consist of polyaromatic hydrocarbons and of oxides and salts of metals such as arsenic, with particle sizes of the order of 0.1 μm. These will be partially removed by the fly ash forming a filter bed on the baghouse filters. The size of the particles forming the bed in the baghouse filter is of the order of 10 μm. In analysing data on collection efficiency, dimensional analysis is used in correlating the data and in determining the most important variables. Since diffusion dominates the dynamics of fine airborne particles, determination of the diffusion coefficient (Hutching *et al.*, 1984) becomes important. A variation of the QELS (quasi-elastic light scattering) technique (Hutching *et al.*, 1984) has been used. The Brownian motion of aerosol particles can be monitored by a laser Doppler velocimeter. From the correlation function of the scattered light intensity the diffusion coefficient is determined.

Flangan and Seinfeld (1988) define the collection efficiency, η, as $\eta = 1 - N_{out}/N_{in}$, where N_{out} is the number of particles per m^3 of gas out and N_{in} is the number of particles per m^3 of gas in. The number of particles can be determined using an optical microscope if a membrane filter is used for collecting them. An expression of collection efficiency based on the mass of particles is $\eta_m = 1 - M_{out}/M_{in}$, where M_{out} is the mass of particles out and M_{in} is the mass of particles in. The collection efficiency based on the mass is easier to measure experimentally than the collection efficiency based on the particle number. In industry, fabric filtration is used for air cleaning.

The pressure drop across the filter is caused by the combined resistance of the fibres and the collected particles. For the fibrous filter, Davies (1973) gave the pressure drop based on the total drag force of all fibres as $\Delta p = \eta t u f(\alpha)/D_f^2$, where $f(\alpha) = 64\alpha^{1.5}(4 + 56\alpha^3)$ and $0.006 < \alpha < 0.3$. Here t is time. From the above equation we see that the pressure drop is directly proportional to time and face velocity. We also see the pressure drop is D_f^2. Using the equation, the effective fibre diameter can be determined by measurements of Δp, t and α. Thus the collection efficiency can be determined. The fibre size has an effect on the fibre efficiency. The relationship between the filter quality and fibre diameter can be described by the following equation where q_f, defined as filter quality, is $q_f = 4\alpha D_f \eta_s/\pi\mu u f(\alpha)$ or $q_f = \gamma t/\Delta p$, where η_s is the single-fibre efficiency, γ is the surface tension and u is the face velocity of filter. When the fraction $1 - \alpha$ is held constant and pressure drop is kept constant as the pore size is decreased, the efficiency is increased. With the void fraction constant, as the fibre diameter decreases the pore size also decreases;

thus the filter efficiency increases with decreasing fibre diameter, the filter quality increases with decreasing fibre size.

For aerosol sampling membrane filters are important. The pore size ranges from 0.01 to 10 μm and pore densities vary from 10^6 to 10^8 cm^{-2}. The solid volume fraction ranges from 0.15 to 0.30. The most common material for membrane filters is a cellulose ester. Membrane filters are also made of polycarbonate films having 10^5–10^9 capillary pores per cm^2, with the pore size from 0.03 to 12 μm. The overall collection efficiency of the capillary pore membrane filters is low for particles smaller than the pore size. For very small particles, there is a minimum efficiency around 0.05 μm.

If an adsorbent is introduced to capture toxin particles, the capturing process can be described by the pore tree model. The tree model describes the tree-like pore structure of carbon char. Each pore is considered as the trunk of a tree. The number of pores is described by the continuous branching and is a function of the skewed distance into the pore tree. The process of gas adsorption onto a solid particle involves four steps: (1) transport of gaseous components from the bulk stream to the particle surface, (2) transport of gaseous components in the particle pores, (3) adsorption of the gaseous components, (4) surface chemical reaction of the adsorbed atoms or molecules with the solid.

Simons (1992) derived equations to describe the effects of bulk diffusion to the particle exterior surface, diffusion in the larger pores, and diffusion in the smaller pores. The collection efficiency of any absorbent depends on its structure. The structure influences gas diffusion and reaction process. The equations derived by Simons can be used to predict the capture ability of CMA ash.

6.3   Materials and Methods

The materials used in this study were CMA, CMC and ferrous sulphate. The CMA can usually be obtained from the reaction of calcium and magnesium carbonate (or oxide) with acetic acid. The cost of acetic acid mainly determines the cost of CMA production, because dolomite, the source of calcium and magnesium, is relatively inexpensive (DeVito *et al.*, 1993). Low-cost acetic acid can be derived from anaerobic fermentation of glucose in biomass feedstocks; sewage sludge, woody biomass, and whey permeate are suitable feedstocks for fermentation (Cadle, 1975; Zumwalde and Dement, 1977; Field *et al.*, 1990). The CMA needs to be burned in or just after the combustion chamber to obtain CMA ash with the desired porous constructure. We used a vertical furnace to form the CMA ash. To get a larger surface area with a smaller particle size, an electric mill is used to grind the CMA powder, with a particle size from greater than 60 μm, to less than 20 μm. The CMC was ground by hand using a mortar and pestle to less than 45 μm. Ferrous sulphate crystals dissolved in distilled water were used in a nebulizer to form a ferrous sulphate aerosol to simulate fine toxin particles. The concentration of ferrous sulphate in solution was 0.001 g ml^{-1}.

The equipment shown in Figure 6.2 simulates an industrial baghouse filtration process. It consists of glassware, including the tube and o-rings (Thomas Scientific Company, NJ), rotameter (0–4 l min^{-1}; Dwyer Instruments, Inc.), nebulizer (10 ml volume, Lane Health Center Northeastern University) used to produce the fine particles, two heating tapes (Thermolyne Brisk heat flexible electric heating tape, 120 volts, length of 2.4 m and width 2.5 cm) used to heat the air flowing through the glass column, CMA ash or CMC ash on two membrane filters (Micron Separations, Inc., 90 cm in diameter) in series to capture the fine particles. The pore size of the membrane filters was 0.1 μm. The filter paper covered with wet CMA ash was dried for at least 24 h in the oven to remove the moisture and the initial weight of filter paper measured. The dried filter paper was placed between two o-rings, clamped and sealed.

The CMA was burned in a vertical furnace in the simulated post-combustion environment. The gas flow rate was 3 l min^{-1}. It contained 2000 ppm SO_2, 2100 ppm NO_x, and 3 % O_2. The temperature in the furnace tube was 1000°C. The experimental conditions above were chosen based on previous work (Shuckerow, 1995) and gave good results. The gas flow was metered. CMA in a test tube was fed pneumatically to the furnace, at a rate to give the desired ratio of calcium to sulphur. The residence time for the gas in the furnace tube was 1 s. Filter paper was used to collect the CMA ash coming out of the bottom of the furnace. An optical microscope was used to measure the particle size of the CMA ash. The CMA ash particles were coloured and had a porous structure, with a particle size of 5–35 μm. The bulk density of the CMA and CMA ash was determined by weight and volume measurements. Shuckerow (1995) had previously measured the surface area of CMA ash by nitrogen adsorption using the Brunauer–Emmett–Teller multilayer theory. The porosity of CMA ash had been determined by mercury intrusion pore symmetry. The structure of CMA ash could be observed from scanning electron microscope photographs: the particles are cenospheres of thin, porous walls with blowholes. Micrographs revealed that CMC particles were not spherical and did not change structure greatly on passing through the furnace.

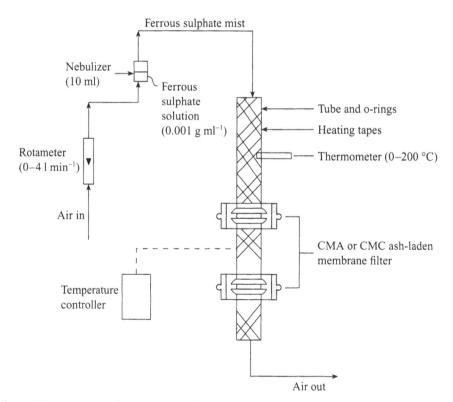

Figure 6.2   Apparatus for capture of air toxins

Either 0.655 g or 0.802 g of CMA ash were slurried in water and then placed on the filter paper using a funnel. Four different air flow rates were chosen: 2, 2.5, 3.5, or 4 l min⁻¹. The ferrous sulphate solution was placed in the nebulizer where, as shown in Figure 6.2, metered air from the laboratory supply produced a ferrous sulphate mist. The mist entered the glass column, the temperature of which was controlled by the heating tape at 108°C in order to dry the ferrous sulphate mist completely and form fine particles. A mercury thermometer was used to measure the temperature of the air. Some of the dry ferrous sulphate particles were captured by the ash on the top filter paper, some went through the top filter paper and were collected by ash on the second filter paper, and some went out of the glass column. The amount of ferrous sulphate collected by each filter was found from the weight of filter paper plus ash before and after each run. The ferrous sulphate particle sizes were determined from electron micrographs.

Nebulizers are widely used for producing the aerosol particles. There are four major kinds (Zumwalde and Dement, 1977): the evaporation–condensation monodisperse aerosol generator, the fluidized-bed aerosol generator, the vibrating orifice aerosol generator, and the DeVilbiss model 40 nebulizer, which we used. The DeVilbiss nebulizer blows compressed air from a small tube at high velocity through the liquid to entrain liquid droplets. These larger droplets are redeposited into the bulk liquid when the spray stream impacts a surface. The particle concentration produced by the nebulizer is 10^6–10^7 per cm³. The air stream was then heated to evaporate the water in the droplets, leaving an aerosol of solid ferrous sulphate particles. The particle size can be controlled by varying the concentration in the liquid.

6.4 Results and Discussion

The ferrous sulphate particle size was measured directly from scanning electron micrographs. The particles were spherical, with sizes from 0.016 to 0.5 μm and with an average particle size 0.066 μm. Figure 6.3 shows the particle size distribution. The size distribution of the CMA ash was found from optical micrographs and is given in

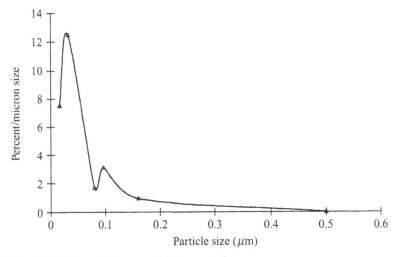

Figure 6.3 Distribution of ferrous sulphate particles

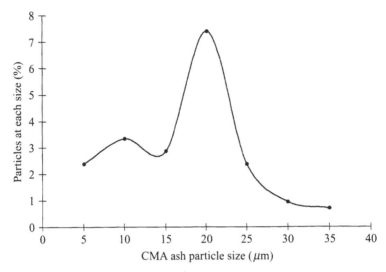

Figure 6.4   Distribution of CMA ash particles

Table 6.1   Collection efficiencies of CMA ash-laden filter paper

Air flow (l min^{-1})	Filter efficiency (%)	
	0.6 g ash per filter	0.8 g ash per filter
2.0	72.1	66.3
2.5	68.3	63.6
3.5	57.6	61.3
4.0	52.0	38.8

Table 6.2   Collection efficiencies of CMC ash-laden filter paper

Air flow (l min^{-1})	Filter efficiency (%)	
	0.6 g ash per filter	0.8 g ash per filter
2.0	14.01	13.25
2.5	13.60	12.20
3.5	12.80	10.10
4.0	11.10	7.43

Figure 6.4. The bulk density of CMA decreased during burning from 0.432 g ml^{-1} for CMA to 0.342 g ml^{-1} for CMA ash. In burning, the CMA was pyrolyzed to CaO and MgO, and CO_2 gas is produced by combustion. The formation of CO_2 may be the main cause of the CMA structure change. The reactions are:

Table 6.3 Collection efficiencies of one CMA ash-laden filter

Air flow (l min^{-1})	Efficiency (%)
2.0	60.6
2.5	52.7
3.5	48.7
4.0	34.7

$$CaMg_2(CH_3CO_2)_6 \rightarrow CaCO_3 + 2MgCO_3 + 3CH_3 - CO - CH_3$$

$$CaCO_3 \rightarrow CaO + CO_2$$

$$MgCO_3 \rightarrow MgO + CO_2$$

$$CH_3 - CO - CH_3 \rightarrow H_2C = C = CH_2 + H_2O$$

$$H_2C = C = CH_2 + O_2 \rightarrow CO_2 + H_2O$$

Twelve trials were carried out to measure the collection efficiency of CMA ash (0.6 g on one filter only, 0.6 or 0.8 g on each filter; flow rates of 2.0, 2.5, 3.5 and 4.0 l min^{-1}), and other trials were carried out with the CMC ash. The temperature and the concentration of ferrous sulphate solution were fixed.

As presented in Tables 6.1–6.3, the highest mass collection efficiency that CMA ash achieved was 72 %. The amount of CMA ash on the filters affected the collection efficiency in one case; in all but the collection efficiency was lower at higher ash weights, although this may just be a measure of the scatter of the data. The CMC collection efficiencies were much lower than those of CMA ash. Although the surface area of the CMC ash increased during burning it was still much less than the area of the CMA ash. Here again the collection efficiency decreased with increasing levels of ash.

The collection efficiency for the two layers of CMA ash is greater than for a single layer. Figure 6.5 shows the collection efficiencies of CMA ash and CMC ash as a function of the gas velocity: the collection efficiency of CMA ash decreases with increasing air velocity. The collection efficiency changed little with ash weight used. From Figure 6.6 we see that CMC ash-laden filter paper had a much lower collection efficiency than the CMA ash-laden filter paper. As with CMA ash-laden filter paper, the collection efficiency decreases with increasing ash weight.

The particle Reynolds number, $D_p\rho u$, was calculated on the mean and CMA ash size, and the collection efficiencies are plotted against in Figure 6.7. The efficiencies are also plotted against the Peclet numbers, $D_p u/D_0$. Both of these are essentially just proportional to the air velocity. Figure 6.7 shows the collection efficiencies of CMA ash-laden filter paper decreasing with increasing Reynolds and Peclet numbers. Essentially, the two graphs show that the collection efficiencies decrease as gas velocity increases.

In Figure 6.8 the collection efficiency of one and two layers of CMA ash-laden filter paper at various gas velocities. Two layers of CMA ash-laden filter paper have, as expected, a higher collection efficiency than one layer. We would expect the top filter to remove most of the larger particles, and so the collection efficiency of the second filter, which would try to catch the remaining smaller particles, would be less. From Figure 6.8(b) and (c), however, we see that in most cases the bottom filter has a higher collection efficiency than the top filter. In some cases the collection efficiency was the same, or that of the top filter was a little higher than that of bottom filter.

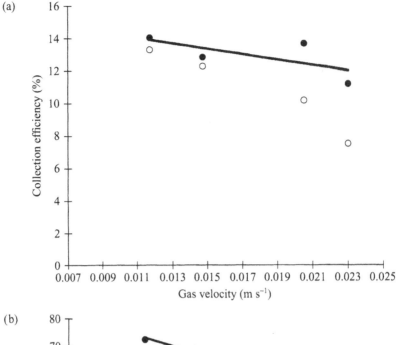

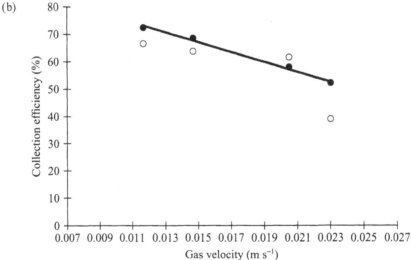

Figure 6.5 (a) Correlation of efficiency of CMA ash (0.6 g; ●) or CMC ash (0.8 g; ○) with gas velocity; (b) Correlation of efficiency of weight of CMA ash (0.6 g; ● or 0.8 g; ○) with gas velocity

We also calculated the Stokes number based on velocity and particle diameter. Figure 6.9 shows that the collection efficiency decreases as Stokes number increases. As the Stokes number here increases only because the air velocity increases, this only means that collection efficiency decreases with increasing velocity.

The pressure drop across the ash-laden filter paper should increase with time because more and more ferrous sulphate particles are captured by the ash. From the experimental data of pressure drop versus time, however, we did not see this. The pressure drop measurement had a large experimental error and could be used only as a relative value. In

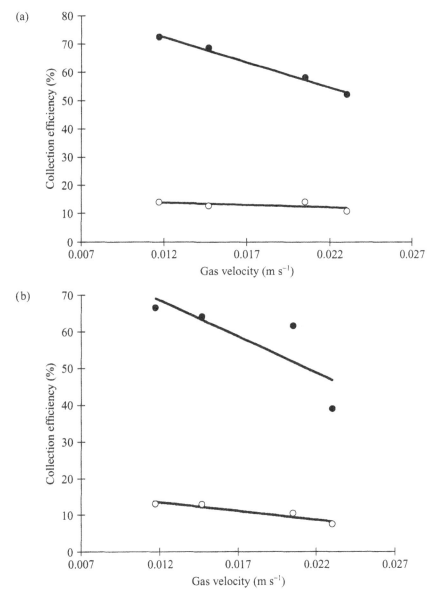

Figure 6.6 Correlation of efficiency of CMA (●) and CMC (○) ash with gas velocity. (a) 0.6 g ash per filter; (b) 0.8 g ash per filter

theory, the pressure drop is a function of the amount of ash, its particle size, and its porosity. The particle size and porosity of the ash may have varied from sample to sample.

6.5 Conclusions and Recommendations

The collection efficiency of calcium magnesium acetate ash was about 70 % for capturing fine particles of an average size of 0.66 μm, at an air velocity of 0.009 m s^{-1}. At higher

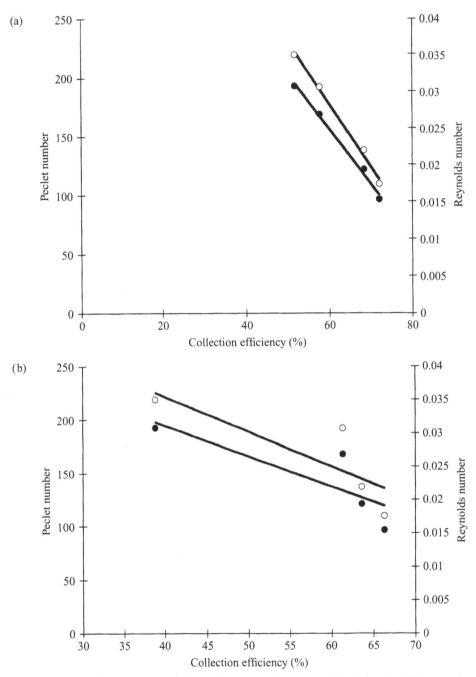

Figure 6.7 Correlation of Reynolds number (●) and Peclet number (○) with efficiency of CMA ash (a) 0.6 g per filter and (b) 0.8 g per filter

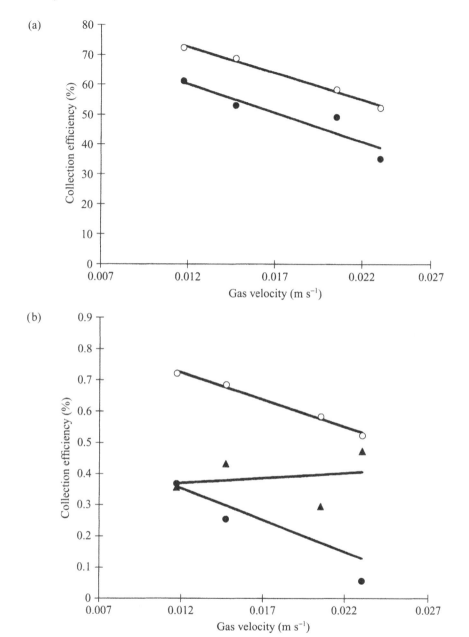

Figure 6.8 (a) efficiency of one (●) or two (○) layers of CMA ash-laden filter paper (0.6 g ash per filter); (b) efficiency of top filter (●), bottom filter (▲) and overall efficiency of two filters (○) containing 0.6 g CMA ash per filter

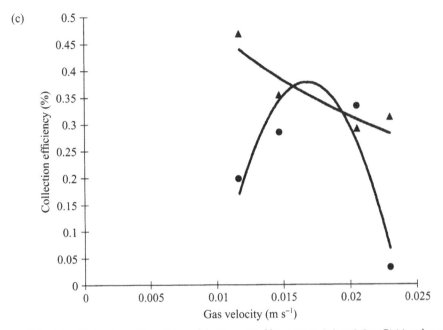

Figure 6.8   (c) efficiencies of top (●) and bottom (●) filters containing 0.8 g CMA ash per filter

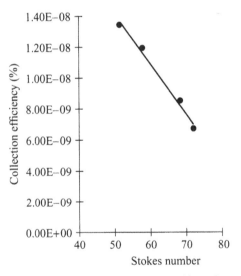

Figure 6.9   Efficiency of CMA ash-laden filter (0.6 g per filter) decreases as Stokes number increases (where $1.00E - 08$ is 1×10^{-8}, etc.)

velocities it was less. The collection efficiency of calcium magnesium carbonate ash was just 14 % at the same air velocity. Calcium magnesium carbonate is not an effective additive for capturing fine particles.

This study shows the potential of calcium magnesium acetate ash in capturing very fine particles. The main parameter changed in this study was the air velocity. Further studies should be done on the other parameters, such as temperature and the particle size, so that a mathematical model may be developed.

References

AKERS, D.J., 1996, Coal Cleaning Controls HAP Emissions, *Power Engineering*, 7, 33–34.

AKERS, D. and HUDYNCIA, M., 1990, Laboratory Guidelines and Procedures: Trace elements in coal, industrial report from the Electric Power Research Institute, *EPRI CS-5644*, 5, Palo Alto, CA, USA.

BISHOP, W.J., WALLIS, M.J. and WITHERPPON, J.R., 1992, Air Emission Control Technology, *Options for POTWs*, Electric Power Research Institute, Palo Alto, CA, USA, pp.51–52.

CADLE, R.D., 1975, *The Measurement of Airborne Particles*, New York: Wiley, 1975.

CALVERT, S., 1984, Particle control by scrubbing, in Calvert, S. and England, H.M. (eds) *Handbook of Air Pollution Technology*, New York: Wiley, p.39.

CHANGE, R. and OFFEN, G., 1995, Mercury Emissions Control Technologies: An EPRI Synopsis, *Power Engineering*, November, Electric Power Research Institute, Palo Alto, CA, USA.

DAVIES, C.N., 1973, *Air Filtration*, London: Academic Press.

DEVITO, M.L., ROSENDALE, J.B. and CONRAD V., 1993, Comparison of Trace Element Contents of Raw and Clean Commercial Coals, presented at the DOE Workshop on Trace Elements in Coal-Fired Power Systems, April, 1993.

FIELD, R.A., WU, D. and MCFARLAND, R., 1990, Evaluation of continuum regime theories for bipolar charging of particles in the 0.3–1.3 μm diameter size range, *IEEE Transactions on Industry Applications*, 12, 523–527.

FLANGAN, R.C., 1979, Submicron Aerosols from Pulverized Coal Combustion, in Seventeenth Symposium on Combustion, The Combustion Institute, Pittsburgh, PA, pp. 97–104.

FLANGAN, R.C. and SEINFELD, J.H., 1988, *Fundamentals of Air Pollution Engineering*, New Jersey, Prentice Hall.

GENNRICH, T., 1993, Filter bags help meet particulate control standards, *Power Engineering*, 13, 37–39.

GOMAA, H. and ALLAWI, A., 1994, Minimize air-toxic emissions, *Environmental Focus*, **August**, 121–122.

GOREN, S.L. *et al.*, 1968, Model for predicting pressure drop and filtration efficiency in fibrous media, *Environmental Science and Technology*, 2, 279–286.

GRADON, L., 1987, Influence of electrostatic interactions and slip effect on aerosol filtration efficiency in fiber filters, *Indust. Eng. Chem. Res.*, 26, 306–311.

HAN, H. and ZIEGLER, E.N., 1984, The performance of high efficiency electrostatic precipitators, *Environment Progress*, 3, 201–206.

HENRY, R.F., PODOLSKI, W.F. and SAXENA, S.C., 1982, Electrostatically Augmented Gas Cleanup Devices for Particulate Removal presented at the IEEE IAS Annual Meeting, San Francisco, CA, 1982.

HIGLEY, J.K. and JOFFE, M.A., 1996, Airborne Molecular Contamination: Cleanroom Control Strategies, Electric Power Research Institute, Palo Alto, CA, USA, pp.211–212, July, 1996.

HINDS, W.C., 1982, *Aerosol Technology: Properties, Behavior, and Measurement of Airborne Particles*, New York: John Wiley & Sons.

HURD, P.K. and MULLINS, J., 1962, Aerosol size distribution from mobility, *Journal of Colloid Science*, 17, 91–100.

HUTCHING, D.K., GREGORY, D.D. and SIMS, R.A., 1984, Simultaneous determination of the diffusion coefficient and electrical mobility of single aerosol particles, *The Institute of Physics*, 875–879.

KAO, A.S., 1994, Formation and removal reactions of hazardous air pollutants, *Air & Waste*, 7, 43–59.

KUDLAC, G.A. and FARTHING, G.A., 1992, SNRB catalytic baghouse laboratory pilot testing, *Environmental Progress*, **11**, 33–43.

LEITH, D. and LICHT, W., 1972, The collection efficiency of cyclone type particle collectors – a new theoretical approach, *AICHE Symposium Series* No. 126, Vol. 68, pp.196–206.

LEVENDIS, Y.A. and WISE, D.L., 1994a, US Patent 5,312,605.

LEVENDIS, Y.A. and WISE, D.L., 1994b, US Patent 5,352,423.

LIU, B.Y.H. and YEH, H.C., 1997, On the theory of charging of aerosol particles in an electric regime, *Journal of Colloid Interface Science*, **58**, 142–149.

LOFTUS, P.J., STICKLER, D.B. and DIEHL, R.C., 1992, A confined vortex scrubber for fine particulate removal from flue gases, *Environmental Progress*, **39**, 27–32.

'Microscopic Killers', *The New York Times*, 12 May 1996.

MURADOV, N.Z., RAISSI, A., MUZZEY, D., PAINTER, C.R. and KEEME, M.R., 1996, Selective photocatalytic destruction of airborne VOCs, *Solar Energy*, **56**, 445–449.

OH, S.H., MacDONALD, J.S., VANEMAN, G.L. and HEGEDUS, L.L., 1981, Mathematical modeling of fibrous filters for diesel particulate-theory and experiment, SAE Technical, **13**, 23–27.

PALMER, H.B. and CULLIS, C.E., 1965, The formation of carbon from gases, in Walker, P.I. (ed.) *Chemistry and Physics of Carbon*, Vol. 1, New York: Marcel Dekker, 265–325.

PERRY, R.H. *et al.*, 1984, *Perry's Chemical Engineering Handbook*, 6th ed., New York: McGraw-Hill.

RAABE, O.G., 1976, The generation of aerosols of fine particles, in Liu, B.Y.H. (ed.) *Fine Particles*, New York: Academic Press, pp.37–53.

SENIOR, C.L. and FLANGAN, R.C., 1982, Ash vaporization and condensation during the combustion on a suspended coal particles, *Aerosol Science and Technology*, **1**, 371–383.

SHUCKEROW, J.I., 1995, M.Sc. Thesis, Northeastern University, Boston, MA.

SIMONS, G.A.,1992, Predictions of CMA utilization for *in-situ* SO_2 removal in utility boilers, *Resources, Conservation and Recycling*, **23**, 161–168.

SINCLAIR, D.C., COUNTESS, R.J., LIU, B.Y.H. and PUI, H.L., 1979, *DYH in Aerosol Measurement*, Gainesville: University of Florida.

SINGH, S., 1993, Pulse-jet dust collectors, *Chemical Engineering*, **2**, 125–127.

STECIAK, J., LEVENDIS, Y.A., WISE, D.L. and SIMONS, G.A., 1995, Dual SO_2–NO_x concentration reduction by calcium salts of carboxylic Acids, *Journal of Environmental Engineering*, **12**, 595–604.

TARDOS, G.I., SNADDON, R.W.L., 1984, Electrical charge measurements on fine airborne particles, *IEE Transactions on Industry Applications*, November/December, **17**, 1578–1582.

YEH, H.C. and LIU, B.Y.H., 1974, Aerosol filtration by fibrous filter, *Journal of Aerosol Science*, **5**, 191–204.

ZUMWALDE, R.D. and DEMENT, J.M., 1977, *Review and Evaluation of Analytical Membrane Filter Method for Evaluating Airborne Asbestos Fibers*, DHEW (NIOSH) Pub. No. 77–204.

PART TWO

Recycling and Treatment of Organic Wastes

7

Duckweed-Based Wastewater Treatment for Rational Resource Recovery and Reuse

HUUB J. GIJZEN AND SIEMEN VEENSTRA

7.1 Introduction

7.1.1 *World-wide Need for Low-cost Wastewater Treatment*

Rapid industrialization and urbanization over the past decades have generated increasing amounts of wastewater, resulting in environmental deterioration and pressure on reliable water sources in many countries. Treatment of these waste flows prior to disposal, therefore, is of urgent concern worldwide. In countries with a high gross national product (GNP) substantial investments were made over a period of several decades in high-cost sewerage and centralized treatment facilities. These investments have contributed to a reduction of the waste load to receiving water bodies. However, few of the treatment systems installed have a capacity to remove nitrogen and phosphorus from the effluent. This, together with the increased use of fertilizer over the past 30 years, is frequently causing eutrophication of surface water and groundwater contamination. The situation in countries with a low GNP is even more pressing. Developing nations house the largest part of the world population, take a 90 % share of the world-wide population growth, and show strong urbanization patterns. There is at present hardly any infrastructure for the effective treatment of wastewaters in these countries. Most developing countries have only recently initiated the development of environmental legislation regarding effluent standards and it is expected that the effective implementation of such legislation will need substantial efforts and investments in the coming decades.

Although industrialized nations may have the economical capacity to deal with environmental problems via high-cost technologies, these may not provide an immediate solution for countries with a low GNP. Briscoe (1993) stated that due to the huge cost associated with centralized urban sewerage in Latin America just 2 % of the sewage flows are at present properly treated. Grau (1994) calculated that the period of time required to meet EU effluent standards in several central eastern European countries exceeds by far the economic lifespan of treatment plants and sometimes even that of sewers. The challenge ahead, therefore, is to develop reliable and appropriate technology that is within the economical and technological capabilities of developing countries. The

Table 7.1　Nitrogen and phosphorus uptake by floating macrophytes

Location	Macrophyte	Daily uptake ($g\ m^{-2}$ per day)		Reference
		N	P	
Florida, USA	Water hyacinth	1.30 (0.25)*	0.24 (0.05)	Reddy and DeBusk, 1987
Florida, USA	Water lettuce	0.99 (0.26)	0.22 (0.07)	Reddy and DeBusk, 1987
Florida, USA	Pennywort	0.37 (0.37)	0.09 (0.08)	Reddy and DeBusk, 1987
USA	*Lemna sp.*	1.67	0.22	Zirschky and Reed, 1988
India	*Lemna sp.*	0.50–0.59	0.14–0.30	Tripathi *et al.*, 1991
Louisiana, USA	Duckweed	0.47	0.16	Culley and Meyers, 1980
Bangladesh	*Spirodela polyrrhiza*	0.26	0.05	Alaerts *et al.*, 1996

* Values in parentheses were obtained during winter season

non-availability of such technology has in fact been one of the reasons for the limited application of wastewater treatment in many urban and rural situations.

This chapter will discuss the role of duckweed-based wastewater treatment in an integrated concept of recycling and re-use of valuable waste components. The focus on duckweed as a key step in waste recycling is due to the fact that it forms the central unit of a recycling engine driven by photosynthesis and therefore the process is sustainable, energy efficient, cost efficient and applicable under a wide variety of rural and urban conditions.

7.1.2　*The Use of Aquatic Macrophytes*

The objective of wastewater treatment is to remove or convert contaminants that are considered detrimental to human health or to the environment. Historically, attention has focused on removal of suspended solids, organic matter and pathogens, but more recently also toxic compounds and nutrient removal are considered to be important issues. Aquatic macrophytes have been studied for their use as effective scavengers of nutrients from wastewaters. Their use has been suggested as a low-cost option for the purification of wastewater and simultaneous production of plant biomass (Reddy and DeBusk, 1985, 1987; Araujo, 1987; Brix and Schierup, 1989; Brix, 1991; Buddhavarapu and Hancock, 1991; Skillicorn *et al.*, 1993; Oron, 1994; Gijzen, 1996). When applying aquatic plants in shallow ponds, a combination of secondary and tertiary treatment may be realized. In addition, aquatic macrophytes assimilate nutrients into a high-quality biomass that may have an economic value. This contrasts with advanced costly nitrification–denitrification, where nitrogen is converted into atmospheric N_2 and therefore is 'lost' for further reuse. This is rather inefficient and a technology based on nutrient reuse seems to be a more rewarding option from both an energy and resource efficiency point of view.

Various studies have reported the use of water hyacinth (*Eichhornia crassipes*), pennywort (*Hydrocotyle umbellata*), water lettuce (*Pistia stratiotes*) and duckweed (*Lemnaceae*) for the efficient removal of nutrients. The economic potential of each plant species for wastewater treatment depends largely on its efficiency to remove nutrients under a wide range of environmental conditions, its growth and maintenance in a treatment system, and the possible application of plant biomass. Water hyacinth has been used most widely, due to its high nutrient uptake capability (Table 7.1), but no economically attractive application of the generated plant biomass has been identified so far.

7.2   Characteristics of Duckweeds

7.2.1   *What is Duckweed?*

Duckweeds, or *Lemnaceae*, are small floating aquatic plants, which can be found worldwide under widely varying climatic and environmental conditions. The family consists of four genera (*Wolffiella, Wolffia, Lemna, Spirodela*) and at least 37 species have been identified. The plant morphology is simple and structural components are limited to short roots and a frond with a size of only a few millimetres for most species (Landolt and Kandeler, 1987). Parameters affecting duckweed growth in natural environment include temperature, light intensity, other climatic conditions, pH and availability of nutrients in the water. These factors will also be important for duckweed growth on sewage, but in this case also the potential of growth inhibition due to high concentrations of sulphides and ammonia need to be considered (Veenstra *et al.*, 1995). Optimum values for the above-mentioned climatic and environmental parameters will differ not only between species but most probably also between varieties of the same species. It is therefore important to use locally available duckweed varieties, since these may prove to give better performance under local conditions.

7.2.2   *Growth Rate and Composition*

Since structural components like stems and leaves are missing, most of the tissue of duckweed plants is actively involved in photosynthesis. It is not surprising, therefore, that *Lemnaceae* are considered to be among the most vigorously growing plant species on Earth. Duckweed yields reported in literature vary significantly. Low productivity, of 11 t dry matter (DM)/ha/year, was observed by Edwards (1987) when growing *Lemna perpusilla* on septic tank effluent, while values as high as 55 t DM/ha/year were reported by Oron *et al.* (1986) for the growth of *Spirodela polyrrhiza* on domestic wastewater. Highest production rates are usually observed in trials at laboratory scale or small pilot scale. Average biomass yields in full scale ponds will probably be in the range of 15-30 t DM/ha/year.

   Protein content of duckweed depends largely on growth rate and growth conditions. Culley and Epps (1973) reported protein content of 14–26 % in duckweed growing on natural water bodies, whereas 29–41 % protein was observed in duckweed cultivated in waste stabilization ponds. Duckweed has an enormous potential in protein production, far higher than that of any other crop (Table 7.2). The relatively high protein content of duckweed is generating interest in animal feed applications of this aquatic plant. *Wolffia arrhiza* is even used for human nutrition in Burma, Laos and northern Thailand. Annual yields of about 2000 kg ha^{-1} have been obtained from very simple cultivation methods in ponds without fertilizer being applied.

Table 7.2 Annual productivity of duckweed and selected crops in terms of dry matter and protein

Plant/crop	Production (t DM/ha/year)	Crude protein (% dry matter)	Protein production (kg/ha/year)
Duckweed	17.60	37	6510
Soybean	1.59	41.7	660
Cottonseed	0.76	24.9	190
Peanuts	1.60–3.12	23.6	380–740
Alfalfa hay	4.37–15.69	15.9–17.0	690–2670

Adapted from Hillman and Culley, 1978

7.3 Duckweed and Domestic Wastewater Treatment

7.3.1 *Advantages of Duckweed*

The use of duckweed for the efficient and low-cost treatment of domestic wastewaters in rural or urban contexts is recommended by various authors (Reddy and DeBusk, 1987; Zirschky and Reed, 1988; Skillicorn *et al.*, 1993; Oron, 1994; Alaerts *et al.*, 1996; Gijzen, 1997) because of a number of advantages this technology offers over conventional treatment technology:

- Duckweed grows rapidly and shows a high nutrient uptake rate when exposed to sewage. Duckweed growth is less sensitive to low sewage temperature high nutrient levels, pH fluctuations, and pest and diseases than most other aquatic plants.
- Duckweed can tolerate the rather high concentrations of detergent present in domestic wastewater.
- Duckweed is capable of absorbing, accumulating and/or disintegrating a wide variety of substances, which are not easily biodegraded in conventional wastewater treatment plants (Landolt and Kandeler, 1987). Accumulation of heavy metals may be disadvantageous when aiming at feed applications of duckweed biomass.
- Because of its high protein and relatively low fibre content duckweed is highly nutritious and digestible for a wide variety of animals including pigs, ruminants, poultry and fish.
- Harvesting of duckweed plants from the water surface is far less complicated than the harvesting of other macrophytes, which are commonly interconnected over large distances.
- A complete duckweed cover on the wastewater may effectively prevent the development of algae in the water body and provides quiescent conditions, which contribute to a more clear effluent with lower total suspended solids (TSS).
- The presence of a duckweed cover has been reported to decrease the development of mosquitoes in the water body because it prevents mosquito breeding.
- Duckweed positively affects the reduction of the release of odorous compounds.
- In (semi-)arid regions of the world, which are increasingly confronted with water shortage, it is of great relevance to know that water losses due to evapo(transpi-)ration rates can be reduced when covering a water body with duckweed.

Table 7.3 Duckweed production and protein content

Species	Cultivation	Production (DM/ha/year)	Protein (% DM)	References
S. polyrrhiza	Domestic sewage	17–32	–	Alaerts *et al.* (1996)
L. minor	UASB – effluent	10.7	28.9	Weller and Vroon (1995)
L. gibba	Pretreated sewage	55	30	Oron (1994)
L. gibba	Domestic sewage	10.9–54.8	30–40	Oron *et al.* (1986)
S. polyrrhiza, *L. perpusilla* and *W. arrhiza*	Septage from septic tank	9.2–21.4	24–28	Edwards *et al.* (1992)
Lemna spp.	Domestic sewage	27	37	Zirschky and Reed (1988)
S. polyrrhiza	Domestic sewage	17.6–31.5	30	Gijzen (1996)

These advantages have triggered duckweed research and application. Several full-scale applications of duckweed-based wastewater treatment systems exist, for example in the USA, Bangladesh and China (Zirschky and Reed, 1988). Duckweed systems have been studied for dairy waste lagoons (Culley *et al.*, 1981), domestic sewage (Skillicorn *et al.*, 1993; Oron, 1994; Alaerts *et al.*, 1996), secondary effluent (Harvey and Fox, 1973; Sutton and Ornes, 1977), waste stabilization pond effluents (Wolverton, 1979) and fish culture systems (Rakocy and Allison, 1981; Porath and Pollock, 1982). Even in moderate climatic conditions duckweed-based systems are being advocated for treatment of domestic wastewater from small communities in Belgium and Poland (Nielsen and Ngo, 1995). In Siberia the USA-based Lemna Corporation is currently installing a system with an estimated cost of US$50 million, whereas a $30 million agreement was recently signed with China.

7.3.2 *Major Treatment Parameters Affected by Duckweed*

Aquatic plants used in wastewater treatment are generally considered to not contribute directly to biodegradation processes. Their role is more of a supportive nature, providing surface area for attached biofilms, providing oxygen for aerobic metabolism, and stimulating quiescent conditions in the water phase which enhance settling processes. With more information becoming available, the supporting role of macrophytes and their contribution to overall treatment efficiency is increasingly being recognized.

Nutrients

Lemna species show high nutrient uptake rates in domestic sewage, characterized by abundance in the macronutrients nitrogen, phosphorus and potassium (N, P and K). Typical daily nitrogen and phosphorus uptake rates for duckweed are 0.4 g N m^{-2} and 0.03 g P m^{-2} (Table 7.3). Culley *et al.* (1981) calculated daily nutrient uptake rates per m^2 by *Lemnaceae* to amount to 0.4 g N, 0.1 g P, and 0.1 g K. The results from different studies are not comparable, because different species have been used and different climatic and operational conditions applied.

Daily nitrogen removal via microbiological nitrification–denitrification rates may be in excess of 1 g m^{-2}. Culley *et al.* (1978) demonstrated a 20–40 % lower nitrogen content in pond effluents covered with *Lemnaceae* than in algal-based pond effluents without aquatic plants.

Pathogen removal

Pathogen removal from duckweed-covered lagoons treating domestic wastewater is important in order produce an effluent quality that allows for further reuse in agriculture and/or aquaculture. Therefore the effluent guidelines of < 1000 coliforms/100 ml and helminth eggs < 1/l (Mara and Cairncross, 1989) should be met to ensure safe effluent re-use. Other health-related concerns are related to the manual harvesting of duckweed and its application as animal feed. These concerns need to be studied and evaluated carefully because they may directly affect the feasibility of duckweed application to domestic wastewater treatment.

Islam *et al.* (1996) monitored faecal coliform levels in a plug-flow duckweed lagoon and found a reduction from 4.5×10^4 per 100 ml in the influent to values below 100 per 100 ml in the effluent. These results are very positive, but do not provide information on the relative contribution of duckweed on coliform removal or survival. Dewedar and Bahgat (1995) reported that the decay rate observed for faecal coliforms exposed to direct sunlight was 0.177 h^{-1}, whereas faecal coliform numbers under a dense cover of *Lemna gibba* did not decline over a period of 5 days. These results suggest that elimination of exposure to direct sunlight by a duckweed cover may result in extended survival for faecal coliforms in treatment ponds. In addition, Culley and Epps (1973) showed that the presence of a duckweed cover reduces the dissolved oxygen level in the water phase and thus diminishes the disinfection effects of free oxygen radicals produced in the water phase.

Efficient removal of coliforms has been reported for conventional algal-based pond systems. Direct effects of sunlight penetration, combined with indirect stimulation of algal growth resulting in fluctuations of pH and dissolved oxygen, stimulate coliform reduction. Van Haandel and Catunda (1997) demonstrated a positive correlation between pond depth and decay rate for *Escherichia coli*, stressing the effect of sunlight penetration into the water phase. With deeper ponds and presence of a duckweed cover disinfection via solar radiation will decrease and coliform die-off rates will be reduced. Pathogenic parasites usually settle in ponds. Duckweed might enhance this process by improving the hydraulic stability of the water column.

BOD removal

The role of macrophytes in biochemical oxygen demand (BOD) removal is not fully understood and probably varies between different species. The introduction of duckweed to a pond limits the free diffusion of oxygen from the air into the water phase. A compensating effect may be the diffusion of produced oxygen through the duckweed roots. Oxygen will induce aerobic mineralization of organic compounds by the heterotrophic microbial biomass adhering to the plant surface. In this concept, ponds with a dense cover of aquatic plants could be considered as fixed film reactors (Stowell *et al.*, 1981; Tchobanoglous, 1987).

Hillman (1961) reported that duckweed may contribute to the removal of simple organic compounds via direct uptake. Körner *et al.* (1998) could not confirm this, but suggested that aerobic conditions at micro-level in the root zone might induce high BOD conversion. Zirschky and Reed (1988) reported that duckweed supports attached biofilm development to the roots, but its quantitative contribution to overall BOD removal is not clear. Alaerts *et al.* (1996) reported BOD removal efficiencies of 95–99 % in a 0.7 ha sewage lagoon which was operated with a cover of *S. polyrrhiza* at a hydraulic retention time of about 20 days. Daily BOD loading rate in the lagoon varied between 48 and 60 kg ha^{-1}. DeBusk and Reddy (1987) observed BOD removal rates as high as 300–400 kg/ha/day under very high BOD loading conditions in treatment systems with water hyacinth. However, such highly loaded systems usually produce an effluent BOD level of over 60 mg^{-1}.

Rather than looking for ways to enhance BOD removal in duckweed ponds it is much more rewarding to consider the use of anaerobic technology before duckweed ponds. Via anaerobic treatment technology the organic matter is reduced by 65–85 %. The resulting effluent is effectively stripped from TSS and BOD, but has conserved all other contaminants such as nitrogen, phosphorus, and faecal coliforms. The application of post treatment by macrophyte-based ponds seems therefore most logical and feasible (van Haandel and Catunda, 1997; Gijzen, 1997).

TSS removal

The removal of TSS in waste stabilization ponds is mainly determined by the rate of organic particle biodegradation, particle settling rates, and the potential production of algal-related TSS. In particular, in developing countries high concentrations of TSS are encountered in the raw sewage, which might enhance flocculent settling and contribute to substantial BOD and TSS removal in anaerobic ponds or reactors (Veenstra *et al.*, 1995). The development of algae results in an increase of effluent TSS to over 200 mg l^{-1} (Arceivala, 1986). Since algal growth will be largely suppressed in duckweed-covered ponds TSS removal will be efficient and allow effluent distribution to agriculture via drip irrigation systems without rapid emitter fouling (Taylor *et al.*, 1995).

Removal of heavy metals

Heavy metal removal mechanisms in macrophyte-based treatment systems include plant uptake, chemical precipitation and adsorption. *Lemnaceae* show a fair degree of tolerance to heavy metals and posses a great ability to accumulate them. Landholt and Kandeler (1987) suggested that duckweed could effectively be used to remove metals from wastewater flows. Gaur *et al.* (1994) found that the accumulation of cadmium, chromium, cobalt, copper, nickel, lead and zinc by *S. polyrrhiza* and *Azolla pinnata* was related to the concentration of metals applied.

The accumulation factor for heavy metals in *Lemnaceae* depends largely on its concentration, the presence of other metals, and the duckweed species (Landolt and Kandeler, 1987). For lead and cadmium accumulation factors of over 1000 have been reported. *Lemna valdiviana* accumulates copper by a factor of 500–54 000, depending on concentration offered, whereas *Lemna minor* showed accumulation factors from 80–8000. Extremely high accumulation factors of 660 000 and 850 000 were reported for aluminium and manganese, respectively. Gellini and Piccardi (1981) reported that *L. minor* causes a 75 % reduction in copper content of water containing 5 mg l^{-1} within 48 h. This suggests that duckweed can be effectively used for heavy metal stripping. However, the duckweed produced under these conditions should be considered as a chemical waste, and not be used in any feed application.

7.3.3 *Duckweed Ponds in Combination with Anaerobic Technology*

The application of anaerobic technology results in a substantial reduction of organic matter and suspended solids from wastewater. In cases where sewage is concentrated and has temperatures above 15°C, anaerobic technology can be most rewarding. The technology does not require complicated and costly mechanical equipment or intensive energy input. On the contrary, recovery and valorization of the produced high caloric methane gas may positively affect the overall economy of the process. Anaerobic technology shows very poor nitrogen and phosphorus removal rates, whereas typical removal rates for coliforms via attachment to and settling of suspended solids is only some 50–90 %. Helminth eggs are more susceptible to removal by filtration, entrapment and settling within the anaerobic reactor. As duckweed ponds focus on the reduction of pathogens and dissolved nitrogen and phosphorus compounds, a combination with anaerobic pretreatment seems to be most effective (van Haandel and Catunda, 1997; Gijzen, 1996, 1997). Advantages of such a combination are:

- The design of duckweed ponds can entirely be based on nutrient removal and pathogen reduction rather than on the biodegradation of organic matter and the subsequent transfer of substantial amounts of oxygen per m^2 per day. This may lead to a significant reduction of HRT and pond area required to achieve effective treatment results.

- Accelerated nutrient uptake rates can be achieved if substantial chemical oxygen demand (COD) reduction is realized in the anaerobic reactor prior to the duckweed pond (Oron *et al.*, 1987).

- Anaerobic pretreatment contributes to the liquefaction of suspended organic particles. In this process organically bound nutrients will be mineralized and will become available for duckweed growth.

- Anaerobic technology as well as duckweed pond technology can be easily scaled down to small capacity and can thus be applied to decentralized sewage treatment in urban areas in developing countries, or to small communities world-wide.

In particular for highly concentrated wastewaters from the agro-industrial sector anaerobic treatment is economically feasible (van Haandel and Lettinga, 1994). Breweries, slaughterhouses, dairies and other food industries already make use of this technology in order to treat their effluents. As long as no strict criteria are imposed on effluents via legislation there might be no driving force to add post-treatment technology to the anaerobic reactors. However, it could be an economic incentive to initiate the application of duckweed pond technology to these effluents on a commercial basis in those areas where duckweed has economic potential. In the rural setting it will be interesting to check whether duckweed technology can successfully be applied to anaerobically digested manure and thus contribute to the recycling of energy and nutrients.

7.3.4 *Other Sources of Waste*

Food processing industry

Almost no information is available regarding the treatment of specific industrial wastewater in duckweed ponds. Duckweed may be successfully applied for the treatment of non-domestic wastewater with a high nutrient content such as effluents from food-processing businesses, slaughterhouses and chemical industries involved in the production of nitrogen and phosphorus compounds. Wastewaters with extremely high nitrogen levels in combination with high pH may be toxic to duckweed (Wang, 1991). Generally no inhibitory effects are observed at concentrations up to 50 ppm ammonium at pH levels below 8. The sensitivity to ammonium may differ for different duckweed species and may be decreased by gradual adaptation to higher ammonium concentrations.

Algal culture effluent

Koles *et al.* (1987) demonstrated that duckweed can be used for the efficient removal of ammonia, phosphorus and TSS in the effluent from an algae culture unit. This application is unique, since the effluent has a high pH and contains high concentrations of phosphorus and ammonium ions. The authors reported an average duckweed production of 87 g fresh weight m^{-2}/day, with a maximum value of 280 g m^{-2}/day (about 50 t DM ha^{-1}/year). Another possible application of duckweed has been reported for the treatment and recycling of water from intensive fish cultivation (Porath and Pollock, 1982).

Xenobiotics

The removal of various xenobiotic compounds such as phenol, naphthalene and chlorodibromomethane from wastewater by a water hyacinth-covered system in a pilot-scale study has been demonstrated by Conn and Langworthy (1984). Federle and Schwab (1989) reported the efficient biodegradation of alcohol ethoxylate and mixed amino acids by the microbiota associated with the duckweed *L. minor*. The results of these studies indicate that macrophytes do accelerate the biodegradation of xenobiotics, but the mechanisms involved need to be further studied.

Manure

In the rural context, other sources of waste materials can be considered and their treatment in duckweed ponds could contribute to the recycling of valuable components (Rodriquez and Preston, 1996). In principle, any organic material that is readily biodegradable and has a high nutrient content could be treated in duckweed ponds. Examples of such waste materials include all kinds of animal manure, latrine wastes, effluents from biogas plants, kitchen waste, composted agricultural wastes and market wastes. Treatment of latrine septage in a duckweed system could provide basic sanitation and simultaneously stimulate nutrient recycling in the rural areas of low GNP countries. At the same time, the duckweed ponds could be fed with other waste streams, such as digested manure and composted agricultural and domestic wastes.

7.4 Integrated Concepts

7.4.1 *Integrated Systems for the Rural and Urban Environment*

Integrated systems have been defined as a combination of processes and practices where optimum use of resources is achieved via waste recycling aimed at the recovery and reuse of energy, nutrients and possibly other components. The conversion processes for different sources of waste are arranged in such a way that a minimum input of external

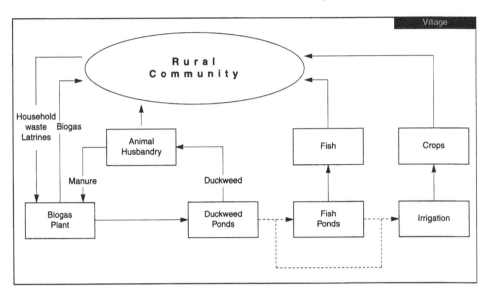

Figure 7.1   Community-based rural integrated waste recycling system using duckweed ponds

energy and raw materials is required and maximum self-sufficiency is achieved. In rural Asia, integrated systems form an old concept that has been applied for hundreds or even thousands of years. In China there are huge farms, which are almost completely self-sufficient in terms of energy and nutrients because of effective recycling of their waste streams. The application of integrated concepts provides a good balance between re-source utilization, reuse and environmental protection. Such balance is completely absent in densely populated urban areas, with high consumption rates and where concentrated waste production is sustained by the large-scale importation of energy and nutrients into the urban environment. The absence of reuse processes for energy, nutrients and other valuable components in urban waste has generated serious environmental and public health concerns. It is suggested therefore that urgent attention be given to the develop-ment of rational reuse strategies in the urban context. This approach, in combination with waste minimization schemes, will provide a more sustainable solution than the present approach of costly end of pipe treatment. The challenge for the coming years therefore will be to develop integrated concepts and processes for the minimization, recovery and reuse of waste materials in both rural and urban environments in high and low GNP countries. If effective programmes and action plans could be defined by relevant organ-izations, such as the World Bank, United Nations, national universities and research centres, this challenge could be met within a reasonable time with the help of modern science and technology.

Duckweed ponds could provide an important role in recycling and reuse schemes in both rural and urban areas. The process steps and products of an integrated duckweed-based treatment system for rural and urban recycling of waste streams are presented in Figures 7.1 and 7.2 respectively. Initially anaerobic technology is advocated to reduce the bulk of organic and suspended matter. The energy produced, either in rural biogas digesters or urban high-rate reactors (e.g. upflow anaerobic sludge blanket; UASB), can be used by the community (rural context) or for the operation of subsequent treatment steps (urban application), thereby reducing treatment costs. Various full-scale anaerobic

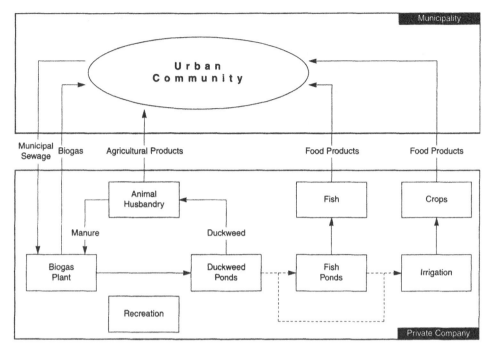

Figure 7.2   Urban integrated waste recycling system using duckweed ponds

reactors have been installed to treat domestic wastewater in India, Colombia, and Brazil (Draaijer *et al.*, 1992).

7.4.2   Production and Uses of Duckweed

Duckweed production and nutritive value

Various authors recommend the application of duckweed biomass as a source of high-quality protein in animal nutrition (Hillman and Culley, 1978; Stephenson *et al.*, 1980; Culley *et al.*, 1981; Edwards, 1990). In Section 7.3 the advantages that make duckweed particularly attractive as an economic biomass resource were listed. Besides these positive characteristics, there are also a number of negative factors that need to be studied when considering duckweed as a potential resource:

• Owing to the efficient absorption of heavy metals and possibly other toxic compounds, duckweed should be cultivated using effluents containing low concentrations of such compounds. No information is available on the possible transmission of pathogens if duckweed harvested from domestic wastewater treatment ponds is used as an animal feed.

• Duckweed has a relatively high moisture content (about 95 %), which will affect the cost of handling, transportation and drying. This characteristic will be less important in an integrated system, where duckweed is used on site.

• The genera *Lemna* and *Spirodela* may contain high amounts of calcium oxalate. The presence of this component may limit the use of certain duckweed species for

Table 7.4 Feed conversion ratios for duckweed to fish

Duckweed species	Fish species	FCR	Reference
Lemna	Tilapia	1.6–3.3	Hassan and Edwards, 1992
Unknown	Grass carp 3 g	1.6	Shireman *et al.*, 1978
Unknown	Grass carp 63 g	2.7	Shireman *et al.*, 1978
Unknown	Unknown	1.6–4.1	Baur and Buck, 1980
Unknown	Unknown	3.1	Hajra and Tripathy, 1985
Spirodela	Polyculture	1.2–3.3	PRISM, Bangladesh

non-ruminant and human nutrition. When hit is properly mixed with other feed, constituents no harmful effects are expected.

- Production of the crop is limited to areas with a minimum of 20 cm of standing water throughout the year or during a major part of the year.

Biomass productivity and nutritive value are important factors when considering duckweed as an animal feed. The nutritive value largely depends on the type of animal, digestibility of major plant components, and protein content. Frye and Culley (1981) reported a lower limit of 20–30 mg nitrogen per litre in the medium in order to achieve high protein content in duckweed biomass.

Duckweed as an animal feed

Most feeding trials with duckweed have been done with fish cultivation. The *Lemnaceae* appear to be suitable for a wide range of fish species and may also be applied in polycultures of fish. Duckweed may be fed fresh as a sole feed or in combination with other feed components. In addition, can be dried and offered in pellets. Intensive fish production with duckweed as a predominant feed constituent has been reported by a number of authors (van Dyke and Sutton, 1977; Robinette *et al.*, 1980; Gaigher *et al.*, 1984; several references cited in Landolt and Kandeler, 1987; Hassan and Edwards, 1992). Promising results have been obtained with the grass carp *Ctenopharyngodon idella* (Porath and Koton, 1977; Shireman *et al.*, 1977, 1978; Baur and Buck, 1980; Edwards, 1980; Hajra and Tripathy, 1985). The feed conversion ratio (FCR = g DM duckweed per gram fish fresh weight) is lower in young grass carps than in mature fish (Table 7.4). The reported FCR values for duckweed compare very favourably with FCR values reported for other fish feeds such as catfish chow or rye grass (Shireman *et al.*, 1978).

PRISM, a local NGO in Bangladesh, has experimented with duckweed-based wastewater treatment and fish production for more than 9 years. The ponds yield over 12 tons ha^{-1}/ year of fish, making the overall treatment process profitable with an estimated net annual revenue of US\$3000 ha^{-1} (Gijzen, 1997). Excellent results have been obtained for both duckweed and polyculture fish production (Skillicorn *et al.*, 1993). *Lemnaceae* may also be used for shrimp cultivation (Heckmann *et al.*, 1984) or to grow zooplankton, which subsequently is used by carnivore fish species.

For higher animals (ruminants, pigs and chicken) preliminary trials show that small portions of duckweed in the diet stimulate their growth, whereas high proportions of duckweed tend to decrease weight gain or create digestion problems (Leng *et al.*, 1995).

7.4.3 *Effluent Reuse*

In view of the rapidly growing shortage of renewable water resources in many parts of the world there is a growing interest in the use of treated effluents from wastewater treatment plants. These effluents not only serve as water source but also provide nutritious input for fish and crops in aquaculture or agricultural irrigation schemes. In most developing countries the agricultural sector is by far the largest consumer of water resources (typically over 90 %) compared to the domestic and industrial sector. In many countries the conventional 'supply-driven' water consumption has already led to uncontrolled abstraction and substantial groundwater depletion in many arid and semi-arid areas. Reuse of wastewater effluents could definitely provide some relief on the use of scarce water resources. Duckweed ponds could condition the sewage effluent and make it fit for reuse. When looking at the water quality criteria required for reuse of wastewater in agriculture (Pescod, 1992) duckweed ponds definitely have good potential for effluent cleansing.

The effluent of duckweed ponds may be useful in drip irrigation schemes applied in arid regions of the world as the duckweed cover helps to minimize water losses, reduces the salinity of the effluent by the uptake of dissolved ionic compounds, and prevents the occurrence of algal blooms. Emitter fouling can thus be minimized as the amount of suspended solids in the irrigation water will be low (Taylor *et al.*, 1995).

There seems to be a contradiction when promoting the use of sewage nutrients for duckweed growth while claiming at the same time that the final effluent can supply nutrients for crop production. Domestic sewage in developing countries is characterized by high nutrient concentration. In order to meet the crop water and nutrient requirements the concentration of nutrients in the effluent needs to be carefully matched to the water demand, which frequently urges the need for effluent dilution with fresh water. A better option is to partially reduce the nutrient concentration before agricultural application. Another reason for applying duckweed ponds in integrated waste recycling systems is that excessively high nitrogen levels in effluents might be toxic to crops or fish, or create substantial groundwater pollution when recharged into the soil in an uncontrolled way.

7.4.4 *Recreation*

A major limitation for the application of duckweed technology in urban centres is the non-availability of space at a reasonable cost. However, outside the city, land cost will be substantially lower and duckweed ponds may be feasible. Anaerobic treatment facilities may be planned at convenient locations in or near the city, requiring only limited space. The effluent of the anaerobic reactors can be channelled away from the city to duckweed pond facilities. The duckweed harvested at regular intervals can be used to cultivate fish in adjacent ponds, while the effluent can be made available for irrigation. The space requirement for the duckweed pond is estimated at 0.5–1 m^2 per capita in (sub-)tropical climatic regions. Although this is significantly lower than for conventional stabilization ponds, considerable land area will be required.

Owing to the rapid growth of cities, sufficient land area has to be reserved for recreational purposes. In mega-cities with over 1 million inhabitants large-scale recreational facilities can only be planned outside the city at convenient locations. In recognition of the fact that the green duckweed-covered ponds provide a pleasant ecological appearance, one might think of a multifunctional use of space, combining wastewater management with a recreational destination. The possible production of unpleasant odours from the wastewater is effectively reduced by the presence of a duckweed cover on the pond surface. This multifunctional use of the available space may greatly enhance the economical feasibility of the entire system. Together with the income from the products generated (energy, fish, irrigation water), the proposed integrated system has the potential to become a commercial enterprise generating substantial revenues.

7.5 Conclusions

- Conventional treatment technology does not provide a sustainable solution for wastewater management in countries with relatively low GNP. The objective of developing sustainable solutions for waste treatment could be achieved if waste treatment is considered in an integrated concept together with cleaner production and reuse.

- We can learn from the processes and efficiencies of natural systems how efficiently biomass is constantly produced and recycled. Such natural processes have been optimized over millions of years of evolution and it would therefore be wise to simulate and stimulate these processes when developing waste treatment and reuse technology. The duckweed-based integrated concept as presented in this chapter is basically a simulation of natural processes and systems. During the development and testing of such technology the best results can be expected if (sanitary) engineers work closely together with ecologists, biologists and chemists in an interdisciplinary approach.

- The combination of duckweed-based wastewater treatment with cultivation of herbivorous fish could be a most rewarding low-cost technology for waste recycling that can be applied to both rural and urban waste and wastewater. In the urban context, such a process could be combined further with a recreational function, which will positively affect its economical feasibility. An integrated approach has the potential to generate net profit from the treatment of waste. The implementation of reuse-oriented treatment systems therefore could be taken up by self-sufficient and even profitable enterprises. The operation of the integrated system will also generate employment, which is a high priority in most countries.

- Anaerobic pretreatment of waste and wastewater may have several important advantages, including production of valuable energy and reduction of land area required for subsequent treatment steps. In combination with duckweed-covered ponds, anaerobic technology further reduces the chances of production of bad odours and therefore offers the possibility of combining the treatment function with other functions (recreation, fish production etc.).

- The technical and economical feasibility of selected integrated treatment systems should be demonstrated via pilot and full-scale demonstration projects at various locations under different conditions of wastewater composition, climate, culture etc. The World Bank, together with other (inter)national actors, could play a most important role in the planning and financing of such initiatives which could induce a worldwide breakthrough in waste treatment and reuse.

References

ALAERTS, G.J., MAHBUBAR, M.R. and KELDERMAN, P., 1996, Performance of a full-scale duckweed-covered sewage lagoon, *Water Res.*, **30**, 843–852.

ARAUJO, M.C., 1987, Use of water hyacinth in tertiary treatment of domestic sewage, *Water Sci. Technol.*, **19**, 11–17.

ARCEIVALA, S.J., 1986, *Wastewater Treatment for Pollution Control*, New Delhi, India: Tata McGraw-Hill.

BAUR, R.J. and BUCK, D.H., 1980, Active research on the use of duckweeds in the culture of grass carp, tilapia, and fresh water prawns, *Natural History Survey*, R.R.I. Kimmundy, 111.

BRISCOE, J., 1993, When the cup is half full. Improving water and sanitation services in the developing world, *Environment*, **4**, 7–37.

BRIX, H., 1991, The use of macrophytes in wastewater treatment: biological features, in Madoni, P. (ed.) *Proceedings of the International Symposium on Biological Approaches to Sewage Treatment Process: Current Status and Perspectives, Perugia, Italy, 15–17 October 1990*, pp.321–328.

BRIX, H. and SCHIERUP, H.H., 1989, The use of aquatic macrophytes in water pollution control, *Ambio*, **18**, 100–107.

BUDDHAVARAPU, L.R. and HANCOCK, S.J., 1991, Advanced treatment for lagoons using duckweed, *Water Environ. Technol.*, **3**, 41–44.

CONN, W.M. and LANGWORTHY, A.C., 1984, Practical operation of a small aquaculture, in *Proceedings of Conference on Water Reuse II, vol. 2*, American Water Works Association.

CULLEY, D.D. and EPPS, E.A., 1973, Use of duckweed for waste treatment and animal feed. *J. WPCF*, **45**, 337–347.

CULLEY, D.D. and MEYERS, 1980, *Effect of Harvest rate on Duckweed Yield and Nutrient Extraction in Dairy Waste Lagoon*, Baton Rouge, LA:

CULLEY, D.D., REJMANKOVA, E., KVET, J. and FRYE, J.B., 1981, Production, chemical quality and use of duckweeds (*Lemnaceae*) in aquaculture, waste management, and animal feeds, J. World Maricult. *Soc.*, **12**, 27–49.

CULLEY, D.D., GHOLSON, J.H., CHISHOLM, T.S., STANDIFER, L.C. and EPPS, E.A., 1978, *Water Quality Renovation of Animal Waste Lagoons Utilising Aquatic Plants*. Ada, Oklahoma: US Environmental Production Agency.

DEBUSK, T.A. and REDDY, K.R., 1987, BOD removal in floating aquatic macrophyte-based wastewater treatment systems, *Water Sci. Technol.*, **19**, 273–279.

DEWEDAR, A. and BAHGAT, M., 1995. Fate of faecal coliform bacteria in a wastewater retention reservoir containing *Lemna gibba* L, *Water Res.*, **29**, 2598–2600.

DRAAIJER, H., MAAS, J.A.W., SCHAAPMAN, J.E. and KHAN, A., 1992, Performance of the 5 mld UASB reactor for sewage treatment at Kanpur India, *Water Sci. Technol.*, **7**, 123–133.

EDWARDS, P., 1980, *Food Potential of Aquatic Macrophytes*, Manila, Philippines: ICLARM.

EDWARDS, P., 1987, Use of terrestrial vegetation and aquatic macrophytes in aquaculture, in Moriarty, D.J.W. and Pullin, R.S.V. (eds) *Proceedings of the Conference on Detrital Systems for Aquaculture, August 1985, Bellagio, Italy*, pp.311–334.

EDWARDS, P., 1990, An alternative excreta re-use strategy for aquaculture: the production of high protein animal feed, in Edwards, P. and Pullin, R.S.V. (eds) *Proceedings of an International Seminar on Wastewater Reclamation and Re-use for Aquaculture, Calcutta, India, December 1988*, pp.209–221.

EDWARDS, P., HASSAN., M.S., CAO, C.H. and PACHARAPRAKITI, C., 1992, Cultivation of duckweeds in septage-loaded earthen ponds, *Biores. Technol.*, **40**, 109–117.

FEDERLE, T.W. and SCHWAB, B.S., 1989, Mineralization of surfactants by microbiota of aquatic plants, **55**, 2092–2094.

GAIGHER, I.G., PORATH, D. and GRANOTH, G., 1984, Evaluation of duckweed (*Lemna gibba*) as feed for Tilapia (*Oreochromis niloticus* X.O. Aureus) in a recirculating unit. *Aquaculture*, **41**, 235–244.

GAUR, J.P., NORAHO, N. and CHAUHAN, Y.S., 1994, Relationship between heavy metal accumulation and toxicity in *Spirodela polyrrhiza* (L.) Schleid. and *Azolla pinnata* R.Br. *Aquatic Botany*, **49**, 183–192.

GELLINI, R. and PICCARDI, E.B., 1981, Sulla possibilita di assorbimento di metalli pesanti da parte di alcune piante acquatiche, *Inquinamento*, **23**, 45–46.

GIJZEN, H.J., 1996, Anaerobic wastewater treatment: an important step in rational re-use strategies of nutrients and energy, in Rojas, O. and Acevado, L. (eds) *Proceedings of IV Seminario-Taller Latino Americano sobre tratamiento anaerobio de aguas residuales, Bucaramanga, 19–22 November 1996*, Bucaramanga, Colombia: Universidad Industrial de Santander, pp.537–548.

GIJZEN, H.J., 1997, Duckweed based wastewater treatment for rational resource recovery, in *Proceedings of Seventh National Congress on Biotechnology and Bioengineering, Mazatlan, Mexico, 8–12 September 1997*, pp.39–40.

GRAU, P., 1994, What next? *Water Quality International*, **4**, 29–32.

HAJRA, A. and TRIPATHY, S.D., 1985, Nutritive value of aquatic weed, Spirodella polyrhiza (Linn.) in grass carp, Ind. *J. Anim. Sci.*, **55**, 702–705.

HARVEY, R.M. and FOX, J.L., 1973, Nutrient removal using *Lemna minor. J. Water Poll. Control Fed.*, **45**, 1928–1938.

HASSAN, M.S. and EDWARDS, P., 1992, Evaluation of duckweed (*Lemna perpusilla* and *Spirodela polyrhiza*) as feed for Nile tilapia (*Oreochromis niloticus*). *Aquaculture*, **104**, 315–326.

HECKMANN, R.A., WINGET, R.N., INFANGER, R.C., MICKELSEN, R.W. and HENDERSEN, J.M., 1984, Warm water aquaculture using waste heat and water from zero discharge power plants in the Great Basin. *Great Basin Nat.*, **44**, 75–82.

HILLMAN, W.S., 1961, The *Lemnaceae*, or duckweeds, A review of the descriptive and experimental literature, *Botanical Review*, **27**, 221–287.

HILLMAN, W.S. and CULLEY, D.D., 1978, The uses of duckweed. *American Scientist*, **66**, 442–451.

ISLAM, M.S., ALAM, M.J., SHAHID, N.S., HASAN, K.Z., IKRAMULLAH, M., SACK, R.B. and ALBERT, M.J., 1996, Faecal contamination of a fish-culture farm where duckweed is grown in hospital wastewater and used as a fish-feed, Abstract, in *Proceedings of the Fifth Annual Scientific Conference, ASCON V*, Dhaka.

KOLES, S.M., PETRELL, BAGNALL, L.O., 1987, Duckweed culture for reduction of ammonia, phosphorus and suspended solids from algal-rich water, in Reddy, K.R., Smith, W.H. (eds) *Aquatic Plants for Wastewater Treatment and Resource Recovery*, Orlando, FL: Magnolia, pp.769–774.

KÖRNER, S. *et al.*, 1998, *Water Res.*, **32**, 3092–3098.

LANDOLT, E. and KANDELER, R., 1987, *The family of Lemnaceae: a monographic study.* Veroffentlichungen Des Geobotanischen Institutes der Edgenossenschatfliches Technische Hochschule, Stiftung Ruebel, Zurich.

LENG, R.A., STAMBOLIE, J.H. and BELL, R., 1995, Duckweed – A potential high-protein feed resource for domestic animals and fish, in *Proceedings of the AAAP Conference, Bali*, pp.103–114.

MARA, D.D. and CAIRNCROSS, S., 1989, *Guidelines for the Safe Use of Wastewater and Excreta in Agriculture and Aquaculture. Measures for Public Health Protection.* Geneva, Switzerland: World Health Organization.

NIELSEN, H. and NGO, V., 1995, A natural solution for reliable, cost-effective wastewater treatment, *Wastewater Int.*, **10**, 20–21.

ORON, G., 1994, Duckweed culture for wastewater renovation and biomass production, *Agricult. Water Man.*, **26**, 27–40.

ORON, G., PORATH, D. and WILDSCHUT, L.R., 1986, Wastewater treatment and renovation by different duckweed species, *J. Environ Eng.*, **112**, 247–261.

ORON, G., PORATH, D. and JANSEN, H., 1987, Performance of the duckweed species *Lemna gibba* on municipal wastewater for effluent renovation and protein production. *Biotechnol. Bioeng.*, **29**, 258–268.

PESCOD, M.B., 1992, Wastewater treatment and use in agriculture. FAO Irrigation and Drainage Paper no 47, Rome, Italy.

PORATH, D. and KOTON, A., 1977, Enhancement of protein production in fish ponds with duckweed (Lemnaceae), *Isr. J. Botany*, **26**, 51.

PORATH, D. and POLLOCK, J., 1982, Ammonia stripping by duckweed and its feasibility in circulating aquaculture, *Aquat. Botany*, **13**, 125–131.

RAKOCY, J.E. and ALLISON, R., 1981, Evaluation of a closed recirculating system for the culture of tilapia and aquatic macrophytes, *Bio-Eng. Symp. Fish Cult.*, **1**, 296–307.

REDDY, K.R. and DEBUSK, T.A., 1985, Nutrient removal potential of selected aquatic macrophytes, *J. Environ. Qual.*, **14**, 459–462.

REDDY, K.R. and DEBUSK, T.A., 1987, State of the art utilisation of aquatic plants in water pollution control, *Water Sci. Technol.*, **19**, 61–79.

ROBINETTE, H.R., BRUNSON, M.W. and DAY, E.J., 1980, Use of duckweed in diets of channel catfish, in *Proceedings of the 13th Annual Conference of the SE Association on Fish and Wildlife Age*, pp.108–114.

RODRIQUEZ, L. and PRESTON, T.R., 1996, Use of effluent from low cost plastic biodigesters as fertilizer for duckweed ponds (published on Internet).

SHIREMAN, J.V., COLLE, D.E. and ROTTMANN, R.W., 1977, Intensive culture of grass carp Ctenopharyngodon idella in circular tanks, *J. Fish Biol.*, **11**, 267–272.

SHIREMAN, J.V., COLLE, D.E. and ROTTMANN, R.W., 1978, Growth of grass carp fed natural and prepared diets under intensive culture, *J. Fish. Biol.*, **12**, 457–464.

SKILLICORN, P., SPIRA, W. and JOURNEY, W., 1993, *Duckweed Aquaculture, A New Aquatic Farming System for Developing Countries*, Washington, DC: The World Bank.

STEPHENSON, M., TURNER, G., POPE, P., COLT, J., KNIGHT, A. and THOBANOGLOUS, G., 1980, The use and potential of aquatic species for wastewater treatment, Appendix A. The Environmental requirements of aquatic plants. *California state water Resources Control Board*, **65**, 291–440.

STOWELL, R.M., LUDWIG, R., COLT, J. and TCHOBANOGLOUS, G., 1981, Concepts in aquatic treatment system design, *J. Envir. Eng. Proc.*, **107**, 919–940.

SUTTON, D.L. and ORNES, W.H., 1977, Growth of *Spirodela polyrrhiza* in static sewage effluent, *Aquat. Botany*, **3**, 231–237.

TAYLOR, H.D., BASTOS, R.K.X., PEARSON, H.W., and MARA, D.D., 1995, Drip irrigation with waste stabilisation pond effluents: solving the problem of emitter fouling, *Water Sci. Technol.*, **12**, 417–424.

TCHOBANOGLOUS, G., 1987, Aquatic plant systems for water treatment: Engineering considerations, in Reddy, K.R. and Smith, W.H. (eds) *Aquatic Plants for Water Treatment and Resource Recovery*, Orlando, FL: Magnolia.

TRIPATHI, B.D., SRIVASTAVA, J. and MISRA, K., 1991, Nitrogen and phosphorus removal-capacity of four chosen aquatic macrophytes in tropical freshwater ponds; *J. Envir. Conserv.*, **18**, 143–147.

VAN DYKE, J.M. and SUTTON, D.L., 1977, Digestion of duckweed (*Lemna* spp.) by the grass carp (*Ctenopharyngodon idella*), *J. Fish. Biol.*, **11**, 273–278.

VAN HAANDEL, A. and LETTINGA, G., 1994, *Anaerobic Sewage Treatment, A practical Guide for Regions with a Hot Climate*. Chichester: John Wiley & Sons.

VAN HAANDEL, A. and CATUNDA, P.F.C., 1997, Anaerobic digestion of municipal sewage and post treatment in stabilisation ponds, in *Proceedings of Seventh National Congress on Biotechnology and Bioengineering, Mazatlan, Mexico, 8–12 September 1997*.

VEENSTRA, S., AL-NOZAILY, F.A. and ALAERTS, G.J., 1995, Purple non-sulfur bacteria and their influence on waste stabilisation pond performance in the Yemen Republic, **12**, 141–149.

WANG, W., 1991, Ammonia toxicity to macrophytes (common duckweed and rice) using static and renewal methods, *Envir. Toxicol. Chem.*, **10**, 1173–1177.

WELLER, B. and VROON, R., 1995, Treatment of domestic wastewater in a combined UASB reactor – duckweed pond system, MSc thesis, Agricultural University Wageningen.

WOLVERTON, B.C., 1979, Engineering design data for small vascular aquatic plants wastewater treatment systems, in *Proceedings of an EPA Seminar on Aquaculture Systems for Wastewater Treatment*, EPA 430/9-80-006. Washington, D.C.

ZIRSCHKY, J. and REED, S.C., 1988, The use of duckweed for wastewater treatment. *Journal WPCF*, **7**, 1253–1258.

8

Anaerobic Treatment of Tequila Vinasse

KUPPUSAMY ILANGOVAN, JOSEFINA LINERIO, ROBERTO BRIONES
AND ADALBERTO NOYOLA

8.1 Introduction

Alcoholic beverages are produced from various feedstocks such as sugar cane molasses, sugar cane juice, sugar beet molasses, grapes, malt and rice. Tequila is one of the most important traditional alcoholic beverages of Mexico, manufactured from *Agave tequilana* weber (Cedeño, 1995). In Mexico, there are 35 tequila distilleries, producing annually 70 millions litres of tequila and generating approximately 700 000 m^3 of vinasse, a liquid waste left after distillation of fermented agave juice that contains high concentrations of organic matter, salts, and yeast cells. Most tequila distilleries use the vinasse for irrigation and cultivation of maize or agave but long-term irrigation with untreated vinasse may affect the soil fertility and texture. In recent years the Upflow Anaerobic Sludge Blanket (UASB) technology has been applied world-wide in breweries, distilleries, the pulp and paper industry, food processing industry and certain petrochemical plants (Macarie *et al.*, 1995). Vinasse from distilleries can be treated cost-effectively by anaerobic digestion in order to reduce pollution load and to produce methane, which can be used as fuel. The aim of the present investigation was to study the feasibility of the anaerobic treatment of tequila-vinasse using laboratory-scale UASB reactors under different operating conditions.

8.2 Materials and Methods

Two UASB reactors with liquid volume of 2.3 l (8 cm internal diameter, 45 cm high) were used in this study (Figure 8.1). Influent feedstock from tequila distillation was collected from Centinela Distillery, Arandas, State of Jalisco, Mexico and stored at 5(±2)°C. The wastewater contained different concentrations of agave fibres and other particulate materials. In order to have a homogeneous influent to the reactor, wastewater was filtered (using polyester rapid filter paper porosity of 25 μm) before feeding. Seed granular sludge (700 ml), used to start up the UASB 1 and UASB 2, was received from a full-scale UASB reactor treating malting industry wastewater.

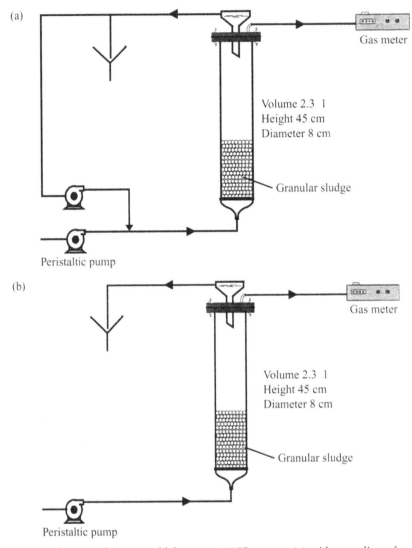

Figure 8.1 Schematic diagrams of laboratory UASB reactor (a) with recycling of treated effluent (UASB 1) and (b) without recycling of treated effluent (UASB 2)

The effluent recycling ($Q:Q_r$, 1:1 and 1:8 ratios) was performed in one of the UASB reactors (UASB-1) in order to increase the upflow velocity and to recover internal alkalinity. The influent pH was adjusted to 4.5–5 throughout the start-up period with sodium bicarbonate. Both UASB 1 and UASB 2 were continuously fed with diluted vinasse with an organic load of 7.5 kg COD m^{-3}/day up to 60 days of start-up period and gradually increased to 19 kg COD m^{-3}/day for UASB 2 (without recycling of effluent) whereas organic load up to 25 kg COD m^{-3}/day was applied to UASB 1 (with recycling). Tap water was used for dilution. Recycling of effluent in UASB 1 was performed on the 60th day of operation onwards. The UASB reactors were placed in a temperature-controlled room (32 ± 2°C) and operated at a constant hydraulic retention time of 3 days over a period of 140 days.

Biogas production was measured using a water displacement gas meter and composition of biogas was analysed using gas chromatography (Fisher Partitioner Model 1200). Total solids, volatile solids, total suspended solids, volatile suspended solids, pH, total and soluble COD, sulphate, chloride, sodium, iron, magnesium and calcium were analysed using *Standard Methods* (APHA, 1989). Alkalinity was measured following the method of Jenkins *et al*. (1991) by simple alkalimetric titration at pH 5.75 (bicarbonate alkalinity) and pH 4.3 (total alkalinity). The alkalinity ratio (α) between bicarbonate and total alkalinity was calculated by

(alk.4.3 − alk.5.75)/alk.4.3

and used as control parameter for operating both reactors, the desired value being below 0.4. Volatile fatty acids (VFA) were analysed using an FID gas chromatograph (SRI 1800). Metal concentrations of influent and effluent were analysed using atomic absorption spectrometer (Perkin Elmer 1100).

8.3 Results and Discussion

The physicochemical characteristics of the reactor feed contained 66.26 g l^{-1} total COD, of which 40.49 g l^{-1} were soluble COD, pH 3.4, total solids 42.84 g l^{-1}, volatile solids 39.46 g l^{-1}, total suspended solids 9.1 g l^{-1}, volatile suspended solids 8.8 g l^{-1}, sodium 152 mg, calcium 356 mg l^{-1}, iron 35 mg l^{-1}, magnesium 217 mg l^{-1}, potassium 290 mg l^{-1}, chloride 1.66 g l^{-1} and sulphate 0.88 g l^{-1}. Unlike stillage of sugar cane molasses, tequila-vinasse is more easily biodegradable due to its low levels of sulphate, potassium and chlorides. Elevated concentrations of potassium and chlorides in sugar cane molasses stillage inhibit the anaerobic microbial activity. The presence of 9 g l^{-1} of potassium in molasses stillage removed the exchangeable form of micronutrients from the sludge matrix (Ilangovan and Noyola, 1993).

Influent and effluent pH profiles are shown in Figure 8.2. Effluent pH was virtually constant throughout the period of operation. During the period of operation of UASB 2 the influent pH was raised with sodium bicarbonate (3.5–5 g l^{-1}), whereas in UASB 1 neutralization of influent was carried out only during the start-up period (60 days) and from day 70 of operation recycling was performed in UASB 1.

In order to avoid acidification in UASB 1, influent pH was adjusted for a short period (days 81–85 and 107–112 of operation) when a different batch of raw vinasse was fed. However, the alkalinity of UASB 1 remained stable without exogenous addition of bicarbonates and indicate that effluent bicarbonate alkalinity might be sufficient to buffer the pH of the influent (Figure 8.3a). Profiles of alkalinity data shown in this figure indicate the good performance of the reactor. Alkalinity profiles of UASB 2 reflect instability after 60 days of operation (Figure 8.3b).

Both reactors at start-up stage gradually increased to 80 % COD removal (Figure 8.4) with 7.5 kg COD m^{-3}/day. However, UASB 2 was clearly affected when fed with 20 kg COD m^{-3}/day, while UASB 1 showed improvements in performance and ability to accept higher organic load (up to 25 kg COD m^{-3}/day) with 80 % COD removal efficiency and increasing biogas production. Effluent recycling in UASB 1 enhanced biogas production (11.5 l biogas/litre reactor/day) over UASB 2 without recycling (10 l biogas/litre reactor/day). Acetate and propionate accumulation (Figure 8.5) was noticed in UASB 2 (acetate 0.3 g l^{-1}; propionate 0.35 g l^{-1}) at the 60th day of operation, whereas in UASB

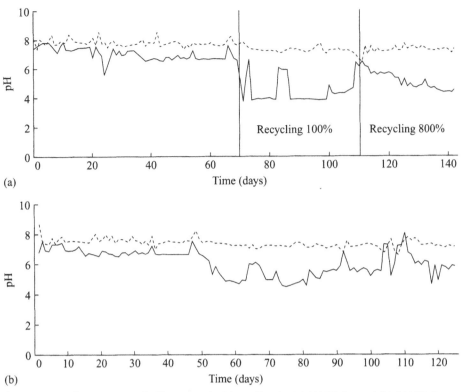

(a)

(b)

Figure 8.2 Influent (—) and effluent (---) pH profiles in (a) UASB 1, and (b) UASB 2

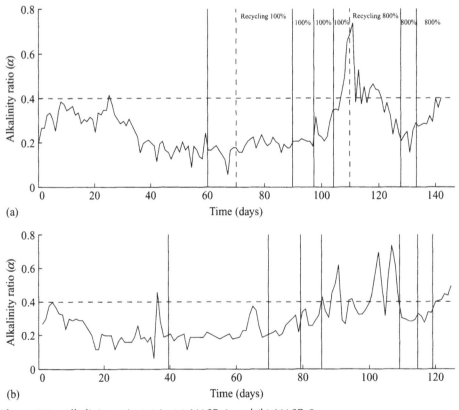

(a)

(b)

Figure 8.3 Alkalinity ratio (α) in (a) UASB 1 and (b) UASB 2

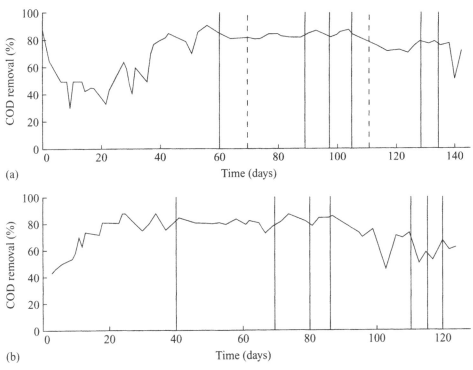

Figure 8.4 COD removal in (a) UASB 1 and (b) UASB 2

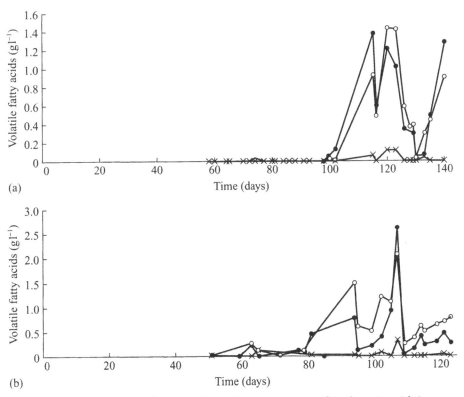

Figure 8.5 Volatile fatty acids (•, acetic acid; ○, propionic acid; ×, butyric acid) in (a) UASB 1 and (b) UASB 2

1, a limited accumulation of VFA (acetate 0.05 g l^{-1}, propionate 0.03 g l^{-1}) was noticed until day 100 of operation.

Therefore, effluent recycling favoured the degradation of accumulated VFA and as a consequence enhanced the biogas production. As a result of an increase in organic load (15–20 kg COD m^{-3}/day) acetic and propionic acid concentration at day 105 of operation in UASB 1 increased to 1.4 and 0.9 g l^{-1} respectively, whereas UASB 2 showed higher concentration of acetate (2.5 g l^{-1}) and propionate (1.5 g l^{-1}). Concentration of butyric acid remained negligible throughout the period of operation in both reactors, with the exception of limited increases during instability periods. From these results, it is evident that recirculation significantly favoured VFA degradation and maintained satisfactorily the performance of the reactor. Recycling operation showed that the internally generated bicarbonate alkalinity could be enough to achieve good performance of the reactor without requirement for further reagent. The high COD removal and moderate effluent VFA concentration indicate the stable process performance of UASB 1 reactor even at high organic loads.

From the experimental results it was concluded that it is feasible to treat undiluted tequila-vinasse with a UASB reactor with recycling of effluent and achieve 80 % of COD removal. Recycling of effluent eliminates the need for exogenous addition of bicarbonate to the reactor. Performance data indicate that undiluted tequila-vinasse was easily biodegradable with UASB system with 3.0 days of hydraulic retention time. A UASB system coupled with appropriate post treatment could yield higher disposable effluent quality standards.

Acknowledgements

The assistance of Centinela Distillery is greatly appreciated. This research was jointly supported by DGAPA (Project Number: IN502094) and Institute of Engineering, UNAM. The authors are grateful to E. Alvarez and A. Espinosa for their technical assistance.

References

APHA, 1989, *Standard Methods for the Examination of Water and Wastewater*, 17th edition, Washington DC, USA: APHA, AWWA, WPCF.

CEDEÑO, M., 1995, Tequila production, *Critical Reviews in Biotechnology*, **15**(1), 1–11.

ILANGOVAN, K. and NOYOLA, A., 1993, Availability of micronutrients during anaerobic digestion of molasses stillage using an upflow anaerobic sludge blanket (UASB) reactor, *Environmental Technology*, **14**, 795–799.

JENKINS, S.R., MORGAN, J.M. and ZHANG, X., 1991, Measuring the usable carbonate alkalinity of operating anaerobic digestors. *Journal of the Water Pollution Control Federation*, **63**(1), 28–34.

MACARIE, H., NOYOLA, A., GUYOT, J.P. and MONROY, O., 1995, Anaerobic digestion of petrochemical wastes: the case of terephthalic waste water, in *Proceedings of the Second International Minisymposium on Removal of Contaminants from Water and soil*. UNAM, Mexico D:F. 2: 35–51.

9

Immobilization of Living Microalgae and their Use for Inorganic Nitrogen and Phosphorus Removal from Water

CARLOS GARBISU, ITZIAR ALKORTA, MARÍA J. LLAMA
AND JUAN L. SERRA

9.1 Introduction

Although nitrates occur naturally, they have become important pollutants as a consequence of human activities (Winteringham, 1974). In fact, due to the use of fertilizers in excess and the increasing accumulation rate of wastes from human and animal populations, many sources of drinking water contain high levels of nitrate ion (Kurt et al., 1986).

The major sources of nitrogenous wastes can conveniently be considered in two groups, chemical and biological. The main chemical sources are fertilizer production, explosives manufacture, coal conversion and other industries such as the manufacture of sodium carbonate, petroleum refining, refrigeration plants using ammonia in scouring and cleaning operations, certain synthetic fibre, plants, etc. Biological sources of nitrogenous pollutants are sewage, animal husbandry, food and fermentation industries, natural sources (decaying vegetation, animal and bird droppings).

The best-known effect of fertilizer run-off and the other forms of nitrate pollution is eutrophication of water, which in extreme cases can result in water becoming clogged with algae, killing most other plants and animals (Dudley, 1990). Regarding nitrogen pollution, most concern stems from the possible health hazards that have been attributed to nitrate, either directly as a causative factor of methaemoglobinaemia, or indirectly as the source of carcinogenic nitrosamines (Abel, 1989; Hagmar et al., 1991). However, the possible damage to the ozone layer of the Earth's atmosphere by gaseous oxides of nitrogen, released from nitrate ion in the soil, is also considered an important aspect of air pollution (Royal Commission on Environmental Pollution, 1979).

Although to date it does not seem to present any problem for human health, phosphorus removal from municipal and industrial wastewaters is required to protect water bodies from eutrophication, especially in lakes and enclosed bays with stagnant water (Comeau et al., 1987). Municipal, agricultural and forest wastes are each responsible for about 30 % of the total known sources of phosphates, with detergents contributing about 60 % of the municipal phosphate effluent.

9.2   Microalgae and Cyanobacteria

Photosynthetic microorganisms can be grouped into two categories: photosynthetic bacteria and microalgae. Photosynthetic bacteria perform anoxygenic photosynthesis and possess bacteriochlorophyll, a light-absorbing pigment chemically distinct from the chlorophyll *a* present in all other photosynthetic organisms (algae and higher plants).

Cyanobacteria (also known as blue–green algae or cyanophyceae) constitute the largest, most diverse and most widely distributed group of photosynthetic prokaryotes (Stanier and Cohen-Bazire, 1977). They hold an intermediate position between photosynthetic bacteria and eukaryotic algae, but they do not possess either bacteriochlorophyll or chlorophyll *b*. In contrast to photosynthetic bacteria, cyanobacteria perform oxygenic photosynthesis similar to that of higher plants.

From a biotechnological point of view, the term 'microalgae' (which has no taxonomic meaning) includes those microorganisms that possess chlorophyll *a* and other photosynthetic pigments capable of performing oxygenic photosynthesis (Abalde *et al.*, 1995). Therefore, two different cellular types are combined within such a term: cyanobacteria and eukaryotic microalgae.

The large-scale cultivation of microalgae was probably first considered seriously in Germany during World War II, with the aim of producing lipids by means of growing *Chlorella pyrenoidosa* and *Nitzschia palea* (Abalde *et al.*, 1995). This initial research was taken up by a group of scientists at the Carnegie Institution of Washington in order to use the green alga *Chlorella* for large-scale production of food. Since then, many research teams in several countries have been successfully developing techniques for the cultivation of microalgae on a large scale, mostly freshwater ones such as *Chlorella*, *Scenedesmus*, *Coelastrum*, *Dunaliella* and *Spirulina*. Some cyanobacteria, such as *Spirulina*, are among the most commonly utilized organisms in microalgal biotechnology.

Although there are unmistakable indications that photosynthesis is the principal mode of energy metabolism in these organisms, in the natural environment they regularly experience dark diurnal periods and some can even survive long periods in complete darkness. Consequently, they possess a respiratory metabolism capable of providing maintenance energy in the dark (Fry and Peschek, 1988).

Cyanobacteria have a world-wide distribution and most species are cosmopolitan (Fogg *et al.*, 1973). They may occur in extreme habitats such as hot springs, in desert rocks and below the Antarctic ice and are known to be amongst the earliest colonizers of arid land. They were probably largely responsible for our oxygen-enriched atmosphere and the roots of the cyanobacterial lineage extended far into the geological past. Owing to the longevity of the lineage and its influence on the Earth's early environment, cyanobacteria must rank as among the most successful and ecologically significant of the myriad forms of life to have appeared over the history of biological evolution (Schopf and Walter, 1982).

9.3   Biological Wastewater Treatment

Biological removal of nitrogen appears a valid option and offers some advantages over chemical and physicochemical treatments (Proulx and de la Noüe, 1988): it can operate under relatively wide pH variations (at least in the case of using microalgae), it does not create secondary pollution problems, and the biomass can be exploited commercially.

The low capital and operational costs of biological processes for removal of phosphorus have also been attracting attention in recent years (Somiya *et al.*, 1988). The biological process offers the advantages of not requiring addition of chemicals and of reducing the volumes of sludge produced. Recently, instead of the chemical precipitation methods, the biological phosphorus removal capability of the anaerobic/oxic process has been substantiated experimentally (Hashimoto *et al.*, 1987).

The application of high-rate algal ponds in wastewater treatment appears to have the greatest potential of all biotechnologies based on microalgae, if exploited fully as a multipurpose system (Richmond, 1986). Although the most widely utilized biological method for wastewater treatment has focused mainly on removal by bacteria, over the last 20–25 years increasing attention has been paid to the possibility of using microalgae (Tam and Wong, 1989). In fact, wastewaters from domestic, industrial or agricultural sources are a favourable microalgal culture medium (Callegari, 1989). The idea of using microalgae for this purpose, as initially proposed and experimentally tested by Caldwell (1946) and, some 10 years later, by Oswald and Gotaas (1957), has gained momentum and a fair number of papers have dealt with such systems (Lavoie and de la Noüe, 1985; Abeliovich, 1986; de la Noüe and de Pauw, 1988; Oswald, 1988; Hashimoto and Furukawa, 1989; Pouliot *et al.*, 1989; Tam and Wong, 1989; Ozaki *et al.*, 1991; Garbisu *et al.*, 1991, 1992, 1993, 1994).

At present, one of the promising fields using microalgae in wastewater treatment is the utilization of cyanobacteria. Through the appropriate use of the nutrient uptake capabilities of these prokaryotes, nutrients in the secondary effluent of wastewater treatment plants can be removed and used in the growth of algal biomass, which can then become a source of biomass (Hashimoto and Furukawa, 1989). Such light-driven biotechnology may provide an effective means of removing inorganic nitrogen (ammonium, nitrate, nitrite) and phosphorus (phosphate) in primary and secondary urban wastewater treatment processes where the ions are implicated in the eutrophication of the receiving bodies of water (Shelef and Soeder, 1980).

In countries with a high number of hours of sunlight there has been considerable development of combined algal and bacterial systems for wastewater treatments (Smith, 1988). However, treatment of wastewaters with microalgae is possible under harsh conditions by using greenhouse technology (Pouliot and de la Noüe, 1985). At present, algal treatment systems operate with high efficiency in various parts of the world (Fallowfield and Garrett, 1985; Tam and Wong, 1990; Schramm, 1991).

The cost of microalgal cultures can be at least partially overcome by selling the biomass or extracting high added-value products (Metting and Pyne, 1986). However, large-scale solar biotechnology is still confronted with the expense of harvesting microalgal biomass (de la Noüe and Proulx, 1988a,b).

9.4   Utilization of Inorganic Nitrogen and Phosphorus by Cyanobacteria

In general, cyanobacteria can use nitrate or ammonium as the sole nitrogen source for growth, and may have the additional ability to fix atmospheric nitrogen (N_2). In the utilization of any form of inorganic nitrogen, first, the nitrogenous compound in the outer medium should enter the cell and, second, since ammonium is the only nitrogen form directly incorporated into amino acids, dinitrogen and nitrate must be reduced to ammonium. Finally, ammonium is incorporated as an α-amino group into carbon skeletons (Guerrero and Lara, 1987). Nitrate is probably the most common source of nitrogen for utilization by microalgae and cyanobacteria in the environment and only the ionized form is encountered at physiological or ecological pH values (Kerby *et al.*, 1989). In addition, nitrate assimilation is affected by a number of environmental and nutritional factors such as light, temperature, pH and carbon source availability (Serra *et al.*, 1990). Under autotrophic conditions, the utilization of any form of inorganic nitrogen by cyanobacteria is strictly dependent on the availability of light and CO_2. Light is needed to synthesize the ATP and reductant required for the assimilation processes involved, whereas the CO_2 requirement results from a set of interactions between carbon and nitrogen metabolism (Guerrero and Lara, 1987).

Orthophosphate is believed to be the form of phosphorus commonly taken up by cyanobacteria (Marco and Orús, 1988). Phosphate flow into the cell is closely linked to photosynthetic ATP formation and thus depends strongly on the overall phosphorylation potential of the cell (Falkner *et al.*, 1984). The transport of phosphorus through the cell membrane appears to be the rate-limiting step in its incorporation (Falkner *et al.*, 1980).

The intracellular accumulation of phosphate is energy dependent, being higher in the light than in the dark, and depends strictly on the pH of cytoplasm and not on the energy conversion at the thylakoid membrane, which is responsible for the energy supplies.

Finally, the major phosphate reserve of cyanobacteria is polyphosphate, which accumulates as discrete granules in the cytoplasm of the cell when phosphate is in excess. Polyphosphate is mobilized during periods of nutrient shortage, representing a valuable pool of activated phosphorus, which can be used in a variety of metabolic processes (Okamoto *et al.*, 1986).

9.5   Immobilization Techniques

One of the major problems in use of microalgae for the biological treatment of wastewaters is their recovery from the treated effluent. In order to solve the problem of harvesting the biomass, many systems have been proposed or tried (Borowitzka, 1986; Metting and Pyne, 1986; de la Noüe and de Pauw, 1988; Mohn, 1988; Oswald, 1988), but only two appear feasible: chemical flocculation, especially with chitosan (Lavoie and de la Noüe, 1983, 1984) and bioflocculation (Eisenberg *et al.*, 1981). Among the most recent ways of bypassing this problem are immobilization techniques applied to algal cells (de la Noüe and Proulx, 1988a,b). In fact, in the case of photosynthetic cells, over the last 20–25 years considerable progress has been achieved in the immobilization of photobiological organisms and organelles such as cyanobacteria, photosynthetic bacteria, algae and chloroplasts (Hall and Rao, 1989).

9.5.1   *Cell Immobilization*

Rosevear (1984) defined immobilization as a technique which confines a catalytically active enzyme or cell within a reactor system and prevents its entry into the mobile phase, which carries the substrate or product. According to Fukui and Tanaka (1982), two different types of immobilized cells can be distinguished:

- Immobilized, treated cells: cells utilized in a dead state being subjected to an appropriate treatment before or after the immobilization. However, the desired enzymes are in an active and stable form.
- Immobilized, living cells: either resting or growing in gel matrices.

Although the advantages (and drawbacks) of using immobilized rather than free-living cells depend on the intrinsic properties of the cells and the purpose of their use, some general observations can be made. Immobilized cells may offer certain specific advantages over batch or continuous culture fermentation where free cells are used, such as (Tampion and Tampion, 1987; Hall and Rao, 1989): accelerated reaction rates due to increased cell density per unit volume, increased cell metabolism and cell wall permeability, no wash-out of cells, high operational stability and better control of catalytic processes, separation and reuse of catalyst, reduction in cost due to easier separation of cells and excreted product.

There are three major techniques to be used for cell immobilization: entrapment, adsorption and coupling. The immobilization techniques which most resemble the circumstances in which cells find themselves in nature, are their entrapment within gels and adsorption to surfaces. In fact, many organisms normally exist adsorbed to surfaces while entrapped in a gel or slime of their own making (Bucke, 1986).

Entrapment

In the case of entrapment, the cell is merely restricted in its movements and confined to a small volume of liquid within a defined microenvironment. The matrix must be constructed *de novo* around the cells rather than adsorbing the cells onto a preformed material. One of the major problems found when entrapping cells in a gel is the diffusional resistance of the gel to the substrate and products. Another important problem that arises when using entrapped cells lies in their ability to divide and eventually break the support (Shi, 1987). The gelation of alginic acid by polyvalent metal ions is one of the most important methods of cell immobilization. The technique involves the drop-wise addition of cells suspended in sodium alginate onto a solution of calcium chloride (Kierstan and Bucke, 1977) to form a very stable gel where the cells are entrapped.

A number of polyurethane prepolymers are commercially available and have been used successfully to entrap cells. Urethane is formed by the reaction between an isocyanate and a hydroxyl group. Condensation of the urethanes with other isocyanates, alcohols or amines produces a cross-linked polymer. By varying the temperature, prepolymer structure, condensation reagent, among others, polyurethane foams of different porosity, strength and translucency can be produced. Addition of cells to prepolymers before condensation can result in a uniform distribution of the biomass in the foam (Rao and Hall, 1984).

Adsorption

Adsorption of cells onto a solid surface is probably the mildest of the cell immobilization techniques. Many organisms are capable of naturally adhering to surfaces; in fact, whenever a suspension of cells is brought into contact with a surface there will almost always be a certain amount of adsorption to the solid.

Adsorption thus describes an immobilization method in which the biocatalyst is attached to a surface by non-covalent interactions. The type of adsorption can range from non-specific binding dependent on weak surface forces through the strong interaction of ionic bonds to the highly specific, multipoint binding of natural biological affinities. Cells are capable of producing extracellular polysaccharides once adsorbed on a surface and this may further strengthen the hold of a biocatalyst to the matrix.

Polyurethane and polyvinyl foams have been used as supports for immobilized cells (Mavituna and Park, 1985; Garbisu *et al.*, 1991, 1992, 1993, 1994).

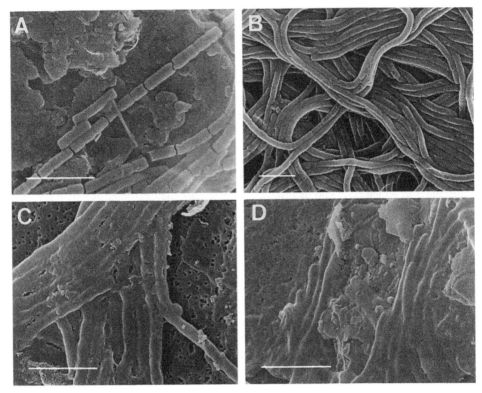

Figure 9.1 Scanning electron micrographs of free-living *Phormidium laminosum* cells at different physiological states: after (A) 2 days and (B) 10 days of growth in a culture medium with nitrate, or after (C) 1 day or (D) 3 days of starvation in a nitrogen-free medium. Scale bars = 5 μm in all cases

Coupling

Covalent coupling is perhaps the most popular technique when immobilizing enzymes, but few systems using this technique on cells have been reported. This is because it is a generally harsh technique, which results in loss of viability of cells. In addition, where viable cells can be immobilized by covalent coupling, any cell division is likely to result in cell leakage from the support, as the daughter cells will be probably less strongly attached.

Coupling is based on covalent bond formation between activated support and cells. Another possibility is to covalently cross-link the cells to one another, providing greater stability to the aggregates that can be achieved by flocculation. The most widely used coupling agent is probably glutaraldehyde although carbodiimine, isocyanate and aminosilane have also been used (Tampion and Tampion, 1987).

Microalgae have been used successfully for the removal of nitrogen and phosphorus pollutants from water. Chevalier and de la Noüe (1985a,b) have shown that cells of *Scenedesmus* sp. immobilized on κ-carrageenan beads are as efficient as free cells in taking up ammonium and orthophosphate from secondary urban effluents. *Phormidium* sp. cells were attached to chitosan flakes and used for removing nitrogen (ammonium, nitrate, nitrite) and orthophosphate from urban secondary effluents (de la Noüe and Proulx, 1988b). Robinson *et al.* (1989) found that *Chlorella emersonii* entrapped in calcium

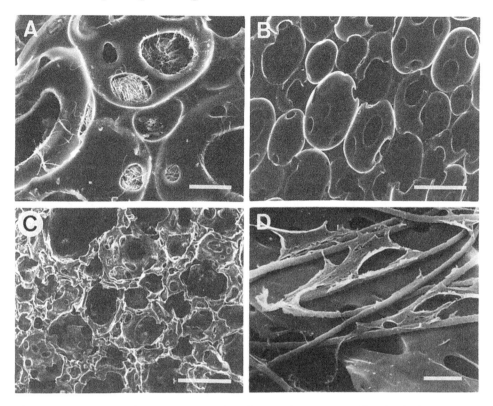

Figure 9.2 Scanning electron micrographs of: (A) *P. laminosum* cells adsorbed onto polyurethane foams after 45 days of colonization (bar = 100 μm); (B) polyurethane foam (bar = 500 μm); (C) polyvinyl foam (bar = 500 μm); (D) *P. laminosum* filaments adsorbed onto polyvinyl foams (bar = 5 μm)

alginate beads was able to remove phosphorus from secondary treated effluent with acceptable efficiencies.

Garbisu *et al.* (1991, 1992, 1993, 1994) reported on the utilization of the filamentous thermophilic cyanobacterium *Phormidium laminosum* immobilized in polymer foams for the removal of nitrate, nitrite and phosphate from water (Figures 9.1 and 9.2). In view of several studies (de la Noüe and Proulx, 1988a,b; Garbisu *et al.*, 1991, 1992, 1993, 1994), it appears that the filamentous cyanobacteria *Phormidium* spp. may be among the most promising microalgal genera for the processes of biological tertiary treatment used in the depollution of water. In fact, *P. laminosum* shows some characteristics that, in an immobilized state, make it a suitable choice for the removal of inorganic nitrogen and phosphorus from water: it is a non-N_2-fixing cyanobacterium that can utilize nitrate, nitrite, or ammonium ions as the sole nitrogen source for growth; it can stand wide variations of pH (6–11) and temperature (15–50°C) and can grow at relatively low light intensities; its filamentous nature is an advantage when immobilizing the cells by entrapment or by adsorption into matrices; its hydrophobicity and tendency to attach to surfaces gives it a good potential in processes involving immobilized cells; and its metabolism can be readily altered by nutritional stress.

Consequently, we carried out some studies to study the feasibility of using polymer-immobilized *P. laminosum* for the removal of inorganic nitrogen and phosphorus from water (Garbisu *et al.*, 1991, 1992, 1993, 1994). In order to do so, batch and continuous-flow

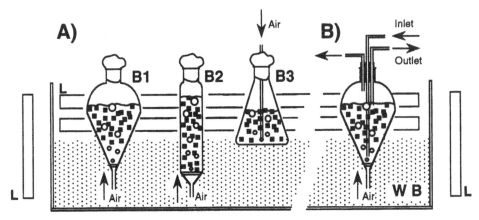

Figure 9.3 Schematic representation of the air-agitated bioreactors used: B1, funnel shape; B2, column; B3, Erlenmeyer flask; the liquid phase volumes were 250, 125 and 250 ml, respectively. (A) Batch fluidized-bed; (B) Continuous-flow air agitated. L, Fluorescent lamp; WB, water bath. The liquid phase consisted of ultrapure water supplemented with 50 mg l^{-1} NaNO$_3$ or 10 mg l^{-1} Na$_2$HPO$_4$. Preboiled, washed, and dried 5-mm foam cubes were placed in a cell suspension and immobilization by adsorption was carried out. Once the foam cubes were fully colonized, they were introduced into the bioreactors. Effluent samples were collected at timed intervals and the ion concentration determined. The reactor was continuously illuminated with cool white fluorescent lamps at a light intensity of 100 μmol photon m^{-2} s^{-1}. Algal leakage from the reactor was identified by the presence of free cells in the effluent and was measured by absorbance at 678 nm. The B1 bioreactor was also tested in a continuous-flow mode (B) by connecting it to a supply tank containing ultrapure water supplemented with the same ion concentrations. The water was supplied at the top of the bioreactor using an on-line variable-speed peristaltic pump at different flow rates. Uptake efficiency (UE) in the continuous-flow bioreactor was defined as: UE = [(I − E)/I] × 100, where *I* and *E* were the influent and effluent concentrations of the ions, respectively. An efficiency value of 100 % was obtained when no ion appeared in the effluent (i.e. when *E* = 0)

air-agitated and packed-bed bioreactors (Figures 9.3 and 9.4) were used to study the removal of nitrate, nitrite and phosphate from ultrapure water by polyvinyl-immobilized *P. laminosum* cells as a preliminary approach to the applicability of these systems for the removal of ions from polluted potential drinking water.

The results showed that *P. laminosum* cells immobilized in polyvinyl foam and packed in a column reactor took up nitrate continuously for at least 3 months. During this time, the cells remained active and nitrate uptake efficiencies in excess of 90 % were achieved.

We studied the effects of light intensity and CO$_2$ concentration on this packed-bed bioreactor performance with respect to the nitrate uptake efficiency of the system (Garbisu *et al.*, 1991). From these studies, it was concluded that not only CO$_2$ concentration but also light intensity affects nitrate uptake, although the effect of the latter must be probably associated with the CO$_2$ fixation rate. In fact, the limiting factor in nitrate utilization appears to be the availability of CO$_2$ fixation products, the light dependence of the process being a reflection of the light dependence exhibited by CO$_2$ fixation. Thus, at limiting light intensity, a moderate competition between nitrate utilization and CO$_2$ fixation occurs as both processes require reducing power and/or ATP, resulting in a decrease in the rate of CO$_2$ fixation in the presence of nitrate. However, at high light intensities and saturating CO$_2$ concentrations, no competition for assimilatory power between nitrate

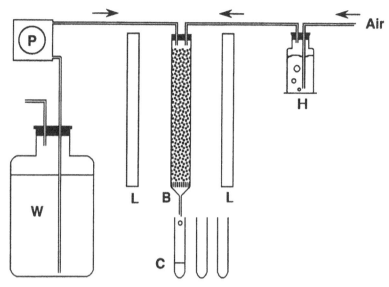

Figure 9.4 Schematic diagram of the continuous-flow packed-bed bioreactor. L, Fluorescent lamp; B, bioreactor; H, humidifier flask; C, collector; W, water (feeding tank); P, pump. A glass column (2.7 cm inside diameter, 50 cm long) was used. Previously colonized foam cubes were packed in the bioreactor and the column was connected to a feed tank containing ultrapure water supplemented with 50 mg l^{-1} $NaNO_3$ or 10 mg l^{-1} Na_2HPO_4. The medium was supplied at the top of the column using a peristaltic pump at different flow rates. Column effluent samples were collected at timed intervals and the ion concentration determined. The column was continuously illuminated with cool white fluorescent lamps at a light intensity of 100 μmol photon m^{-2} s^{-1}. Uptake efficiency and cell leakage were identified as in the continuous-flow air-agitated bioreactor

assimilation and CO_2 fixation occurs because the photosynthetic apparatus is able to generate enough assimilatory power for both simultaneous processes (Garbisu *et al.*, 1991).

Continuous-flow air-agitated bioreactors showed nitrogen removal efficiencies of up to 90 % for residence times of 14 h in short-term experiments. Although nitrogen-starved cells showed higher inorganic nitrogen-uptake rates than nitrogen-sufficient ones, the photosynthetic activities of the cells decreased progressively with the time of nitrogen starvation. Nitrogen deficiency can be easily induced by incubating the cells for a given time in a nitrogen-free medium (Serra *et al.*, 1990). Although the incubation of cells in a nitrogen-free medium up to about 70 h resulted in higher uptake rates, longer incubations led to lower uptake rates, probably due to the excessive degeneration of the cell structures. These nitrogen-starved cells produced high amounts of exopolysaccharides, which appear to assist the immobilization process (Garbisu *et al.*, 1992). The fact that after approximately 60 h of nitrogen starvation none of the photosynthetic activities (photosystem I, II or I + II) could be detected might be due either to the photosystems having been rendered inactive or to the high amounts of exopolysaccharides produced by the nitrogen-starved cells which could act as an additional diffusion barrier to the entrance of the electron mediators used to measure these activities into the cells and to gas exchange (Garbisu *et al.*, 1992).

If nitrogen-starved cells were to be used practically for the removal of nitrogen and phosphorus from water, several bioreactors would need to be set up in parallel, alternating

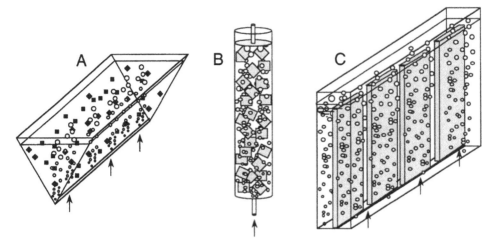

Figure 9.5 Schematic diagram of the continuous-flow air-agitated photobioreactors of different shapes used for the removal of nitrate from water. (A) Triangular reactor; (B) column reactor; (C) rectangular reactor. Air was supplied to the cultures through the bottom of the bioreactors (arrows)

cycles of nitrogen starvation and supply, to keep the system operating continuously with starved cells.

Inorganic nitrogen uptake by nitrogen-starved cells occurred in both light and dark under aerobic conditions. In anaerobiosis light was required for the uptake, confirming that the necessary energy might perhaps be derived from the respiratory electron transport chain under aerobiosis. The uptake of nitrate and nitrite in the dark and in the absence of an added carbon source by nitrogen-starved cells could be due to these cells accumulating high carbohydrate reserves during the starvation period, which can apparently substitute for photosynthetic CO_2 fixation products in stimulating ion uptake (Guerrero and Lara, 1987).

Ammonium inhibited nitrate uptake but did not affect the uptake of nitrite. Nitrate uptake appears to be regulated by a negative feedback control exerted by certain compounds produced during nitrate assimilation and ammonium metabolism. The effect of ammonium upon nitrate uptake may be seen as an exaggeration of this feedback control continuously modulating the uptake rate (Guerrero and Lara, 1987). The fact that nitrite uptake was not inhibited by ammonium appears to support the idea that two different transport systems (permeases) operate in the uptake of nitrate and nitrite ions (Garbisu *et al.*, 1992).

Blanco *et al.* (1993) also studied the nitrate-removal capacity from water of immobilized microalgae and found results similar to those reported by Garbisu *et al.* (1991, 1992, 1993, 1994). In their studies, as well as *P. laminosum* cells they utilized other microalgae such as *Phormidium uncinatum*, *Scenedesmus obliquus* and *Phormidium* sp. in different air-agitated photobioreactors of various shapes (Figure 9.5). In all those works, and using residence times of 3–5 h and light intensities of 100 μmol photon m^{-2} s^{-1}, they found nitrate uptake efficiencies of up to 90 % of the concentration of this ion supplied in the influents to the bioreactors (50 mg l^{-1} nitrate).

By contrast, *P. laminosum* cells immobilized in polymer foams did not show high phosphate uptake efficiencies (Garbisu *et al.*, 1993, 1994). Nitrogen-starved cells were also examined in relation to their phosphate uptake characteristics and it was shown that starvation led to lower uptake rates. The addition of nitrate to nitrogen-starved cells

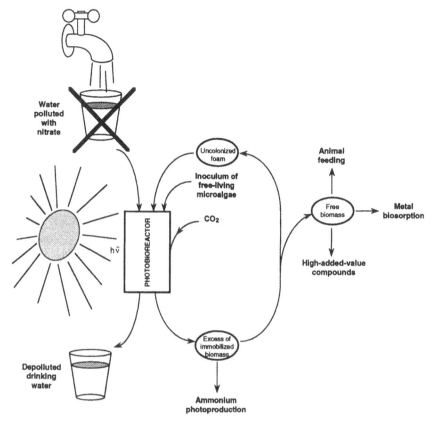

Figure 9.6   Depollution of nitrate-containing water using cyanobacteria (*Phormidium* sp.) immobilized in polymeric matrices

markedly increased phosphate uptake. Phosphate uptake was inhibited in the dark. Although our data showed that nitrogen-sufficient *P. laminosum* had significant phosphate uptake only when light was available, this should not be an insuperable obstacle when a bioreactor is designed to deal with the practical problem of removing phosphate from polluted water. In fact, given the likelihood that light will not penetrate very far into the columns of immobilized algae, various ways have been used to solve this light penetration problem (which always arises when using photobioreactors). Approaches studied are: matrices with internal reflections, organisms which do not require much light (e.g. cyanobacteria and nutrient-deficient organisms), fluidized-bed systems in the case of foam matrices, or different bioreactor designs which enhance light penetration (e.g. tubular photobioreactors) (Garbisu *et al.*, 1993).

When considering the possibility of using these immobilized systems for the removal of ions from polluted water, the fact that no significant leakage of cells was observed in the effluent during reactor operation is of vital importance. Lack of cell leakage overcomes one of the major problems in the utilization of microalgae for the biological treatment of water, i.e. the recovery of microalgae from the treated effluent. Once the supports are fully colonized and the cells begin to leak out from the foams into the effluent, the bioreactor can be stopped and most of the immobilized cells easily removed from the foam matrices by squeezing and the foams subsequently reused without re-inoculation (Figure 9.6). If cells cannot be recycled due to their age or metabolic state, they can be utilized for different purposes, e.g. feed (fish), extraction of high-added-value compounds (phycobilin pigments), or removal of heavy metals by means of chemically treated biomass (see Chapter 11).

From all these studies, it was concluded that *P. laminosum* immobilized on polymer foams is of potential value for biological removal of inorganic nitrogen and phosphorus from water in continuous-flow bioreactor systems.

9.6 Concluding Remarks

Biological removal of nitrogen and phosphorus from water appears to offer some advantages over chemical and physicochemical treatments and, therefore, has been attracting attention in recent years. In this context, although the most utilized biological methods for wastewater treatment have focused mainly on removal by bacteria, during the last 20–25 years numerous investigations have been carried out on the possibility of using microalgae. Wastewaters from domestic, industrial or agricultural sources are a favourable microalgal culture medium. Through the appropriate use of the nutrient uptake capabilities of these organisms, nutrients (such as inorganic nitrogen and phosphorus) in the secondary effluents of wastewater treatment plants can be removed and used in the growth of algal biomass.

So far, one of the major problems in the use of microalgae for the biological treatment of wastewaters is their recovery from the treated effluent. Among the most recent ways of bypassing the problem of harvesting the biomass are immobilization techniques. Consequently, various studies have been carried out to study the feasibility of using immobilized microalgae (mainly cyanobacteria immobilized in different matrices) in batch and continuous-flow bioreactors for the removal of inorganic nitrogen and phosphorus from water. From all these studies, it was concluded that some polymer immobilized cyanobacteria are of potential value for biological removal of inorganic nitrogen and phosphorus from water in continuous-flow bioreactor systems.

Acknowledgements

This work was partially supported by grants from Acciones Integradas Hispano-Británicas, Spanish Ministry of Education (PB88-0300), Commission of the European Communities DG XII-E, (STEP CT91-0126) and the University of the Basque Country (UPV 042.310-TC111/92).

References

ABALDE, J., CID, A., FIDALGO, P., TORRES, E. and HERRERO, C., 1995, *Microalgas: Cultivo y Aplicaciones*, La Coruña: Servicio de Publicaciones Universidad de La Coruña.

ABEL, P.D., 1989, *Water Pollution Biology. Ellis Horwood Series in Wastewater Technology*, Chichester: John Wiley & Sons.

ABELIOVICH, A., 1986, Algae in wastewater oxidation ponds, in Richmond, A. (ed.) *CRC Handbook of Microalgal Mass Culture*, Boca Raton: CRC Press, pp.331–338.

BLANCO, F., GOROSTIZA, I., DE LAS FUENTES, L., GIL, J.M., URRUTIA, I., GARBISU, C. *et al.*, 1993, Eliminación de nitrato de aguas potencialmente potables por microalgas inmovilizadas, in *III Congreso de Ingeniería Ambiental-Proma*, Bilbao: Feria Internacional de Bilbao, pp.187–195.

BOROWIZTKA, M.A., 1986, Microalgae as sources of fine chemicals, *Microbiologica Sciences*, **3**, 372–375.

BUCKE, C., 1986, Methods of immobilising cells, in Webb, C., Black, G.M. and Atkinson, B. (eds), Process Engineering Aspects of Immobilized Cell Systems, Warwickshire: The Institution of Chemical Engineers, pp.20–34.

CALDWELL, D.H., 1946, Sewage oxidation ponds. Performance, operation and design, *Sewage Work Journal*, **18**, 433.

CALLEGARI, J.P., 1989, Feu vert pour les microalgues, *Biofutur*, **76**, 25–40.

CHEVALIER, P. and DE LA NOÜE, J., 1985a, Wastewater nutrient removal with microalgae immobilized in carrageenan, *Enzyme and Microbial Technology*, **7**, 621–624.

CHEVALIER, P. and DE LA NOÜE, J., 1985b, Efficiency of immobilized hyperconcentrated algae for ammonium and orthophosphate removal from wastewaters, *Biotechnology Letters*, **7**, 395–400.

COMEAU, Y., RABIONWITZ, B., HALL, K.J. and OLDHAM, W.K., 1987, Phosphate release and uptake in enhanced biological phosphorus removal from wastewater, *Journal of Water Pollution Control Federation*, **59**, 707–715.

DE LA NOÜE, J. and DE PAUW, N., 1988, The potential of microalgal technology: A review of production and uses of microalgae, *Biotechnological Advances*, **6**, 725–770.

DE LA NOÜE, J. and PROULX, D., 1988a, Biological tertiary treatment of urban wastewaters with chitosan-immobilized *Phormidium, Applied Microbiology and Biotechnology*, **29**, 292–297.

DE LA NOÜE, J. and PROULX, D., 1988b, Tertiary treatment of urban wastewaters by chitosan-immobilized *Phormidium* sp, in Stadler, T., Mollion, J., Verdus, M.C., Karamanos, Y., Morvan, H. and Christiaen, D. (eds) *Algal Biotechnology*, New York: Elsevier Applied Science, pp.159–168.

DUDLEY, N., 1990, *Nitrates: The Threat to Food and Water*, London: Green Print.

EISENBERG, D.M., KOOPMAN, B., BENEMAN, J.R. and OSWALD, W.J., 1981, Algal bioflocculation and energy conservation in microalgal sewage ponds, *Biotechnology and Bioengineering Symposium*, **11**, 429–448.

FALKNER, G., HORNER, F. and SIMONIS, W., 1980, The regulation of the energy-dependent phosphate uptake by the blue-green algae *Anacystis nidulans, Planta*, **149**, 138–143.

FALKNER, G., FALKNER, R., GRAFFIUS, D. and STRASSER, P., 1984, Bioenergetic and ecological aspects of phosphate uptake by blue–green algae, *Archives of Hydrobiology*, **101**, 89–99.

FALLOWFIELD, H.J. and GARRETT, M.K., 1985, The treatment of wastes by algal culture, *Journal of Applied Bacteriology, Symposium Supplement*, 187S–205S.

FOGG, G.E., STEWART, W.D.P., FAY, P. and WALSBY, A.E., 1973, *The Blue-Green Algae*, London: Academic Press.

FRY, I.V. and PESCHEK, G.A., 1988, Electron paramagnetic resonance-detectable Cu^{2+} in *Synechococcus* 6301 and 6311: aa_3-type cytochrome-*c* oxidase of cytoplasmic membrane, *Methods in Enzymology*, **167**, 450–459.

FUKUI, A. and TANAKA, A., 1982, Immobilized microbial cells, *Annual Review of Microbiology*, **36**, 145–172.

GARBISU, C., GIL, J.M., HALL, D.O., BAZIN, M.J. and SERRA, J.L., 1991, Removal of nitrate from water by foam-immobilized *Phormidium laminosum* in batch and continuous-flow bioreactors, *Journal of Applied Phycology*, **3**, 221–234.

GARBISU, C., HALL, D.O. and SERRA, J.L., 1992, Nitrate and nitrite uptake by free-living and immobilized N-starved cells of *Phormidium laminosum, Journal of Applied Phycology*, **4**, 139–148.

GARBISU, C., HALL, D.O. and SERRA, J.L., 1993, Removal of phosphate by foam-immobilized *Phormidium laminosum* in batch and continuous-flow bioreactors, *Journal of Chemical Technology and Biotechnology*, **57**, 181–189.

GARBISU, C., HALL, D.O., LLAMA, M.J. and SERRA, J.L., 1994, Inorganic nitrogen and phosphate removal from water by free-living and polyvinyl-immobilized *Phormidium laminosum* in batch and continuous-flow bioreactors, *Enzyme and Microbial Technology*, **16**, 395–401.

GUERRERO, M.G. and LARA, C., 1987, Assimilation of inorganic nitrogen, in Fay, P. and van Baalen, C. (eds) *The Cyanobacteria*, London: Elsevier Science Publishers B.V. (Biomedical Division), pp.163–186.

HAGMAR, L., BELLANDER, T., ANDERSSON, C., LINDEN, K., ATTEWELL, R. and MOLLER, T., 1991, Cancer morbidity in nitrate fertilizer workers, *International Archives of Occupational and Environmental Health*, **63**, 63–67.

HALL, D.O. and RAO, K.K., 1989, Immobilized photosynthetic membranes and cells for the production of fuels and chemicals, *Chimicaoggi*, **7**, 40–47.

HASHIMOTO, S. and FURUKAWA, K., 1989, Nutrient removal from secondary effluent by filamentous algae, *Journal of Fermentation Bioengineering*, **67**, 62–69.

HASHIMOTO, S., FURUKAWA, K. and SHIOYAMA, M., 1987, Autotrophic denitrification using elemental sulphur, *Journal of Fermentation Technology*, **65**, 683–692.

KERBY, N.W., ROWELL, P. and STEWART, W.D.P., 1989, The transport, assimilation and production of nitrogenous compounds by cyanobacteria and microalgae, in Cresswell, R.C., Rees, T.A.V. and Shah, N. (eds) *Algal and Cyanobacterial Biotechnology*, Harlow: Longman Scientific and Technical, pp.50–90.

KIERSTAN, M. and BUCKE, C., 1977, The immobilization of microbial cells, subcellular organelles and enzymes in calcium alginate gels, *Biotechnology and Bioengineering*, **19**, 387–397.

KURT, M., DUNN, I.J. and BOURNE, J.R., 1986, Biological denitrification of drinking water using autotrophic organisms with H_2 in a fluidized-bed biofilm reactor, *Biotechnology and Bioengineering*, **29**, 493–501.

LAVOIE, A. and DE LA NOüE, J., 1983, Harvesting microalgae with chitosan, *Journal of World Marineculture Society*, **14**, 685.

LAVOIE, A. and DE LA NOüE, J., 1984, Recuperation de microalgues en euax usees: etude comparative de divers agents floculants, *Canadian Journal of Civil Engineering*, **11**, 266–272.

LAVOIE, A. and DE LA NOüE, J., 1985, Hyperconcentrated cultures of *Scenedesmus obliquus*. A new approach for wastewater biological tertiary treatment, *Water Research*, **19**, 1437–1442.

MARCO, E. and ORüS, M.I., 1988, Variation in growth and metabolism with phosphorus nutrition in two cyanobacteria, *Journal of Plant Physiology*, **132**, 339–344.

MAVITUNA, F. and PARK, J.M., 1985, Growth of immobilized plant cells in reticulated polyurethane foam matrices, *Biotechnology Letters*, **7**, 637–640.

METTING, B. and PYNE, J.W., 1986, Biological active compounds from microalgae, *Enzyme and Microbial Technology*, **8**, 386–394.

MOHN, F.H., 1988, Harvesting of microalgal biomass, in Borowizka, M.A. and Borowizka, L.J. (eds) *Microalgal Biotechnology*, Cambridge: Cambridge University Press, pp.395–414.

OKAMOTO, N., TEI, H., MURATA, K. and KIMURA, A., 1986, Phosphate-polymer-dependent phosphorylation of glycolytic substrates by *Brevibacterium* species, *Journal of General Microbiology*, **132**, 1519–1523.

OSWALD, W.J., 1988, Microalgae and wastewater treatment, in Borowizka, M.A. and Borowizka, L.J. (eds) *Microalgal Biotechnology*, Cambridge: Cambridge University Press, pp.305–328.

OSWALD, W. and GOTAAS, H.B., 1957, Photosynthesis in sewage treatment, *Transactions of the American Society of Civil Engineering*, **122**, 73.

OZAKI, H., LIU, Z. and TERASHIMA, Y., 1991, Utilization of microorganisms immobilized with magnetic particles for sewage and wastewater treatment, *Water Science Technology*, **23**, 1125–1136.

POULIOT, Y. and DE LA NOüE, J., 1985, Mise au point d' une usine-pilote d' épuration des eaux usées par production de microalgues, *Revue Française de la Science de l'eau*, **4**, 207–222.

POULIOT, Y., BUELNA, G., RACINE, C. and DE LA NOüE, J., 1989, Culture of cyanobacteria for tertiary wastewater treatment and biomass production, *Biological Wastes*, **29**, 81–91.

PROULX, D. and DE LA NOüE, J., 1988, Removal of macronutrients from wastewater by immobilized algae, in Moo-Young, M. (ed.) *Bioreactor Immobilized Enzymes and Cells: Fundamentals and Applications*, New York: Elsevier Applied Science, pp.301–310.

RAO, K.K. and HALL, D.O., 1984, Photosynthetic production of fuels and chemicals in immobilized systems, *Trends in Biotechnology*, **2**, 124–129.

RICHMOND, A., 1986, *Handbook of Microalgal Mass Culture*, Boca Raton: CRC Press.

ROBINSON, P.K., REEVE, J.O. and GOULDING, K.H., 1989, Phosphorus uptake kinetics of immobilized *Chlorella* in batch and continuous-flow culture, *Enzyme and Microbial Technology*, **11**, 590–596.

ROSEVEAR, A., 1984, Immobilized biocatalysts – a critical review, *Journal of Chemical Technology and Biotechnology*, **34B**, 127–150.

ROYAL COMMISSION ON ENVIRONMENTAL POLLUTION, 1979, Seventh Report. Agriculture and Pollution. London: HMSO, pp. 280 *et seq.*

SCHOPF, J.W. and WALTER, M.R., 1982, Origin and early evolution of cyanobacteria: The geological evidence, in Carr, N.G. and Whitton, B.A. (eds) *The Biology of Cyanobacteria*, Oxford: Blackwell Scientific Publications, pp.543–564.

SCHRAMM, W., 1991, Seaweeds for waste water treatment and recycling of nutrients, in Guiry, M.D. and BLUNDEN, G. (eds) *Seaweed Resources in Europe: Uses and Potential*, Chichester: John Wiley & Sons, pp.149–168.

SERRA, J.L., ARIZMENDI, J.M., BLANCO, F., MARTÍNEZ-BILBAO, M., ALAÑA, A., FRESNEDO, O. *et al.*, 1990, Nitrate assimilation in the non-N_2-fixing cyanobacterium *Phormidium laminosum*, in Ullrich, W.R., Rigano, C., Fuggi, A. and Aparicio, P.J. (eds) *Inorganic Nitrogen in Plants and Microorganisms*, Berlin: Springer-Verlag, pp.196–202.

SHELEF, G. and SOEDER, C.J., 1980, *Algae Biomass: Production and Use*, Amsterdam: Elsevier/North-Holland Biomedical Press.

SHI, D.J., 1987, Energy metabolism and structure of the immobilized cyanobacterium *Anabaena azollae*. PhD thesis, King's College London, University of London.

SMITH, J.E., 1988, *Biotechnology*. 2nd Edition, London: Edward Arnold.

SOMIYA, I., TSUNO, H. and MATSUMOTO, M., 1988, Phosphorus release-storage reaction and organic substrate behaviour in biological phosphorus removal, *Water Research*, **22**, 49–58.

STANIER, R.Y. and COHEN-BAZIRE, G., 1977, Phototropic prokaryotes-cyanobacteria, *Annual Review of Microbiology*, **31**, 225–274.

TAM, N.F.Y. and WONG, Y.S., 1989, Wastewater nutrient removal by *Chlorella pyrenoidosa* and *Scenedesmus* sp., *Environmental Pollution*, **58**, 19–34.

TAM, N.F.Y. and WONG, Y.S., 1990, The comparison of growth and nutrient removal efficiency of *Chorella pyrenoidosa* in settled and activated sewages, *Environmental Pollution*, **65**, 93–108.

TAMPION, J. and TAMPION, M.D., 1987, *Immobilized Cells: Principles and Applications*, Cambridge Studies in Biotechnology 5, Cambridge: Cambridge University Press.

WINTERINGHAM, F.P.W., 1974, Nitrogen residue problems of food and agriculture, in *Effects of Agricultural Production on Nitrates in Food and Water with Particular Reference to Isotope Studies. Proceedings and Report of a Panel of Experts, Vienna*, 4–8 June 1973, organized by the JOINT FAO/IAEA Division of Atomic Energy in Food and Agriculture, International Atomic Energy Agency, Vienna, pp.3–6.

10

Engineered Reed Bed Systems for the Treatment of Dirty Waters

A.J. BIDDLESTONE AND K.R. GRAY

10.1 Introduction

The treatment of aqueous effluents by their application to wetlands for natural purification is a historical practice but over the past 15 years considerable interest has developed in the concept of using constructed wetlands for the treatment of point sources of pollution (Hammer, 1989; Cooper and Findlater, 1990; Biddlestone *et al.*, 1991; Moshiri, 1993; Bavor and Mitchell, 1994; Cooper *et al.*, 1996; Haberl *et al.*, 1997). This interest has arisen because of the need to identify low-cost environmentally friendly techniques for the treatment of dirty waters. Such waters arise from many sources: some contain only a single pollutant, but most contain a wide variety. Materials can be pollutants by adding excess nutrients such as carbon substrate, nitrogen and phosphorus, or by causing toxicity by the addition of heavy metals and pathogenic organisms. The work of Seidel (1973, 1976), closely followed by Kickuth (1984), showed that significant removal of pollutants is possible when contaminated water is passed through beds of reeds planted in soil or gravel. These dirty waters can have a wide range of strengths depending upon their source, as shown in Table 10.1. The major pollutants present in such contaminated waters range from readily biodegradable compounds to heavy metals (Table 10.2).

Table 10.1 Polluting strength of dirty waters

Strength	BOD (mg l^{-1})	Source
Weak	< 45	Effluent from secondary sewage treatment
Medium	45–300	Effluent from primary sewage treatment
Strong	300–3000	Farm dirty water, industrial effluents
Very strong	> 3000	Farm manure slurries, silage liquor, industrial effluents

Dirty water strength can be measured by the biochemical oxygen demand (BOD)

Table 10.2 Major pollutants

Carbon	Readily biodegradable compounds measured as BOD
	Slowly biodegradable plus readily biodegradable compounds
	measured as chemical oxygen demand (COD)
Nitrogen in several forms	Measured as total-N, organic-N, NH_4-N, NO_3 and NO_2-N
Phosphorus in several forms	Measured as total phosphate and orthophosphate
Suspended solids	
Heavy metals	Such as iron, manganese, lead, zinc
Pathogens	Measured in colony forming units (cfu) g^{-1} dry weight or
	wet weight

10.2 Basis of Treatment

The aquatic plants mainly employed in constructed wetlands are the common reed, *Phragmites australis*, the common reedmace, *Typha latifolia*, and the common club-rush, *Schoenoplectus lacustris*. These plants have a root system of rhizomes or thick hollow air passages from which the fine hair roots hang down. The vertical aerial shoots develop upwards from the rhizome. In hot climates the floating water hyacinth *Eichhornia crassipes* has been widely used for water purification.

Oxygen from the leaves passes down through the stems and rhizomes and is exuded from the fine roots so that a thin oxygenated aqueous film surrounds the hairs; as a result this 'root zone' or rhizosphere supports a very large population of aerobic microorganisms, far larger than in comparable soil. In the spaces between the root hairs anaerobic microorganisms predominate, with their slower growing lifestyles.

As a result, water passing through a thickly developed 'root zone' of such aquatic plants encounters alternate aerobic and anaerobic microbial populations which convert the carbonaceous, and to a lesser extent the nitrogenous and phosphorous, contaminants in the water to less polluting materials (Table 10.3). In addition the root growth of the plants causes minor displacement of the media in which the plants are growing and opens up passages for water flow. Above ground the multiplicity of upright stems, often up to 500 per m^2, provides a veritable jungle for water flowing over the bed surface. Microorganisms can form biofilms around the lower stems, which can trap suspended particles in the water by adsorption. Table 10.4 indicates the numbers of microorganisms present in the rhizosphere and on the supporting gravel media.

It shows that microbial populations in the rhizospheres of *Phragmites* and *Typha* are two to three orders of magnitude greater than those in unplanted gravels, indicating the importance of the vegetation, and that the population is dominated by the bacteria. However, Hatano *et al.* (1993) found that although the bacteria are highest in numbers they have relatively low enzymatic activities within the organic substrates tested. As such they would be expected to carry out most of the degradation of the simpler organic materials in wastewaters. In contrast the actinomycetes and fungi, though fewer in number, have a wide range of hydrolysis activities. Many isolates of these organisms showed significant amylase, protease, chitinase, xylanase and cellulase activity. They would be expected to degrade many of the larger molecules in the wastewaters, which contribute to the chemical oxygen demand (COD) level, in addition to the plant litter falling onto the bed surface.

Table 10.3 Removal steps in reed bed treatment systems

Water constituent	Removal steps
Suspended solids	Sedimentation, filtration, adsorption
BOD	Sedimentation and filtration Degradation to CO_2, H_2O and NH_3 by microorganisms attached to plant and sediment surfaces
Nitrogen	Main removal by nitrification–denitrification Ammonia oxidized to nitrite/nitrate by nitrifying bacteria in aerobic zones Nitrates converted to N_2 gas by denitrifying bacteria in anoxic zones A little removal by plant uptake and ammonia volatilization
Phosphorus	Adsorption, complexation, precipitation reactions within the bed matrix, particularly with aluminium, iron, calcium and clay minerals. Very little plant uptake
Heavy metals	Precipitation reactions after pH changes, sedimentation and adsorption onto biomass films on plant stems
Pathogens	Sedimentation and filtration. Competition and natural die-off. Excretion of antibiotics from roots of plants and from composting of plant litter on bed surface

Table 10.4 Comparison of microorganisms found in gravel matrices and rhizospheres of sub-surface flow wetlands (Hatano *et al.*, 1993)

Substrate	Bacteria	Actinomycetes	Fungi
Unplanted gravel[a]	0.6×10^6	3.1×10^4	1.0×10^3
Typha gravel[a]	1.6×10^6	1.4×10^5	4.0×10^3
Typha rhizosphere[b]	3.5×10^9	2.5×10^6	2.8×10^4
Phragmites gravel[a]	1.2×10^7	2.8×10^5	1.1×10^5
Phragmites rhizosphere[b]	0.6×10^9	1.3×10^6	1.6×10^6

[a] Expressed as cfu g^{-1} of dry weight
[b] Expressed as cfu g^{-1} of wet weight

Horizontal subsurface flow of dirty water through the 'root zone' is used for tertiary treatment of sewage and other weak wastewaters (Green and Upton, 1994; Cooper *et al.*, 1996). Downflow through a multi-layered aggregate bed is employed for treating stronger wastes (Burka and Lawrence, 1990; Gray *et al.*, 1990; Cooper *et al.*, 1996). Above-ground flow through the plant stems is used for metal removal and pH change, particularly for acid mine discharges (NRA, 1994). Finally, water transpiration and 'root zone' oxidation is used for stabilizing and drying sludges and slurries (Reed *et al.*, 1995).

In practice the common reed, *Phragmites australis*, has been the aquatic plant mostly used in European applications. The common reedmace, *Typha latifolia*, has been extensively used in the USA and is starting to be employed in schemes in the UK using above-ground flow for metals removal (NRA, 1994). These two species have proved well able to tolerate weak, medium and strong wastewaters; see Table 10.1. So far there have been no major attempts to use reed beds for very strong effluents, as these are likely to put the plants under stress. Moreover, the simple technology of reed beds cannot reasonably be expected to treat this type of material without greatly improved aeration ability.

10.3   Horizontal Flow Beds

Horizontal flow beds consist of either a soil or gravel matrix into which the reeds are planted (Figure 10.1). The matrix is kept flooded, with the water surface less than 5 cm below the top of the bed (Boon, 1985). A basis of design, developed from experience of treating sewage, has been established (Conley *et al.*, 1991; Cooper *et al.*, 1996).

For the tertiary treatment of sewage and for other weak wastewaters the bed length is short, 5–10 m. For the secondary treatment of sewage having a BOD of about 300 mg l^{-1} the bed length can be about 70 m; the bed may need to be subdivided into terraces. The residence time in these horizontal flow beds is several days.

At present the design equations available are based on the work of Kickuth (Cooper *et al.*, 1996). The area of the bed A_h is given by the equation

$$A_h = \frac{Q_d(lnC_o - lnC_t)}{k_{BOD}}$$

where A_h = surface area of bed, m^2; Q_d = daily average wastewater flow rate, m^3/day; C_o = average BOD of inlet wastewater, mg l^{-1}; C_t = required average BOD of outlet water, mg l^{-1}; k_{BOD} = rate constant, m/day.

The value of k_{BOD} depends on the biodegradability of the wastewater and on the bed matrix. Values of this constant are given in Cooper *et al.* (1996). For sewage recent UK experience suggests that an area of 3–5 m^2 per person equivalent is appropriate. The equation is of a similar form to design equations for conventional trickling or percolating 'filters'; it can be derived by considering the bed to be a plug flow reactor and assuming a model for the microbial kinetics.

The cross-sectional area of the bed, A_c, can be calculated from the application of Darcy's law for flow of a fluid through a granular bed:

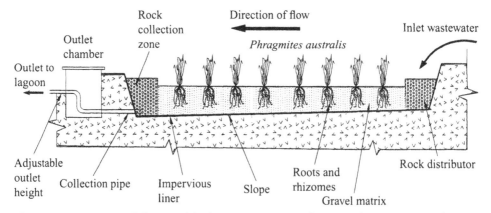

Figure 10.1   Horizontal flow reed bed treatment system showing *Phragmites australis* planted in a matrix of limestone chippings, the inlet wastewater distribution and outlet purified water collection zones of rocks, and the water level control device. The bed is constructed in a shallow excavation with an impervious liner to prevent seepage of polluted water into the subsoil

$$A_c = \frac{Q_s}{k_f dH/ds}$$

where A_c = cross-sectional area m^2; Q_s = average flow rate of wastewater m^{3}s^{-1}; k_f = hydraulic conductivity of the bed matrix, m^{3}m^{-2}s^{-1}(= ms^{-1}); dH/ds = hydraulic gradient (5 %). In the UK, for gravel media a k_f value of 10^{-3} ms^{-1} is used. Kickuth also recommended that the horizontal velocity of the wastewater should be kept below 10^{-4} ms^{-1} in order to prevent erosion and disturbance of the mosaic of aerobic, anoxic and anaerobic zones in the bed.

Research conducted by the University of Birmingham into the use of reed beds for treating farm dirty waters commenced in 1986. Work has been carried out on full-scale reed beds on two farms in conjunction with laboratory work at the University (Gray *et al.*, 1990, Cooper *et al.*, 1996). The work has progressed through three separate experimental periods, with the combined horizontal and downflow reed bed systems at the two farms being identical in form, though differing in size. The trials were conducted under farm conditions with varying influents and strengths. The dirty water used in the trials included septic tank sewage effluent, farmyard run-off from heaps of pig manure and dairy parlour washings. The averaged results for the horizontal flow beds for each experimental period are combined in Figure 10.2, which presents BOD removal against inlet BOD loading. These results represent over 128 separate analyses with inlet BOD values ranging from 1884 down to 88 mg l^{-1}. Table 10.5 gives an indication of the level of pollutant removal in horizontal flow reed beds.

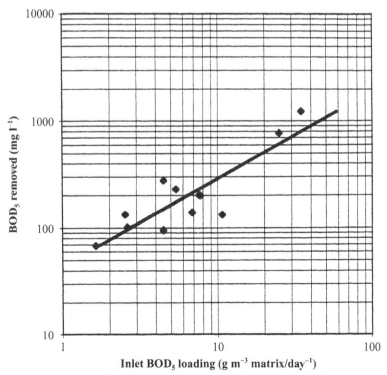

Figure 10.2 BOD$_5$ removal and loading for horizontal flow reed beds ($y = 43.7x^{0.82}$, $R^2 = 0.75$)

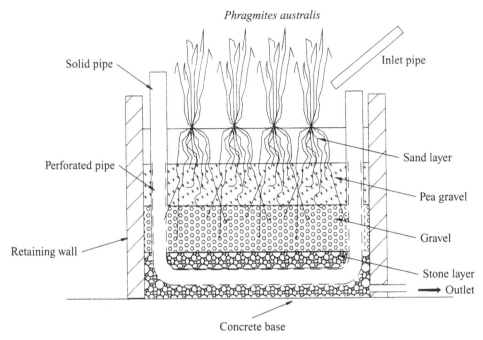

Figure 10.3 Downflow reed bed with *Phragmites australis* planted in a multilayered matrix of various-sized aggregates. A system of perforated pipes allows countercurrent diffusion of oxygen into the bed and carbon dioxide out of it. The wastewater is flushed onto the bed in pulses large enough to allow even distribution over the bed top surface. These beds are often built above ground between retaining walls

10.4   Downflow Beds

Downflow beds, as shown in Figure 10.3, are constructed of sharp sand, pea gravel, gravel and stones in layers of appropriate thickness (Burka and Lawrence, 1990; Gray *et al.*, 1990). Oxygen from the air can diffuse into the bed via perforated pipes set within the layers, as well as from the roots of the reeds. According to the strength of the initial wastewater one, two or three stages of downflow treatment may be required; each stage consists of several identical beds, perhaps six in the first stage, four in the second and three in the third (Biddlestone *et al.*, 1991; Job *et al.*, 1991). These undergo alternating operating and resting periods. During the operating period of several days the surface of the sand top layer gradually becomes choked with fine filtered solids, leading to flooding conditions; by resting the bed for about a week the solids are oxidized off and the bed becomes permeable again. Downflow beds have considerably more potential for oxygenation than have horizontal flow ones. Being self-draining, the residence time in them is short, only a matter of minutes. So far these beds have been designed essentially on hydraulic considerations, to handle the water flow. Experience is slowly being acquired in designing these beds to meet organic loads in terms of kg BOD/day.

The University of Birmingham research on the farm sites has accumulated information on the performance of downflow beds. Figure 10.4 gives data from over 185 determinations with BOD levels between 5075 and 125 mg l^{-1}. This work has shown that some six stages of downflow beds are needed to reduce an influent BOD of 5000 mg l^{-1} down to below 50 mg l^{-1}, the level at which nitrification becomes appreciable. An indication of removal efficiency for various forms of reed bed is given in Table 10.5.

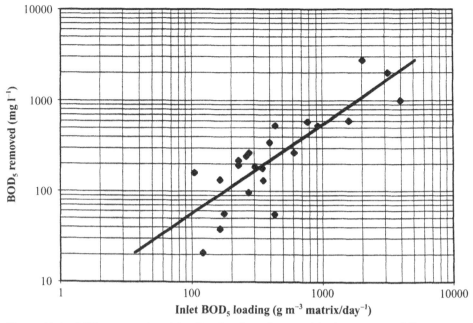

Figure 10.4 BOD$_5$ removal and loading for downflow reed beds ($y = 0.487x^{1.02}$, $R^2 = 0.71$)

Table 10.5 Pollutant removal by reed beds

Bed	Removal	References
Horizontal subsurface flow	Results from a large number of beds in Europe indicate that BOD removal is 80–90 %, with a typical outlet concentration of 20 mg l^{-1}, total nitrogen removal is only 20–30 % and total phosphorus removal is 30–40 %	Cooper (1990), Cooper *et al.* (1996)
Downflow	Using two downflow stages to treat domestic sewage achieved removals of 93 % BOD, 90 % suspended solids, 75 % NH$_4$-N and 37 % orthophosphate	Burka and Lawrence (1990)
Overland flow	Passing the effluent from a lead–zinc mine through a bed of *Typha latifolia* achieved reductions of suspended solids of 99 %, lead of 95 % and zinc 80 %	Lan *et al.* (1990)

BOD: biochemical oxygen demand

10.5 Overland Flow Beds

Overland flow beds aim to precipitate metal salts from solution by pH change and aeration; the precipitates are then adsorbed onto biomass on the plant stems or settle onto the matrix surface (Reed *et al.*, 1995). Sub-surface flow is impossible in these situations because the precipitates would soon choke the bed. These beds are mainly designed on residence time considerations and can achieve substantial reductions (Table 10.5).

10.6 Sludge Treatment Beds

Sludge treatment beds comprise a sand/gravel matrix planted with reeds onto which the sludge/slurry is poured periodically, about once a fortnight (Nielsen, 1990). Dewatering takes place by drainage and evapo-transpiration. If the sludge is carbonaceous in content then oxidation and stabilization can occur. The beds are sized essentially on residence time. As the thickness of solids builds up on the bed the reed and its roots grow up with it. The sludge solids are allowed to accumulate for several years before they are removed; this is easiest done when the reeds have died down in winter.

10.7 Application of Reed Bed Systems

The advantages of constructed wetlands systems include an initial capital cost, which is normally cheaper than that of its mechanical counterpart. For instance, when a combination of septic tank and horizontal flow reed bed is used for secondary sewage treatment in small rural works, Green and Upton (1993) report a capital cost in the range £700–1600 per head of population. This range of costs relates to the size of the unit and to site-specific factors. These costs compare with ones of about £2000 per head of population for the alternative mechanized rotating biological contactor installation.

Where the lie of the land is suitable, gravity flow of wastewater is possible and an electricity supply is not needed. Maintenance and supervision requirements are also low, hence operating costs are far less than those of mechanized units. Mainly because of these cost advantages, reed beds are being constructed in significant numbers for the treatment of weak and medium-strength wastewaters (Table 10.6).

Table 10.6   Applications of reed bed systems

Wastewater type	Application
Municipal sewage	Secondary and tertiary treatment
Sewage treatment following septic tanks	Houses, hotels, stately homes, caravan parks, leisure centres, golf courses
Run-off	From roads, airports and vehicle servicing areas (trains and coaches)
Farm	'Dirty water' from dairy units, slaughterhouses and vegetable washing water
Landfill	Leachate
Industrial	From chemicals manufacture, oil refineries, food and beverage manufacture, paper mills, tanneries
Mine	With metal contamination from operating and derelict coal, metal mines
Sludges, particularly organic ones	Dewatering and stabilization

However, in addition to economic advantages reed bed treatment systems are aesthetic-ally pleasing and provide wildlife habitats for birds and invertebrates, particularly where they are carefully landscaped and include pools (Merritt, 1994).

With this new technology, new applications are still being attempted and problems exposed. Reliable data on flows is often lacking and flows can vary widely in some instances owing to rainfall. The calculation of bed sizes has tended therefore to err heavily on the side of caution; this also helps the bed to cope with occasional overload conditions.

Sewage and farm wastes normally provide adequate plant nutrients for growth of the reeds but with many industrial effluents these nutrients are lacking and need to be added to the inflow.

Reed beds are proving adequate for handling wastewaters with BOD levels up to about 500 mg l^{-1}, using a combination of downflow and horizontal flow beds. They have not yet been proven for wastes with higher BOD levels, such as farm 'dirty water' which has a BOD up to about 3000 mg l^{-1} (Nicholson, 1994). With farm wastes and many other effluents the concentration of ammoniacal nitrogen is also high, often up to 500 mg l^{-1}. This is proving difficult to break down in simple reed bed systems using the normal processes of nitrification/denitrification and is holding up application of wetlands to such wastes (Nicholson, 1994). The evidence gained from horizontal flow beds indicates that, although BOD removal is normally in the range 80–90 %, the removal of total nitrogen can be as low as 20–30 %. After solving the problems of nitrogen removal those of phosphorus removal will become important as this element is heavily implicated in the eutrophication of water masses.

10.8 Conclusions

The technology of engineered reed beds is still relatively new. Their application in horizontal flow beds to the tertiary treatment of sewage has proved technically and economically sound in small rural works. The use of downflow beds in supervised situ-ations for slightly more demanding effluents is showing merit. However, their application to strong wastewaters with BODs up to 3000 mg l^{-1} has not yet been resolved and the removal of ammoniacal nitrogen to satisfactory levels is a major drawback to their em-ployment in farm waste treatment. Use of reed beds for metals removal has great poten-tial and will be greatly needed to cope with seepage from abandoned coal and metal mines. Sludge stabilization and drying on reed beds is starting to make progress. Overall there is still enormous potential for advances in reed bed technology.

References

BAVOR, H.J. and MITCHELL, D.S. (eds), 1994, Wetland systems in water pollution control, *Water Science and Technology*, **29**(4).

BIDDLESTONE, A.J., GRAY, K.R. and THURAIRAJAN, K., 1991, A botanical approach to the treatment of wastewaters, *Journal of Biotechnology*, **17**(3), 209–220.

BOON, A.G., 1985, Report of a visit by members and staff of WRc to Germany to investigate the root zone method for treatment of wastewaters, Report 376-S/1, WRc (Processes).

BURKA, U. and LAWRENCE, P.C., 1990, A new community approach to waste treatment with higher water plants, in Cooper, P.F. and Findlater, B.C. (eds), *Constructed Wetlands in Water Pollution Control*, Oxford: Pergamon Press, pp.359–371.

CONLEY, L.M., DICK, R.I. and LION, L.W., 1991, An assessment of the root zone method of wastewater treatment, *Research Journal Water Pollution Control Federation*, **63**(3), 239–247.

COOPER, P.F., 1990, *European Design and Operations Guidelines for Reed Bed Treatment Systems*, Report UI 17, Water Research Centre, Swindon, UK.

COOPER, P.F. and FINDLATER, B.C. (eds), 1990, *Constructed Wetlands in Water Pollution Control*, Oxford: Pergamon Press.

COOPER, P.F., JOB, G.D., GREEN, M.B. and SHUTES, R.B.E., 1996, *Reed Beds and Contructed Wetlands for Wastewater Treatment*, Swindon: WRc.

GRAY, K.R., BIDDLESTONE, A.J., JOB, G. and GALANOS, E., 1990, The use of reed beds for the treatment of agricultural effluents, in Cooper, P.F. and Findlater, B.C. (eds) *Constructed Wetlands in Water Pollution Control*, Oxford: Pergamon Press, pp.333–346.

GREEN, M.B. and UPTON, J., 1993, Reed bed treatment for small communities – UK experience, in Moshiri, G.A. (ed.) *Constructed Wetlands for Water Quality Improvement*, Boca Raton, Florida: Lewis Publishers, pp.517–524.

GREEN, M.B. and UPTON, J., 1994, Constructed reed beds: a cost effective way to polish wastewater effluents for small communities, *Water Environment Research*, **66**(3), 188–192.

HABERL, R., PERFLER, R., LABER, J. and COOPER, P.F., (eds), 1997, Wetland systems for water pollution control, *Water Science and Technology*, **35**(5).

HAMMER, D.A. (ed.), 1989, *Constructed Wetlands for Wastewater Treatment*: Municipal, Industrial and Agricultural, Chelsea, Michigan: Lewis Publishers.

HATANO, K., TRETTIN, C.C., HOUSE, C.H. and WOLLUM, A.G., 1993, Microbial populations and decomposition activity in three subsurface constructed wetlands, in Moshiri, G.A. (ed.) *Constructed Wetlands for Water Quality Improvement*, Boca Raton, Florida: Lewis Publishers, pp.541–547.

JOB, G.D., BIDDLESTONE, A.J. and GRAY, K.R., 1991, Treatment of high strength agricultural and industrial effluents using reed bed treatment systems, *Transactions of the Institution of Chemical Engineers*, **69**, Part A, 187–189.

KICKUTH, R., 1984, Das wurzelraumvertaren in der praxis, *Landschaft Stadt*, **16**, 145–153.

LAN, C., CHEN, G., LI, L. and WONG, M.H., 1990, Purification of wastewater from a Pb/Zn mine using hydrophytes, in Cooper, P.F. and Findlater, B.C. (eds) *Constructed Wetlands in Water Pollution Control*, Oxford: Pergamon Press, pp.419–427.

MERRITT, A., 1994, *Wetlands, Industry and Wildlife: a Manual of Principles and Practices*, Slimbridge, Gloucester: The Wildfowl and Wetlands Trust.

MOSHIRI, G.A. (ed.), 1993, *Constructed Wetlands for Water Quality Improvement*, Boca Raton, Florida: Lewis Publishers.

NICHOLSON, R.J., 1994, *Treatment of Dilute Effluents by the Root Zone (Reed Bed) Method, WA0501 Final Report*, London: Environmental Protection Division, Ministry of Agriculture Fisheries and Food.

NIELSEN, S.M., 1990, Sludge dewatering and mineralisation in reed bed systems, in Cooper, P.F. and Findlater, B.C. (eds) *Constructed Wetlands in Water Pollution Control*, Oxford: Pergamon Press, pp.245–255.

NRA, 1994, *Wheal Jane – A Clear Way Forward*, Exeter: National Rivers Authority.

REED, S.C., CRITES, R.W. and MIDDLEBROOKS, E.J., 1995, *Natural Systems for Waste Management and Treatment*, 2nd edn, New York: McGraw Hill.

SEIDEL, K., 1973, System for Purification of Polluted Water, US Patent 3 770 623, November 6, 1973.

SEIDEL, K., 1976, Macrophytes and water purification, in Tourbier, J. and Pierson, R.W. (eds) *Biological Control of Water Pollution*, Philadelphia: University of Pennsylvania Press, pp.109–122.

PART THREE

Removal of Recalcitrant Compounds

11

Immobilization of Non-viable Cyanobacteria and their use for Heavy Metal Adsorption from Water

ALICIA BLANCO, MIGUEL ANGEL SAMPEDRO, BEGOÑA SANZ,
MARÍA J. LLAMA AND JUAN L. SERRA

11.1   Introduction

Metals are an integral part of the Earth's crust, mostly present as insoluble precipitates and minerals. Natural concentrations of metal ions in freshwater systems can be attributed to natural geochemical dissolution, weathering and microbial leaching. Nevertheless, the greater load of metal ions enters the environment as a result of industrial activities such as mining, heavy industry and metal refining (Beveridge *et al.*, 1995).

Although most metal ions are essential components of biological systems, all are potentially toxic. The major physiological functions and mechanisms of toxicity result from their electrical charge, coordinating capacities and the possession of multiple oxidation states. Toxic effects include the blocking of functional groups of biologically important molecules – such as enzymes or transport systems for nutrients and ions – the displacement and substitution of essential metal ions from biomolecules and functional cellular components, conformational modification, denaturation and inactivation of enzymes and disruption of cell and organellar membrane integrity. However, because microorganisms possess mechanisms for intracellular accumulation of essential metal ions for growth and metabolism, many other metals having no biological functions can also be accumulated (Gadd and White, 1989).

Toxic metal ion contamination of the environment is a significant world-wide phenomenon. The recognition of toxic effects from minute concentrations of some metal ions has resulted in regulation laws to reduce their presence in the environment to very low levels as ppb (i.e. $\mu g \, l^{-1}$). The major processes used to treat metal-contaminated waters include the use of ion-exchange resins and the addition of lime to precipitate metal ions. Other, less frequently used, methods include activated carbon adsorption, electrodialysis, and reverse osmosis. However, these traditional technologies are often ineffective or very expensive when used to reduce the presence of heavy metal ions to very low concentrations. Thus, new technologies are required to decrease metal ion concentrations to environmentally acceptable levels at affordable costs (Wilde and Benemann, 1993).

Bacteria, cyanobacteria, algae, fungi and yeasts are able to remove metal ions from their surrounding environment by means of physicochemical mechanisms – as adsorption – and/or metabolic-dependent activity – as transport. Some physicochemical interactions may be indirectly dependent on metabolism through the synthesis of certain cell constituents or metabolites that may act as metal chelators or create an environment that induces metal precipitation or deposition. In this way, living and non-viable biomass as well as their by-products can accumulate metal ions (Gadd, 1990).

The extent of metal ions removal depends on the physical and chemical characteristics of the water, such as its pH value, temperature, concentration and state of the metal ions, and a number of biological characteristics as well. These include all those parameters which can modify the cell wall composition, such as growth conditions and the age of the culture. Changes of the cell wall may consist in a variation of the number of functional groups or their states, and thus the capacity of metal accumulation is influenced. Furthermore, when using non-viable biomass, mechanical or chemical treatments after cell growth may increase the maximum metal ion amounts removed by adsorption. This enhancement is based on the use of cell constituents from the interior of the cells or break-up of the cell wall and a diffusion into deeper layers of the cell wall (Glombitza and Iske, 1989).

Native biomass exhibits low mechanical strength, low density, and small particle size. As such, native biomass for metal ions removal must be employed in continuous stirred tank reactors. On the other hand, after metal removal the biomass must be separated from the liquid phase by filtration, sedimentation, or centrifugation. Therefore it is necessary to convert biomass into a form in which it can be used as an ion-exchange resin. Modified biomass must have a particle size similar to that of other commercial resins and possess particle strength, high porosity, hydrophilicity, and resistance to aggressive chemicals. All these properties can be achieved by means of biomass immobilization. Immobilization improves the physical characteristics of biomass for use in reactors, permits continuous operation and protects the microorganisms from microbial degradation (Brierley *et al.*, 1989).

11.2 Microbial Mechanisms for Removal of Metal Ions

Although both living and non-viable cells are able to accumulate metal ions, there may be differences in the mechanisms involved in either case, depending on the extent of their metabolic dependence. Metabolism-independent adsorption of metal ions to cell walls, extracellular polysaccharides, or other materials occurs in living and non-viable cells and is generally rapid. Metabolism-dependent intracellular uptake or transport occurs in living cells and is usually a slower process than adsorption, although greater amounts of metal may be accumulated by this mechanism in some organisms. In growing cultures of microorganisms, metabolism-independent and dependent phases of metal removal can be affected by changes in the medium composition and excretion of metabolites that can act as metal chelators. Thus, in a given microbial system, several mechanisms of metal removal may operate simultaneously or sequentially (Gadd, 1990).

The choice of living or non-viable biomass for metal ions removal depends on each particular case because both options present advantages and disadvantages. In the case of non-viable biomass, the biological contribution is limited to the choice of conditions of growth giving optimal adsorbent qualities. During metal exposure no nutrient feed is necessary and metal toxicity is unimportant; furthermore, the situation is not complicated by the production of metabolic end products that may complex the metals away from the cells. Nevertheless, biomass becomes quickly saturated and metal desorption is necessary before further use. On the other hand, living cells have advantages in their variety of accumulation mechanisms and relative ease of morphological, physiological and genetical manipulation. Additionally, there is far greater potential for a long-term continuous process without the need for desorption, giving continuous biomass growth and replenishment. However, accumulation of metal ions by living cells is inhibited by low temperatures, metabolic inhibitors and the absence of an energy source, and is influenced by the metabolic state of the cells and the composition of the external medium. In addition, metal ions can be present only at low concentration because of their toxicity towards living cells (Macaskie and Dean, 1989).

There are six predominant mechanisms by which microorganisms facilitate removal of soluble metal ions from solution: volatilization, extracellular precipitation, extracellular complexing and subsequent accumulation, intracellular accumulation, oxidation–reduction reactions and adsorption to the cell surface.

11.2.1 *Volatilization*

A number of microorganisms are able to generate volatile organometallic compounds by means of metal methylation reactions. These reactions are thought to be a microbial resistance mechanism to volatilize and thus remove the metal ions from their environment. One well-known example of volatilization is the methylation of selenium, in which the selenate and selenite ions are converted into the volatile selenium compounds dimethyl selenide, dimethyl diselenide and dimethyl selenone (Bender *et al.*, 1991). Volatilization is important in metal ions transformation in the environment; however, because of the toxicity of some methylated metals and the difficulties of capturing them, this mechanism has not been considered for developing a commercial process for treating metal-containing waters (Brierley, 1991).

11.2.2 *Extracellular Precipitation*

Some metabolic products excreted by microorganisms to the environment are capable of immobilizing metals ions. One of the best examples of extracellular precipitation of metal ions is the production of hydrogen sulphide by sulphate-reducing bacteria. These bacteria oxidize organic matter and reduce sulphate to sulphide, which readily reacts with soluble metals to form insoluble metal deposits (Brierley *et al.*, 1989).

Often the site of metal deposition is unspecified, but with *Citrobacter* sp. metal ions bind as metal phosphate on the cell wall. Metal accumulation is mediated by a surface-located acid-type phosphatase. The enzyme cleaves glycerol 1-phosphate or other suitable organic substances and liberates phosphate, which precipitates stoichiometrically with the metal ion to form metal deposits tightly bound at the cell surface (Macaskie and Dean, 1989).

11.2.3 *Extracellular Complexing and Subsequent Accumulation*

Some microorganisms synthesize molecules such as siderophores and extracellular polymers that have a high binding efficiency for metal ions. Many microorganisms release high-affinity iron-binding molecules called siderophores which are catecol or hydroxamate derivatives. These molecules facilitate uptake of iron into the cell; they are also able to bind other metal ions as gallium, nickel, uranium, thorium and copper (Gadd, 1990).

Many capsular and slime exopolymers of microorganisms act as polyanions under natural conditions. The anionic character of polysaccharides is conferred by the carboxylic acid groups, which are partially ionized at neutral to alkaline pH. Polysaccharides also contain an abundance of hydroxyl groups, which tend to interact with metals. It is quite likely that the metal-binding capacity contributed by the protein component of exopolymeric material involves nitrogen-containing groups such as amino and amide. The two most important types of interactions between metal ions and exopolymers are those that involve salt bridges with carboxyl groups on acidic polymers and those that involve weak electrostatic bonds with hydroxyl groups on neutral polymers (Geesy and Jang, 1990). Depending on the nature of the organic metal and the immediate chemical environment, the metal can form metallic aggregates. Aggregate formation begins when a metal is precipitated by hydrolysis, changes its oxidation state, or reacts with counterions in solution. Aggregate growth can then proceed by means of crystal nucleation (Ferris *et al.*, 1989). This kind of metal binding is thought to be the most important mechanism of metal removal in activated sludge (Goldstone *et al.*, 1990).

11.2.4 *Intracellular Accumulation*

Intracellular accumulation of metal ions is an active process depending on cellular metabolism and usually requires an specific transport system (Shravan Kumar *et al.*, 1992), although in some cases intracellular uptake is due to changes in membrane permeability arising from toxic effects (Gadd, 1990). Transport systems encountered in microorganisms are of varying specificity and both essential and non-essential metal ions can be taken up (Skowronski, 1986). Most mechanisms of metal transport appear to rely on the electrochemical proton gradient across the cell membrane, which has a chemical component, the pH gradient, and an electrical component, the membrane potential, although K^+ gradient may also be involved (Gadd and White, 1989). Once inside the cell, metal ions may be located within specific organelles and bound to metal-chelating proteins synthesized to respond to the cytotoxic effects of metal ions (Howe and Merchant, 1992).

11.2.5 *Oxidation–reduction Reactions*

A number of microorganisms are able to actively transform metals by either oxidation or reduction of metal ions, although metal ion reduction can also occur passively after metal binding to reactive sites on and within microbial cells.

Metal oxidation

Chemoautotrophic bacteria obtain energy solely from the oxidation of inorganic substances. One important representative of this group of microorganisms is the bacterium *Thiobacillus ferrooxidans*. Thiobacilli oxidize the sulphides and iron present in their environment to produce sulphuric acid and ferrous ion. The iron-oxidizing bacteria in turn rapidly oxidize the ferrous ion to ferric iron. *T. ferrooxidans* plays a significant role in the formation of acidic drainage and can be used to minimize the environmental impact of acid mine drainage. Acidic drainage must be neutralized in order to precipitate soluble metal ions by raising the pH to between 9.5 and 11. The reason is that ferric iron precipitates as a hydroxide at pH 4.3, as opposed to pH 9.5 for ferrous iron, and thus oxidation of the iron substantially reduces the consumption of the neutralizing agent (Brierley, 1991).

Another example is the case of *Leptothrix* spp., microorganisms that derive energy from the oxidation of organic matter, and actively oxidize manganese and iron, depositing the oxides within the sheath that covers the organisms. However, there is no evidence that these microorganisms obtain energy from the oxidation of these metals (Brierley, 1991).

Metal reduction

Partial metal reductions are thought to be mediated by the microorganism's electron transport system. These partial reductions can transform a metal ion into a form that is less mobile or toxic in the environment. An example is the reduction of Cr(VI) to Cr(III), which is less soluble and toxic, by *Pseudomonas putida* (Brierley *et al.*, 1989).

In some cases metal ion reduction is complete, rendering the elemental metal. Reduction of mercury ions to Hg(0) is a well-known detoxification mechanism associated with the activity of a mercuric reductase enzyme. Hg(0) is volatile and so lost from the medium (Macaskie and Dean, 1989).

Metal reduction can also occur passively. Passive metal reduction occurs when metal ions are bound to an intracellular or extracellular component that functions as a reductant (Brierley *et al.*, 1989). The active reduction of Au(III) to Au(I) and the subsequent passive reduction to Au(0) to form colloidal deposits when Au(I) reacts with the alga *Chlorella vulgaris* has been reported (Greene *et al.*, 1987).

11.2.6 *Adsorption to Cell Surfaces*

Microorganisms can accumulate essential and non-essential metal ions by precipitating or binding the metal ions onto cell walls or cell membranes. The surfaces of microorganisms are composed of macromolecules having an abundance of charged functional groups such as carboxylate, amine, imidazole, sulphydryl, sulphate, hydroxyl, and phosphate. Usually, the net charge of cell surface is negative because of the abundance of carboxylate and phosphate residues (Beveridge and Fyfe, 1985), and cationic metals can passively bind to cell surfaces, a process called *biosorption* (Tsezos and Volesky, 1981). However, because of the presence of amine and imidazole groups, which are positively charged when protonated, cell surfaces may also bind negatively charged metal complexes. Thus, microorganisms contain many polyfunctional metal-binding sites for both cationic and anionic metal complexes.

Because adsorption of metal ions onto cell surfaces is a reversible process, the mechanism probably involved is the electrical bonding between the anionic groups of the cell wall and cationic metals, although van der Waals forces, covalent bonding and redox interactions can also be involved (Brierley, 1991). Electrical attractions usually imply an ion-exchange type of reaction (Greene and Darnall, 1990).

Mechanisms of metal ions binding onto cell surfaces fit the adsorption isotherm types L and S described by Giles and Smith (1974). With type L adsorption there is a progressive decrease in the availability of binding sites with increased metal binding and early saturation. In the type S adsorption the presence of bound metal during the early stages of adsorption promotes an increase in the quantities subsequently adsorbed and can be described by a system of multiple equilibria (Macaskie and Dean, 1989). In this case, whole cells may be necessary for maximal accumulation (Tsezos and Volesky, 1981) but, conversely, with accumulation processes relying on type L adsorption mechanisms, extracted polymer may give a higher specific metal accumulation than whole cells (Rudd *et al.*, 1984).

Cyanobacterial envelopes

The envelopes of cyanobacteria are similar to those of Gram-negative bacteria. They consist of two membrane bilayers (the outer and cytoplasmic membranes) that sandwich a peptidoglycan layer between them. This peptidoglycan layer is responsible for the rigidity of the cell and for its resistance to osmotic lysis. The peptidoglycan is a meshwork of linear *N*-acetylmuramyl-(β-1,4)-*N*-acetylglucosamyl strands that are covalently linked together by bonds between short constituent peptide stems. The end result is a skein of glycan fibres that are linked together to form a three-dimensional macromolecule. The outer membrane is firmly bound to the underlying peptidoglycan by a small lipoprotein. This lipoprotein has its lipid, at one end of the lipoprotein, embedded in the outer membrane, while the other end provides a covalent link to the peptidoglycan. The outer membrane has a lipopolysaccharide in the outer side, which is composed of a polysaccharide and a lipidic moiety called lipid A (Brock and Madigan, 1991). Additionally, numerous cyanobacteria possess an envelope outside of the lipopolysaccharide. This is variously called the sheath, glycocalyx, or capsule – or merely gel, mucilage, or slime, depending on its consistency. The sheaths of cyanobacteria are predominantly polysaccharides, but up to 20 % of the weight may be polypeptides and, depending on the species, many types of sugar residues may be involved (Staley *et al.*, 1989).

The most probable candidates for metal ion binding are the phosphate groups that are resident within the polar head groups of the lipopolysaccharide and phospholipids of the outer membrane, and the carboxylate groups of the peptidoglycan (Beveridge and Fyfe, 1985).

Factors affecting metal ion adsorption

The pH plays a critical role in the binding of metal ions to cell wall functional groups. Most metal ions can be divided into three major categories, depending on how binding to the microorganisms is affected by pH (Greene and Darnall, 1990). One group of metal ions, including Hg(II), Au(III) as $AuCl_4^-$, Ag(I), Pd(II) and Au(I) thiomalate, bind to microorganisms rather independently of pH values between 2 and 7. This behaviour is consistent with the general coordination chemistry of metal ions. These metal ions are all classified as 'soft' and undergo covalent bonding to softer ligand, such as sulphydryl and amine groups (Wulfsberg, 1991). Those bonding interactions are generally minimally affected by ionic interactions and pH.

A second group of metal ions, which binds more strongly to microorganisms as pH increases from 2 to 5, consists of borderline soft and 'hard' metal cations, including Cu(II), Ni(II), Zn(II), Co(II), Pb(II), Cr(III), Cd(II), U(VI), Be(II), and Al(III). A pH dependence of metal cation binding generally occurs when the active metal-binding sites can also bind protons. Thus, metal ions and protons compete for the same binding sites. The increased binding of these ions at high pH is consistent with electrostatic binding to ligands such as carboxylates and phosphates, which can be negatively charged because of deprotonation at high pH values.

The third group of metal ions binds more strongly to microorganisms at pH 2 than at pH 5. These ions include mainly oxoanions and other anionic metal complexes. The increased binding of these ions at low pH is consistent with electrostatic binding to ligands such as amines or imidazoles, which can be positively charged because of protonation at low pH values.

The temperature dependence of metal ions adsorption process is determined by the enthalpy of the reaction. In general, complex formation between metal ions and carboxylate ligands is characterized by a small positive enthalpy, whereas complex formation between metal ions and amine and sulphydryl ligands exhibits a rather large negative enthalpy. Thus, when a ligand containing an amine and a carboxyl group interacts with metal ions, the heat of reaction is dominated by the negative enthalpy due to the amine. Therefore, the magnitude and sign of the enthalpy for a reaction involving metal ions and a substrate that contains mixed functionalities is a weighed average of the contributions of each ligand. The observed enthalpy in a microorganism–metal ion reaction can therefore be positive, negative, or zero depending on the metal to biomass ratio and the relative extent of binding to different sites (Greene and Darnall, 1988).

In general, the presence of two or more metal ions in the medium leads to a decrease in the individual adsorption of each metal ion because of the competition for the same adsorption sites on the cells. However, the observed effect can vary considerably depending on the microorganism, the metal combination, the metal concentration and the order of metal addition due to the existence of different types of adsorption sites on the cell surface, which have preference for binding either hard or soft metal ions (Ting *et al.*, 1991). On the other hand, cations usually present in hard waters as Ca(II) and Mg(II) have no effect on the adsorbent potential for other metal ions, which is an important advantage over commercial ion-exchange resins (Greene and Darnall, 1990).

The metal ions adsorption process is also affected by the presence of organic and inorganic ligands that can act as metal chelators. These ligands compete with the microorganisms for the metal ions. The metal–ligand complexes formed away from the cell are not suitable for adsorption. The extent of the interference depends on the metal–ligand complex stability constants (Macaskie and Dean, 1989).

Desorption of metals from microorganisms

The effect of pH and the presence of competing ligands on metal ion adsorption by microorganisms can be used to reverse the process; to desorb and recover the adsorbed metal ions from cells. For metal ions that show a marked pH dependence in binding to cells, desorption can be accomplished by pH adjustment. Metal ions that show little pH dependence in binding to cells can be successfully desorbed by the addition of specific ligands that form exceptionally stable complexes with these ions (Greene and Darnall, 1990).

Advantages of biosorption over conventional metal ions removal techniques

Several potential advantages are possible with biosorption processes, including: the use of naturally abundant biomaterials that can be cheaply produced; the ability to treat large volumes of metal-contaminated waters because of the rapid kinetics; the high selectivity in terms of removal and recovery of specific metal ions; the ability to handle multiple metal ions and mixed waste; the high affinity, reducing residual metals to below 1 ppb in many cases; the lower requirement for additional expensive process reagents, which typically cause disposal and space problems; the operation over a wide range of physicochemical conditions including temperature, pH, and the presence or other ions as Ca(II) and Mg(II); the relatively low capital investment and low operational costs; the greatly improved recovery of bound metal ions from the biomass; and the greatly reduced volume of hazardous waste produced (Wilde and Benemann, 1993).

11.3   Biomass Immobilization

When practical purposes are considered, free biomass presents serious problems compared with its immobilized counterpart, including those involved in biomass manipulation and separation from the processed effluent. Also, practical application of free biomass in biosorption operations often malfunctions because of pressure drops across a fixed-bed column during down-flow operations; this is caused by cell clumping. Excessive hydrostatic pressure is thus required to generate a suitable flow rate. As some of the employed cells, i.e. algal cells, are rather fragile attrition of the biomass may occur under high pressure (Greene and Darnall, 1990). In summary, free cells are suitable for only a limited number of applications as, for example, discontinuous reactors.

At first, these problems seemed to preclude the practical application of biosorption-based processes. However, the referred difficulties can be overcome in the main by using an appropriate immobilization method for the selected biomass. Thus, biomass immobilization provides good handling and operation characteristics to the biomass. It also permits its use in conventional engineering process designs. Bryers and Characklis (1990) listed some other advantages offered by immobilized whole cells in wastewater treatment. However, a number of disadvantages exist that must be also considered. Apart from higher global costs, immobilization often reduces the efficiency of the biosorbent cells by means of the blockage of some functional surface groups. Biomass immobilization decreases mass transfer due to the presence of lower diffusion rates.

Cell immobilization can be defined as the confinement of whole cells in an insoluble phase, which permits the free exchange of solutes from and towards the biomass but at the same time isolates the cells from their surrounding medium.

There are some considerations about the art of immobilization. Immobilization of biomass in a matrix that is too dense can significantly reduce metal loading. Functional groups can be masked by the binding material, as mentioned previously. In the development of economically and technologically viable biosorption products, there must be a balance between a bead with substantial chemical and physical integrity but reduced metal biosorption capacity, and a bead with high metal-loading capacity but reduced stability.

If the immobilization is too aggressive, the resulting bead lacks porosity and the contact time between the immobilized biomass and the metal-containing water must be increased to allow for the increased diffusion time required for the metal ion to reach the functional group. This can significantly increase column size.

Metals adsorbed onto the immobilized biomass can be eluted in the same way that in the case of free biomass, i.e. by means of the appropriate desorbent agent. Often the optimum eluent consists of a moderately acid solution. The most commonly employed acids are hydrochloric, nitric and sulphuric acids. In some cases, pH reduction is not sufficient, or is simply ineffective, for metal removal from the biomass. Certain metals (Au, Ag, Hg) involve different mechanisms in their binding to the surface active sites. So,

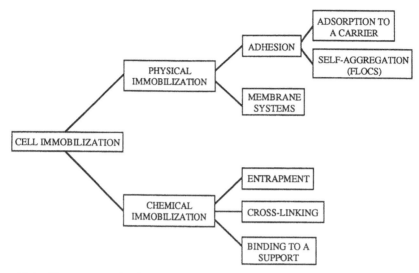

Figure 11.1 The main methods of cell immobilization

chelating compounds, such as EDTA (ethylenediamine tetraacetic acid), or complexing agents, such as thiourea, turn to be necessary for the recovery of such metals from biomass.

11.3.1 *Cell Immobilization Methods*

In recent years the immobilization technology of whole cells has developed greatly. A high number of substances has been examined for the immobilization of different types of biomass. For instance, polyacrylamide, calcium alginate, and silica were used as a support for algal immobilization (Bedell and Darnall, 1990). Other natural materials used as immobilization matrices are agar, agarose, κ-carrageenan and diatomaceous earth. Polyurethane and polyvinyl foams, polyacrylamide, ceramics, epoxy resin and glass beads are among the synthetic materials most frequently used as biosorbent supports.

Cell immobilization techniques can be classified as physical or chemical, depending on the type of binding involved (Figure 11.1). Physical immobilization is based on the natural tendency of cells to form flocs or to adsorb onto inert surfaces. In chemical immobilization three different groups can be distinguished: binding to a supporting surface, cross-linking and entrapment into the porous structure of a polymeric matrix. In the first class, biomass binding to the surface may occur by adsorption, chelation, ionic bonding or covalent bonding. By means of cross-linking, cells become attached to each other covalently. Entrapment can itself be divided into a variety of processes: gelation, polymerization and insolubilization, among others.

In the literature there is abundant information about successful cell immobilization protocols involving both different matrices and biomass from a number of different sources. A broad spectrum of applications is encountered for these immobilized cells. In our laboratory we have improved the methods reported by Ferguson *et al.* (1989) and Klein and Kressdorf (1982) to immobilize non-viable biomass of the thermophilic non-nitrogen-fixing cyanobacterium *Phormidium laminosum* by entrapment in microporous

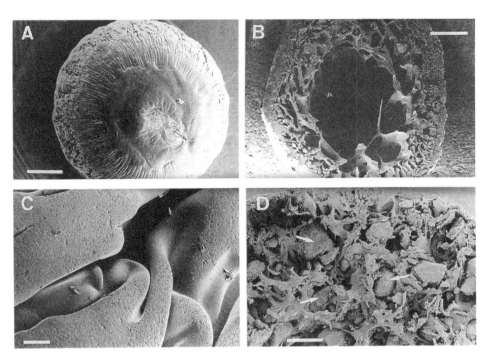

Figure 11.2 Scanning electron micrographs of beads containing *Phormidium laminosum* biomass entrapped in polysulphone. (A) Whole bead; (B) cross-section of the bead; (C) detail of the external bead surface; (D) detail of biomass aggregates inside the pores. Scale bars: 500 μm (A and B), 20 μm (C) and 200 μm (D)

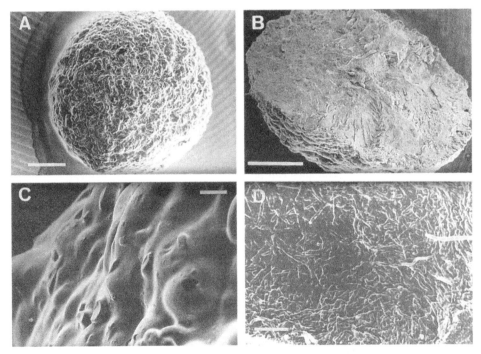

Figure 11.3 Scanning electron micrographs of beads containing *Phormidium laminosum* biomass entrapped in epoxy resin. (A) Whole bead; (B) cross-section of the bead; (C) detail of the external bead surface; (D) detail of biomass aggregates. Scale bars: 500 μm (A and B), 20 μm (C) and 10 μm (D)

beads of polysulphone (Figure 11.2) and epoxy (oxyrane) matrices (Figure 11.3). Non-immobilized biomass of this filamentous cyanobacterium was previously reported to be an excellent heavy metal biosorbent (Sampedro *et al.*, 1995).

11.4   Reactors for the Treatment of Metal-containing Effluents

As mentioned above, retention of the biomass in/on a selected matrix permits its application for practical heavy metal biosorption purposes using conventional engineering systems. Most metal-ion removal processes using non-viable immobilized biomass usually involve either a batch or a column configuration reactor. Briefly:

- *Stirred-tank reactors.* In a batch configuration, immobilized biomass is mixed with the metal ions and, following equilibration, is removed by settling or another suitable mechanism.

- *Packed-bed reactors.* Typically, in a column configuration the metal ions are pumped through a column reactor packed with immobilized biomass. The effluent to be treated is pumped from the top of the column. Then the cleaned fluid is pumped out at the bottom of the reactor. Ideal support for this application should have a low cost and a large surface/volume ratio.

- *Fluidized-bed reactors.* In these reactors, influent flow is pumped through the biomass particles from the bottom of the column with a high enough speed to maintain immobilized cells in a low dense bed. Thus contact area becomes increased, reducing both reactor volume and treatment costs.

- *Dispersed-bed (air-lift) reactors.* Particles of biomass are kept in constant movement by means of air as propellant. As a consequence, particle–solution contact becomes enhanced and so does the global efficiency of the process. Higher costs are the main drawback.

As in ion-exchange equilibria, among the various engineering designs used to incorporate immobilized biomass into water treatment schemes those employing a column configuration offer greater metal-binding capacity and higher efficiency (i.e. high-purity effluents) and are more readily adopted to automation and continuous flow than are batch reactors (Greene and Darnall, 1990).

11.5   Metal Biosorption by Immobilized Biomass

11.5.1   *Factors Affecting Metal Biosorption by Immobilized Biomass*

Apart from the previously mentioned factors affecting metal biosorption when free biomass is considered (pH, surface structure, etc.), a number of other factors can be taken into consideration in the case of immobilized biomass. The most important parameters are contact time, diffusion rate, flow speed, column size and the number of biosorption/elution cycles, among others. These are those normally found in other processes dealing with the same type of reactors.

The kinetics of metal binding to immobilized algae was found to be much slower than the rapid kinetics of metal ion binding by free algal cells. Authors have proposed some ways to improve the binding properties: using a countercurrent feed flow, decreasing the size of the immobilized biomass beads, and increasing the porosity of the beads (Bedell and Darnall, 1990).

The number of cycles a biosorbent can undergo without losing efficiency appears critical for the overall economics of the system.

11.5.2 *Commercially Available Metal Biosorbents*

Several products based on immobilized biomass for detoxification of metal loaded wastewaters or effluents have been patented and commercialized. A variety of types of biomass and supports has been employed in the development of such registered products. In some cases, some of the details remain unknown.

BIOCLAIM®

BIOCLAIM® (Brierley, 1991) is a metal biosorbent including microorganisms, mainly bacteria of the genus *Bacillus*. Before its immobilization, the biomass is washed with an alkaline solution in order to enhance its metal biosorption, and then rinsed with water to remove excess conditioning agent. Microorganisms so treated are then immobilized in polyethyleneimine and glutaraldehyde or other appropriate binders. Sodium hydroxide, potassium hydroxide, alkaline detergents and other reagents are some of the solutions most commonly employed to improve the biosorption capacity. The caustic treatment is said to provoke ruptures on the cell wall of the microorganisms, exposing new sites for metal binding. Also it solubilizes some cell constituents such as lipids, which do not have extensive metal-binding capacity. However, the most important effect of the chemical treatment should be to ensure counterions are present in the surface, decreasing protonation. So, beads containing immobilized biomass retain some residual alkalinity, which increases the pH of the metal solution to be treated. As mentioned above, biosorption capacity increases with pH for some metal ions.

The addition of polyethyleneimine and glutaraldehyde to the prewashed biomass produces a substance with doughlike consistency. This material is readily extrudable through dies of various sizes. This property allows for the production of small (1 mm diameter) beads, which can be rounded using a spherulizing device. The biomass-containing bead, which retains about 60 % moisture, does not need to be dried before use. The bead has a specific gravity of 1.3. The BIOCLAIM bead possesses a greater physical integrity than some chelating cation exchange resins. These characteristics permit the utilization of the product in fixed-bed columns (either up-flow or down-flow), fluidized-bed columns and dispersed-bed operations by air forced motion.

The biomass-loaded particles tend to shrink as the metals are accumulated, resulting in a smaller and denser particle. These dense particles stratify in expanded-bed or dispersed-bed reactors, allowing for selective removal of beads with the higher metal loading. The BIOCLAIM material presents minimal compressibility, thus reducing problems of pressure drop.

Although biosorption materials are relatively non-selective for heavy metal ions, BIOCLAIM beads, and other biosorption products as well, present some metal preference, which is biomass specific. However, when treating mixed-metal wastewaters, the metal with the higher preference did not readily displace metals with lower binding specificities from functional groups.

Metal-loaded BIOCLAIM beads are stripped of metal by means of sulphuric acid, sodium hydroxide for amphoteric metals, or chelating or complexing reagents. The beads

are reactivated with a caustic solution before reusing them to extract metals from contaminated solutions.

AlgaSORB®

AlgaSORB® (Bedell and Darnall, 1990) is a proprietary family of metal biosorbents that
consists of several types of non-living algae and several immobilization matrices. These
products are typically employed in fixed-bed reactors for the removal of soluble, heavy
metal cations from waste streams. A modification of the product has yielded a new
material for the recovery of precious metals.

First-generation AlgaSORB materials were immobilized in silica gel. Like other biomass
materials, AlgaSORB removes metals from solution mainly by ion-exchange processes.
Once loaded with metal ions the algal products can be stripped of metals to produce a
concentrated solution of 10 g l⁻¹. Typically 0.5 M sulphuric acid is used; however, other
reagents such as sodium hydroxide can be used when appropriate. Thus, the material can
undergo a certain number of biosorption–elution cycles. The number of cycles depends
on the conditions employed, and so the economical feasibility of the global process.

Algal biosorbents are particularly useful to remove heavy metals from waters containing high concentrations of $Ca(II)$, $Mg(II)$, $Na(I)$ and $K(I)$, because these ions either do not
bind to the biomass surface or are readily displaced by the metal cations of interest. Like
other biosorption materials, AlgaSORB is unaffected by the presence of organic contaminants in wastewaters.

BIO-FIX®

BIO-FIX® (Ferguson *et al.*, 1989) is a biosorption process that uses biomass immobilized
in polysulfone. The biomass, including sphagnum moss peat, algae, yeast, bacteria, and
aquatic flora, is thermally killed and pulverized. The dried and ground biomass is blended
into a solution containing a given concentration of polysulfone in dimethylformamide.
The mixture is then dropped into water where spherical beads are formed. Nozzle diameter
can be varied to control the bead size. Finally, beads are employed in a moistened state to
proceed with metal biosorption.

The selection of the biomass used in BIO-FIX fabrication is based on the metal
accumulation capacity of the biomass and biomass availability. Sphagnum moss peat, the
marine alga *Ulva* sp., the cyanobacterium *Spirulina* sp., the yeast *Saccharomyces cerevisiae*,
duckweed *Lemna* sp., xanthan and guar gums, and alginate have been evaluated.

BIO-FIX beads have been demonstrated to be most effective in treating wastewaters
with metal concentration ranging from 0.01 to approximately 15 mg l⁻¹ soluble metals.
Moreover, the low affinity for $Ca(II)$ and $Mg(II)$ is an important commercial aspect of
BIO-FIX as well as for other biosorbents. The alkaline earth metals do not readily bind to
the biomass functional groups, even when these metals are present in high concentrations.
When binding does occur, the alkaline earth ions are easily displaced by heavy metal
ones.

Between 75 and 90 % of the soluble metal is biosorbed by the BIO-FIX beads in
the first 20 min of contact. Because the binding of metals to the functional groups on the
biomass is very rapid, the diffusion of metals is probably the rate-limiting step in the
adsorption kinetics.

Metal elution from BIO-FIX beads can be accomplished using sulphuric, nitric, or
hydrochloric acid. EDTA is not as effective as mineral acids for metal elution. Sodium

Table 11.1   Immobilized biomass for metal biosorption processes

Biomass	Immobilization method	Reactor type	Observations	References
Bacillus sp.	Polyethyleneimine and glutaraldehyde	Fixed-bed or fluidized-bed columns	AMT-BIOCLAIM®	Brierley (1991)
Algae	Silica gel	Fixed-bed columns	AlgaSORB®	Bedell and Darnall (1990)
Various types	Polysulfone	Stirred-tank, fixed-bed and fluidized-bed columns	BIO-FIX®	Ferguson et al. (1989)
Streptomyces albus	Polyacrylamide	Stirred-tank reactor	Uranium removal	Nakajima and Sakaguchi (1986)
Persimmon tannin	Formalin	Batch and column	Gold removal	Nakajima and Sakaguchi (1993)
Ascophyllum nodosum	Formaldehyde	Stirred-tank reactor		de Carvalho et al. (1994)
Chlorella homosphaera	Alginate	Fixed-bed reactor		Costa and Leite (1991)
Chlamydomonas reinhardtii and *Selenastrum capricornutum*	Covalent binding to glass (glutaraldehyde)	Minicolumn	Preconcentration for analytical determinations	Elmahadi and Greenway (1994)
Chlorella vulgaris	Polyacrylamide	Column	Selective recovery of metals	Greene et al. (1987)
Zoogloea ramigera	Alginate	Bubble column reactor	Three columns in sequence	Kuhn and Pfister (1989)
Rhizopus arrhizus	Reticulated polyester foam	Column (upward flow)		Lewis and Kiff (1988)
Algae	Silica gel	Column	Trace metal preconcentration	Mahan and Holcombe (1992)
Pseudomonas sp.	Polyacrylamide	Column	Uranium uptake	Pons and Fusté (1993)
Bryoria sp., *Letharia* sp. and *Sargassum* sp.	Silica gel	Column	Preconcentration for analytical applications	Ramelow et al. (1993)
Saccharomyces cerevisiae	Alginate	Batch and continuous flow systems	Heavy metal and radionuclide recovery	De Rome and Gadd (1991)
Rhizopus arrhizus	Alginate, polyacrylamide, expoxy resin and polyvinylfformal	Stirred-tank reactor		Tobin et al. (1993)
Rhizopus arrhizus	Polyvinylformal	Stirred-tank reactor	Uranium removal	Tsezos and Deutschmann (1990)
Rhizopus arrhizus	Reticulated polyester foam	Stirred-tank and upward-flow column		Zhou and Kiff (1991)
Scenedesmus quadricauda	Cross-linking (ethyl acrylate and ethylene glycol dimethacrylate)	Column		Harris and Ramelow (1990)
Saccharomyces cerevisiae	Sand filter	Column (upward-flow)		Huang et al. (1990)
Fungal biomass	Paper filters and polyester fleece	Column		Wales and Sagar (1990)

Table 11.2 Metal adsorption by non-viable biomass of *Phormidium laminosum* immobilized in polysulfone and epoxy resin in packed-bed and fluidized-bed reactors

Reactor	Flow rate (ml min⁻¹)	Metal adsorbed (mg) in beads of							
		Polysulfone				Epoxy resin			
		Cu(II)	Fe(II)	Ni(II)	Zn(II)	Cu(II)	Fe(II)	Ni(II)	Zn(II)
Packed-bed	0.25	5.5	4.62	3.51	6.32	10.23	8.46	7.59	10.39
Packed-bed	0.50	4.59	4.88	4.03	4.94	8.42	7.15	6.46	8.21
Packed-bed	1.00	3.86	3.93	3.54	3.89	6.16	5.56	5.01	6.36
Packed-bed	2.00	3.51	3.24	3.26	3.21	4.69	4.33	3.38	5.04
Fluidized-bed	–	4.28	3.64	4.06	3.75	8.66	6.95	6.01	8.59

Biomass was immobilized by entrapment in polysulfone or in epoxy resin by a modification of the methods reported by Ferguson *et al.* (1989) and Klein and Kressdorf (1982), respectively. Uniform-sized beads of diameter 2.71 (± 0.02) mm and 2.12 (± 0.02) mm, were obtained. The beads (250 mg biomass dry weight) were packed in column reactors operated in continuous mode or in an air-lift reactor operated in batch. 500 ml of 50 ppm metal were either pumped downwards through the column reactors at the indicated flow-rates or incubated for 36 h in the air-lift reactor

carbonate was found to be a good conditioning agent for BIO-FIX after metal elution using nitric acid.

Apart from batch or column reactors, BIO-FIX has also been tested for application in a tough system, which uses the beads contained in porous bags constructed of fine-mesh woven polypropylene filaments. This system is specifically designed for use at remote sites, such as those with acid mine drainages, where minimal maintenance is required and water flows fluctuate.

11.5.3 *Immobilized Biomass-based Biosorption Processes*

Apart from the above-referred products, information from a great number of reported biosorption processes, which are based on a huge diversity of microorganisms, heavy metals and/or conditions, can be found in the literature. What is intended here is not to make an exhaustive review of these studies but only to mention those more remarkable in terms of any particular innovation. Table 11.1 summarizes some of the relevant metal biosorption processes using immobilized biomass.

11.5.4 *Heavy Metal Biosorption by* Phormidium laminosum *Immobilized in Microporous Polymeric Matrices*

In our laboratory we have recently studied the biosorption of a number of heavy metals by non-viable biomass of the cyanobacterium *P. laminosum* entrapped in microporous beads of polysulfone and epoxy resins and in reactors of different design and operation (Table 11.2). The adsorbed metal can be eluted by washing *in situ* with hydrochloric acid,

Table 11.3 Metal adsorption/desorption cycles by *Phormidium laminosum* biomass immobilized in polysulfone beads

Cycle	Adsorbed metal (mg)				Desorbed metal (mg)			
	Cu(II)	Fe(II)	Ni(II)	Zn(II)	Cu(II)	Fe(II)	Ni(II)	Zn(II)
1	3.97	3.34	3.59	4.30	3.76	2.46	2.52	3.52
2	5.07	4.42	4.38	5.39	4.55	2.38	3.79	4.70
3	4.85	4.06	4.32	5.50	4.92	2.37	4.12	4.80
4	4.85	4.30	4.49	5.39	4.80	2.12	3.78	4.65
5	4.70	4.13	4.29	4.73	4.64	1.93	3.67	4.74
6	4.89	4.32	4.28	5.04	4.71	2.10	3.40	4.34
7	4.61	4.19	4.21	5.00	4.66	1.70	3.78	4.35
8	4.84	4.09	4.37	5.06	4.89	1.84	3.45	4.73
9	5.23	4.23	4.19	4.82	4.56	1.70	3.58	4.51
10	4.85	4.19	4.27	4.91	4.70	1.88	3.72	4.47

500 ml of 50 ppm metal were pumped downwards through the column reactors containing 250 mg (dry weight) of immobilized biomass. Bound metal was stripped from biomass with 50 ml of 0.1 M HCl. After each desorption step, the biomass was reconditioned with 30 ml of 0.1 M NaOH. A flow rate of 0.5 ml min^{-1} was used

Table 11.4 Metal adsorption/desorption cycles by *Phormidium laminosum* biomass immobilized in epoxy beads

Cycle	Adsorbed metal (mg)				Desorbed metal (mg)			
	Cu(II)	Fe(II)	Ni(II)	Zn(II)	Cu(II)	Fe(II)	Ni(II)	Zn(II)
1	4.58	5.32	4.17	5.49	3.93	3.45	3.90	4.67
2	3.86	2.07	1.95	2.06	3.77	2.48	2.30	2.23
3	3.74	1.21	2.49	2.20	4.07	2.06	2.05	2.62
4	3.73	1.52	2.79	1.98	3.94	1.73	2.16	2.38
5	3.74	1.27	2.81	2.62	3.82	1.53	2.16	2.70
6	3.18	1.60	2.46	2.56	3.28	1.64	2.23	2.30
7	3.80	1.46	2.67	2.22	3.67	1.42	2.05	2.36
8	3.73	1.13	2.62	2.47	3.56	1.43	2.07	2.26
9	3.95	1.21	1.93	2.36	3.56	1.23	1.95	2.50
10	3.51	1.25	2.27	2.58	3.64	1.10	1.98	2.41

500 ml of 50 ppm metal were pumped downwards through the column reactors containing 250 mg (dry weight) of immobilized biomass at 0.5 ml min^{-1} flow rate. Bound metal was stripped from biomass with 170 ml of 0.1 M HCl pumped at the same flow rate. After each desorption step, the biomass was reconditioned with 30 ml of 0.1 M NaOH at 2 ml min^{-1}.

and after biomass reconditioning with sodium hydroxide the regenerated biosorbent can be used for another metal adsorption cycle (Tables 11.3 and 11.4). This heavy metal biosorption system can be used for at least ten consecutive adsorption/desorption cycles without apparent decrease of efficiency.

11.6 Conclusion

Immobilization of biomass of proved metal-loading capability provides a powerful tool for the detoxification of metal-containing effluents. Biosorption of heavy metal ions must be interpreted in terms of a complementary and emergent technology to conventional wastewater treatments. The broad and diverse number of practical applications of such biosorbents found in the literature gives an idea of the growing importance of this novel technique. However, a certain period of field testing seems to be necessary before this tool can be broadly implemented.

Acknowledgements

This work was partially supported by grants from Acciones Integradas Hispano-Británicas, the Basque Government (PGV9227 and PI95/11), the Commission of the European Communities DG XII-B (CI1*-CT93-0096), the University of the Basque Country (042.310-EC020/95) and Programa de Cooperación con Iberoamérica.

References

BEDELL, G.W. and DARNALL, D.W., 1990, Immobilization of nonviable, biosorbent, algal biomass for the recovery of metal ions, in Volesky, B. (ed.) *Biosorption of heavy metals*, Boca Raton: CRC Press, pp.313–326.

BENDER, J., GOULD, J.P., VATCHARAPIJARN, Y. and SAHA, G., 1991, Uptake, transformation and fixation of Se(VI) by a mixed Selenium-tolerant ecosystem, *Water, Air and Soil Pollution*, **59**, 359–367.

BEVERIDGE, T.J. and FYFE, W.S., 1985, Metal fixation by bacterial cell walls, *Canadian Journal of Earth Sciences*, **22**, 1893–1898.

BEVERIDGE, T.J., SCHULTZE-LAM, S. and THOMPSON, J.B., 1995, Detection of anionic sites on bacterial walls, their ability to bind toxic heavy metals and form sedimentable flocs and their contribution to mineralization in natural freshwater environments, in Allen, H.E., Huang, C.P., Bailey, G.W. and BOWERS, A.R. (eds) *Metal Speciation and Contamination of Soil*, Boca Raton: Lewis Publishers, pp.183–205.

BRIERLEY, C.L., 1991, Bioremediation of metal-contaminated surface and groundwaters, *Geomicrobiology Journal*, **8**, 201–223.

BRIERLEY, C.L., BRIERLEY, J.A. and DAVIDSON M.S., 1989, Applied microbial processes for metals recovery and removal from wastewater, in Beveridge, T.J. and Doyle, R.J. (eds) *Metal Ions and Bacteria*, New York: John Wiley & Sons, pp.359–382.

BROCK, T.D. and MADIGAN, M.T., 1991, *Microbiologia* México: Prentice Hall Hispanoamericana SA, pp.50–68.

BRYERS, J.D. and CHARACKLIS, W.G., 1990, Biofilms in water and wastewater treatment, in Characklis, W.G. and Marshall, K.C. (eds) *Biofilms*, New York: John Wiley & Sons, pp.671–696.

DE CARVALHO, R.P., CHONG, K.-H. and VOLESKY, B., 1994, Effects of leached alginate on metal biosorption, *Biotechnology Letters*, **46**, 875–880.

CHAKRABARTY, A.M., 1976, Plasmids in *Pseudomonas*, *Annual Review of Genetics*, **10**, 7–30.

COSTA, A.C.A. and LEITE, S.G.F., 1991, Metals biosorption by sodium alginate immobilized *Chlorella homosphaera* cells, *Biotechnology Letters*, **13**, 559–562.

ELMAHADI, H.A.M. and GREENWAY, G.M., 1994, Speciation and preconcentration of trace elements with immobilized algae for atomic absorption spectrophotometric detection, *Journal of Analytical Atomic Spectrometry*, **9**, 547–551.

FERGUSON, C.R., PETERSON, M.R. and JEFFERS, T.H., 1989, Removal of metal contaminants from waste waters using biomass immobilized in polysulfone beads, in Scheiner, B.J., Doyle, F.M. and Kawatras, S.K. (eds) *Biotechnology in Minerals and Metal Processing*, Littleton: Society of Mining Engineers, pp.193–199.

FERRIS, F.G., SCHULTZE, S., WITTEN, T.C., FYFE, W.S. and BEVERIDGE, T.J., 1989, Metal interactions with microbial biofilms in acidic and neutral pH environments, *Applied and Environmental Microbiology*, **55**, 1249–1257.

GADD, G.M., 1990, Heavy metal accumulation by bacteria and other microorganisms, *Experientia*, **46**, 834–840.

GADD, G.M. and WHITE, C., 1989, Heavy metal and radionuclide accumulation and toxicity in fungi and yeast, in Poole, R.K. and Gadd, G.M. (eds) *Metal–Microbe Interactions*, Oxford: IRL Press, pp.19–38.

GEESY, G. and JANG, L., 1990, Extracellular polymers for metal binding, in Ehrlich, H.L. and Brierley, C.L. (eds) *Microbial Mineral Recovery*, New York: McGraw Hill Publishers, pp.223–247.

GILES, C.H. and SMITH, D., 1974, A general treatment and classification of the solute adsorption isotherm I. Theoretical, *Journal of Colloid and Interface Science*, **47**, 755–778.

GLOMBITZA, F. and ISKE, U., 1989, Treatment of biomasses for increasing biosorption activity, in Salley, J., McCready, R.G.L. and Wichlacz, P.L. (eds) *Proceedings of the International Symposium on Biohydrometallurgy*, Jackson Hole, Wyoming, pp.329–340.

GOLDSTONE, M.E., KIRK, P.W.W. and LESTER, J.N., 1990, The behaviour of heavy metals during wastewater treatment I. Cadmium, chromium and copper, *Science of the Total Environment*, **95**, 233–252.

GREENE, B. and DARNALL, D.W., 1988, Temperature dependence of metal ion sorption by *Spirulina*, *Biorecovery*, **1**, 27–41.

GREENE, B. and DARNALL, D.W., 1990, Microbial oxygenic photoautotrophs (cyanobacteria and algae) for metal-ion binding, in Ehrlich, H.L. and Brierley, C.L. (eds) *Microbial Mineral Recovery*, New York: McGraw Hill Publishers, pp.277–302.

GREENE, B., MCPHERSON, R. and DARNALL, D.W., 1987, Algal sorbents for selective metal ion recovery, in Patterson, J.W. and PASSINO, R. (eds) *Metal Speciation, Separation and Recovery*, Chelsea, MI: Lewis Publishers, pp.315–332.

HARRIS, P.O. and RAMELOW, G.J., 1990, Binding of metal ions by particulate biomass derived from *Chlorella vulgaris* and *Scenedesmus quadricauda*, *Environmental Science and Technology*, **24**, 220–228.

HOWE, G. and MERCHANT, S., 1992, Heavy metal-activated synthesis of peptides in *Chlamydomonas reinhardtii*, *Plant Physiology*, **98**, 127–136.

HUANG, C., HUANG, C.-P. and MOREHEART, A.L., 1990, The removal of Cu(II) from dilute aqueous solutions by *Saccharomyces cerevisiae*, *Water Research*, **24**, 433–439.

KLEIN, J. and KRESSDORF, B., 1982, Immobilization of living whole cells in a epoxy matrix, *Biotechnology Letters*, **4**, 375–380.

KUHN, S.P. and PFISTER, R.M., 1989, Adsorption of mixed metals and cadmium by calcium-alginate immobilized *Zoogloea ramigera*, *Applied Microbiology and Biotechnology*, **31**, 613–618.

LEWIS, D. and KIFF, R.J., 1988, The removal of heavy metals from aqueous effluents by immobilised fungal biomass, *Environmental Technology Letters*, **9**, 991–998.

MACASKIE, L.E. and DEAN, C.R., 1989, Microbial metabolism, desolubilization, and deposition of heavy metals: metal uptake by immobilized cells and application to the detoxification of liquid wastes, in Mizrahi, A. (ed.) *Biological Waste Treatment*, New York: Alan R. Liss, pp.159–201.

MAHAN, C.A. and HOLCOMBE, J.A., 1992, Immobilization of algae cells on silica gel and their characterization for trace metal preconcentration, *Analytical Chemistry*, **64**, 1933–1939.

NAKAJIMA, A. and SAKAGUCHI, T., 1986, Selective accumulation of heavy metals by microorganisms, *Applied Microbiology and Technology*, **24**, 59–64.

NAKAJIMA, A. and SAKAGUCHI, T., 1993, Uptake and recovery of gold by immobilized persimmon tannin, *Journal of Chemical Technology and Biotechnology*, **57**, 321–326.

PONS, M.P. and FUSTÉ, M.C., 1993, Uranium uptake by immobilized cells of *Pseudomonas* strain EPS 5028, *Applied Microbiology and Biotechnology*, **39**, 661–665.

RAMELOW, G.J., LIU, L., HIMEL, C., FRALICK, D., ZHAO, Y. and TONG, C., 1993, The analysis of dissolved metals in natural waters after preconcentration on biosorbents of immobilized lichen and seaweed biomass in silica, *International Journal of Analytical Chemistry*, **53**, 219–232.

DE ROME, L. and GADD, G.M., 1991, Use of pelleted and immobilized yeast and fungal biomass for heavy metal and radionuclide recovery, *Journal of Industrial Microbiology*, **7**, 97–104.

RUDD, T., STERRIT, R.M. and LESTER, J.N., 1984, Complexation of heavy metals by extracellular polymers in the activated sludge process, *Journal of Water Pollution Control Federation*, **56**, 1260–1268.

SAMPEDRO, M.A., BLANCO, A., LLAMA, M.J. and SERRA, J.L., 1995, Sorption of heavy metals to *Phormidium laminosum* biomass, *Biotechnology and Applied Biochemistry*, **22**, 355–366.

SCOTT, J.A. and PALMER, S.J., 1990, Sites of cadmium uptake in bacteria used for biosorption, *Applied Microbiology and Biotechnology*, **33**, 221–225.

SHRAVAN KUMAR, CH., SIVARAMA SASTRY, K. and MARUTHI MOHAN, P., 1992, Use of wild type and nickel resistant *Neurospora crassa* for removal of Ni^{2+} from aqueous medium, *Biotechnology Letters*, **14**, 1099–1202.

SKOWRONSKI, T., 1986, Influence of some physico-chemical factors on cadmium uptake by the green alga *Stichococcus bacillaris*, *Applied Microbiology and Biotechnology*, **24**, 423–425.

STALEY, J.T., BRYANT, M.P., PFENNING, N. and HOLT, J.G. (eds), 1989, *Bergey's Manual of Systematic Bacteriology*, Vol. 3, Baltimore: Williams & Wilkins, pp.1718–1720.

TING, Y.P., LAWSON, F. and PRINCE, I.G., 1991, Uptake of cadmium and zinc by the alga *Chlorella vulgaris*: II. Multi-ion situation, *Biotechnology and Bioengineering*, **37**, 445–455.

TOBIN, J.M., L'HOMME, B. and ROUX, J.C., 1993, Immobilisation protocols and effects on cadmium uptake by *Rhizopus arrhizus* biosorbents, *Biotechnology Techniques*, **7**, 739–744.

TSEZOS, M. and DEUTSCHMANN, A.A., 1990, An investigation of engineering parameters for the use of immobilized biomass particles in biosorption, *Journal of Chemical Technology and Biotechnology*, **48**, 29–39.

TSEZOS, M. and VOLESKY, B., 1981, Biosorption of uranium and thorium, *Biotechnology and Bioengineering*, **23**, 583–604.

WALES, D.S. and SAGAR, B.F., 1990, Recovery of metal ions by microfungal filters, *Journal of Chemical Technology and Biotechnology*, **49**, 349–355.

WILDE, E.W. and BENEMANN, J.R., 1993, Bioremoval of heavy metals by the use of microalgae, *Biotechnology Advances*, **11**, 781–812.

WULFSBERG, G., 1991, *Principles of Descriptive Inorganic Chemistry*, Mill Valley: University Science Books, pp.266–277.

ZHOU, J.L. and KIFF, R.J., 1991, The uptake of copper from aqueous solution by immobilized fungal biomass, *Journal of Chemical Technology and Biotechnology*, **52**, 317–330.

12

Bioremediation: Clean-up Biotechnologies for Soils and Aquifers

SUSANA SAVAL

12.1 Introduction

Deterioration of soils and aquifers has become evident in the last few years as a result of inappropriate final disposal procedures for all sorts of waste materials. Oil exploitation, uncontrolled fuel spills, poor practices for final disposal of industrial wastes, overuse of pesticides, and operation of sanitary landfills are some causes of pollution of soils and groundwaters.

In the course of the last two decades a wide variety of technologies has been developed for clean-up operations of contaminated soils and aquifers. They can be classified in terms of their principle of operation: physicochemical, thermal and biological. Among the biological technologies bioremediation has evolved as the most promising one because of its economical, safety and environmental features since organic contaminants become actually transformed, and some of them are fully mineralized.

The success of bioremediation techniques is directly related to the metabolic capability of involved microorganisms and can be affected by the surrounding microenvironment. This chapter describes the mechanism of bioremediation; an overview is presented of its virtues and weaknesses.

12.2 The Soil: Where Contaminants and Microorganisms Meet

Soil and subsoil constitute a non-renewable natural resource that plays different roles, as described by Aguilar (1995):

1 Filtering medium during aquifer recharge
2 Protective layer of aquifers
3 Scenario of biogeochemical, hydrologic and food chain processes
4 Natural habitat for biodiversity
5 Space for agricultural and cattle-breeding activities
6 Space for green areas to serve as sources for oxygen regeneration

7   Physical foundation for building construction

8   Sanctuary of the cultural reserve

The first four functions are the most important for the subject of this chapter, and treated as a whole they refer to what is commonly known as 'buffer capacity' of the soil. This buffering phenomenon constantly happens at recharge zones where rainfall migrates vertically downwards towards the aquifers. During seepage a large proportion of solid materials carried by the water are retained at the shallow layers and only water-borne dissolved chemicals seep downwards. Migration time depends on particle size distribution of soil; movement is faster through a fractured medium than through a granular geological material.

Seepage through the subsoil is delayed for compounds that do not migrate at the same rate as water. This frequently occurs in soils with a high organic matter content, such as clays in which organic compounds tend to be retained. There is also another important aspect related to particle size of soils. Clays are characterized by particles of small size (<2 μm); migration is therefore slower and the contact period among exogenic organic compounds and organic matter in the soil is longer, thus favouring the development of the sorption phenomenon. There the wide diversity of heterotrophic microorganisms involved in matter recycling start exerting their metabolic activity by using the existing organic compounds as carbon sources. The longer the substrate/microorganism contact time, the higher the possibility for degradation of the organic matter. The sorption and degradation phenomena of organics in the shallower geological material make it possible for water that continues its migration towards the aquifers to become free from exogenic compounds (Mackay et al., 1985).

As a result of industrial activities spills commonly occur and water-insoluble organic contaminants seep into the soil. These compounds, named NAPLs (non-aqueous phase liquids), have been classified into two types: those lighter than water (known as LNAPLs) and those denser than water (DNAPLs). Typical LNAPLs are petroleum hydrocarbons, their combustible products (such as gasoline, diesel and jet-fuel), benzene, toluene, ethylbenzene and xylenes (BTEX) used as industrial solvents. Other chlorinated industrial solvents, such as tetrachlorethylene, trichlorethylene, chloroform, carbon tetrachloride and methylene chloride, are examples of DNAPLs.

It is important to take into account the classification of contaminants in terms of their density because when LNAPLs reach the water table they tend to float on the water and to spread radially, whereas the DNAPLs continue their downwards path until bedrock is reached to arrest their movement. In terms of contamination effects, DNAPLs are more hazardous because they are capable of polluting the whole aquifer (Mackay and Cherry, 1989).

12.3   Microorganism Survival in Adverse Conditions

Microorganisms possess wide biochemical versatility, which enables them to readily adapt to different microenvironmental conditions of pH, temperature and pressure, even extreme variations. The presence of high pollutant concentrations can be also regarded as extreme conditions. In these cases, contaminants induce a toxic effect on microbial activity to such a degree that the vital functions are inhibited. Microorganisms are capable of developing a certain tolerance to these adverse conditions and to become energy yielding for survival purposes. However, this happens only when microorganisms have the genetic

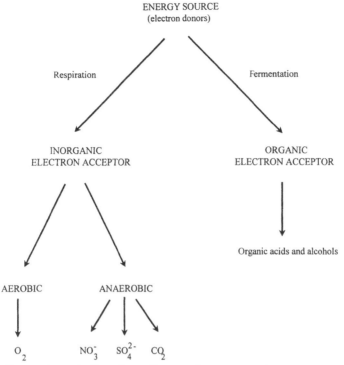

Figure 12.1 Route of metabolism depending on electron acceptor

information available, or if they are capable of developing it, to allow the synthesis of enzymes that participate in the transformation of contaminants.

The organic-type contaminants can be used as carbon sources, thus achieving a redution in their concentration and quite probably complete mineralization, i.e. degradation reaches the generation of carbon dioxide. This is highly desirable, although the presence of a co-substrate might be required to support the energy-yielding activity while biotransformation of contaminants is achieved in different, and hopefully less toxic, chemical entities.

Inorganic contaminants can only be transformed into different molecular entities; some are retained by the cells without becoming degraded. Reduction in the concentration of inorganic contaminants can be observed only when the microbial activity occurs in water where compounds move from the aqueous phase to the inside of the cells.

When reference is made to biodegradation, rather than simply referring to the micro-organisms, it is advisable to consider the enzymes which act as catalysts of the trans-formation reactions induced by the energy-yielding process. For this mechanism to occur the presence of an electron donor and of an electron acceptor is required as well as microenvironmental conditions suitable for synthesis and for the expression of catabolic enzymes. In the case of heterotrophic microorganisms, the electron donor will be the compound used as a source of carbon and energy, most likely the organic contaminant. Its role is to supply energy required by metabolism through electron transfer during the oxidation–reduction reactions to the electron acceptors at the completion of the energy-yielding cycle.

Two types of metabolism exist, depending on the type of electron acceptor: if it has an organic origin, fermentation occurs; for inorganic compounds, the process will be respira-tion (Figure 12.1). In turn, there are two kinds of respiration: aerobic, when the molecular

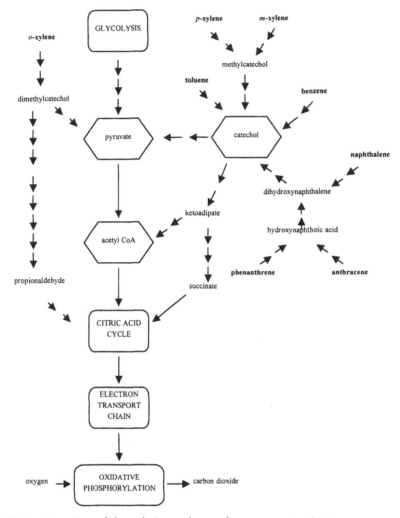

Figure 12.2　Integration of degradative pathways for some contaminants

oxygen becomes the electron acceptor; and anaerobic, when oxidized inorganic compounds such as nitrates, sulphates or carbon dioxide are used. This is why denitrification, sulphate reduction or methanogenesis processes become available.

Traditionally, the initial pathway through which microorganisms start the energy-yielding process is glycolysis, also known as the Emden–Meyerhof–Parnas pathway, which carries glucose or other sugars to an intermediary such as pyruvate. Other compounds with different chemical characteristics (such as amino acids and fatty acids) have different degradation pathways, but all arrive at the same intermediary (acetyl CoA), from which the pathway to follow can be defined. When the respiration pathway is aerobic, three more degradative pathways are then followed: the citric acid cycle, electron transport chain, and oxidative phosphorylation. The presence of molecular oxygen is imperative for the last, which constitutes the most important mechanism for energy supply to the cellular activity; carbon dioxide is generated from the reaction. If this happens, it is assumed that the organic contaminants have become fully mineralized.

For degradation of organic compounds, the presence of enzymes is necessary; however, these are synthesized only when the cells have the specific genetic information and the substrate with which they interact are present in the required concentrations. As an example, mention can be made of the degradation of monoaromatic compounds indicative of contamination, such as gasoline, benzene, toluene and xylene isomers. An intermediary catechol is formed during degradation (Gibson and Subramanian, 1984), which can be degraded through fermentation or respiration, by means of which it can reach other intermediaries of the classic pathway that leads to the generation of carbon dioxide, such as pyruvate or succinate. Figure 12.2 summarizes the integration of degradative pathways for some contaminants with the traditional pathway of glycolysis to oxidative phosphorylation.

Degradation of polyaromatic hydrocarbons (PAHs) such as phenanthrene and anthracene, which are indicators of contamination by diesel, is also shown in Figure 12.2. These molecular entities are formed by three aromatic rings which become transformed through several enzymatic reactions to produce catechol. Naphthalene, which has only two aromatic rings, is also transformed into catechol (Gibson and Subramanian, 1984) and processed by the mineralization pathway.

Actually, most contaminants are found as very complex combinations; for example, gasoline and diesel contain more than 120 different chemical entities (Riser-Roberts, 1992). This situation demands the presence of microbial consortia, i.e. mixed cultures of species that can co-exist without harming each other or which otherwise mutually help each other for the degradation of contaminants. Those native microorganisms which have survived the adverse effects of contamination play a leading role in the biodegradation of contaminants and represent the backbone of bioremediation.

12.4   Advantages of Bioremediation

Bioremediation is a versatile process because it can be adapted to suit the specific needs of each site. Biostimulation can be applied, but only when the addition of nutrients is necessary; bioaugmentation is used when the proportion of degradative microbial flora to contaminant needs to be increased; bioventing is needed when it becomes necessary to supply oxygen from air. Futhermore, bioremediation can be performed off-site when contamination is superficial, but it will have to be *in situ* when contaminants have reached the saturated zone.

One important feature of bioremediation is its low cost compared with other treatment technologies. According to Alper (1993), bioremediation is at least six times cheaper than incineration, and three times cheaper than confinement. It should however be mentioned that all cost comparisons cannot be generalized because they are only applicable to each particular case.

12.5   Knowing the Contaminated Site

Not all contaminated sites are suitable for treatment with bioremediation techniques; it will be necessary to demonstrate their efficacy, reliability and predictability in advance. To this objective site characterization should be performed to obtain information about three closely related aspects: the chemical nature of contamination, the geohydrochemical properties, and the biodegradation potential for the site (Heitzer and Sayler, 1993).

- *Pollutant characterization.* It will be necessary to determine the composition, concentration, toxicity, bioavailability, solubility, sorption and volatilization of all pollutants.
- *Geohydrochemical characterization.* The physical and chemical properties of the geological material should be determined to be able to learn if the microenvironment is suitable for the biodegradative activity. In addition, the geohydrological conditions of the site as well as direction and velocity direction of underground flow are of fundamental interest, particularly when contamination has reached the water table.
- *Microbiological characterization.* It is convenient to analyse the microbial flora in respect to degradative capacity and to the size of the native population with degradative potential.

Integration of the physicochemical and microbiological characterizations should correspond to the results of biofeasibility tests, from which it should be determined whether or not a certain biological treatment is applicable.

Once the site characterization is completed, it is important to proceed almost immediately with the activities leading to its clean-up because contaminants are not static. This is particularly true when pollutants are found in an aquifer.

The characterization of a contaminated site is of utmost importance because a better knowledge of it will facilitate the outlining of an *ad hoc* strategy for its bioremediation (Autry and Ellis, 1992; Rogers *et al.*, 1993). The characterization should be performed in a logical sequence according to a previously established programme, to be able to respond to questions such:

- What chemical compounds are found as contaminants?
- Is the contamination superficial or has it affected the subsoil?
- Are there any records to prove that the contaminants are biodegradable?
- What is the depth and extension of the contaminant plume?
- What is the depth to the water table?
- Is the permeability of the geological material high or low?
- Are there microorganisms capable of degrading the contaminants?
- Is the environment suitable for microbial activity?
- Is it possible to 'build' a bioreactor at the site to be treated?

If answers are affirmative, then bioremediation could be applied, then it will be necessary to carry out biotreatability studies and the evaluation of a bench-scale or pilot-scale project, from which the full-scale process operation will finally be developed.

Something that is commonly encountered in practice is free contaminant in the aquifer; this must be removed before a bioremediation process is applied because the contaminants are toxic to the microorganisms. The latter are capable of tolerating certain concentrations, and some species show a higher tolerance than others, but it is not definitively possible for microorganisms to develop within pure pollutants. This sort of detail should be taken into account when scale-up of the process is being outlined; otherwise a complete failure of bioremediation can be expected.

12.6   Suitability of the Site for Biotreatability Tests

Once information has been gathered on the characteristics of the contaminated site it will be possible to identify its specific requirements. The fact of detecting contaminants at ground surface, in the aquifer or at the mid part of the unsaturated zone, will suggest a strategy of specific bioremediation for each particular case. Therefore, the biotreatability testing procedure will have to be suited to each specific site.

Biotreatability tests are generally performed at a mesocosm level, trying to maintain the environmental conditions that will prevail during treatment in the field. If shallow strata are to be treated with an off-site process, large trays or jars to hold several pounds of soil could be used; it will therefore be necessary to keep humidity and homogeneity constant so that the microbial activity takes place in the whole volume of soil to be treated. If contamination is detected at the water table, columns packed with contaminated soils will be the preferred experimental model for biotreatability tests. It will be necessary in this case to determine the groundwater flow rate to be able to define the operating mechanism of the columns.

When characterization studies indicate that the microbial population with degradative potential is limited or practically zero, it will become important to add exogenous microorganisms. For very practical cases such as bioremediation two alternatives exist. The most common solution is to add commercial compounds; this way is more accessible and faster but it has the least likelihood of success. The other more, interesting, solution is to isolate the rather few degradative microorganisms that were obtained during site characterization and to promote their growth to obtain a culture that can be used for inoculation purposes. This method is safer although it has the shortcoming of requiring a longer time to increase the microbial biomass.

The purpose of biotreatability studies is to predict the behavior of the process and to determine the nutritional requirements for microorganisms to perform biodegradation. The following measurements are thus required:

- Oxygen consumption
- Carbon dioxide generated
- Exhaustion of added nutrients (particularly nitrogen and phosphate sources)
- Contaminant removal

It is necessary to include biotic and abiotic controls to ensure that the contaminant is removed by a microbial activity. When tests are properly carried out it will be possible to predict the behaviour of bioremediation and the time for large-scale application.

For the biotreatability tests to actually represent the field conditions it will be necessary to adopt microenvironmental conditions as close as possible to those encountered at the site to be treated; otherwise the benefits obtained will be minimal.

Because biotreatability tests are time consuming and costly, when bioremediation is applied at a commercial scale these tests are not always performed; nutrients are incorporated empirically, based on past experience. The results are eventually satisfactory but most of the applications are bound to become a complete failure. This should be taken into account because a bioremediation failure may lead to further problems.

12.7 From Laboratory to Field

Even though techniques for growing microorganisms at a commercial level have been perfectly established, additional studies on technologies for bioremediation of soils and aquifers to promote, facilitate or expedite microbial activity in the field are needed. The more relevant aspects are related to bioavailability, the concept of bioreactor, the supply of oxygen, and mass transfer.

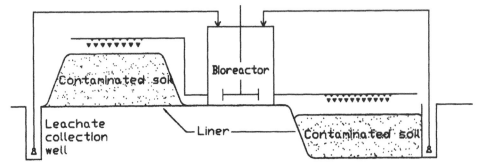

Figure 12.3 Composite diagram for off-site bioremediation techniques: left, a biopile; right, a biocell. Both have collection wells for leachate recirculation

12.7.1 *Bioavailability*

For the biotransformation reactions to be carried out it is necessary for the contaminants to be made ready for attack by the microorganisms (Blackburn and Hafker, 1993). Enzymatic reactions generally occur in aqueous solution but when pollutants are insoluble in water – such as in the case of petroleum hydrocarbons – their energy yielding will be slow due to surface tension between the aqueous and the organic phases. To solve this problem special attention has been paid to the production of surfactants that could be incorporated into the medium and improve the bioavailability. A surfactant acts by 'solubilizing' organic contaminants in the aqueous medium, thus making them accessible to the enzymes responsible for their biodegradation. However, the uncontrolled use of surfactants courses many problems; these are referred to later (Finnerty, 1994).

12.7.2 *The Concept of Bioreactor*

As opposed to conventional biotechnological processes, in which previously built reactors are used, it is necessary for bioremediation purposes to 'build' the bioreactor at the contaminated site. When pollution covers shallow soil layers, it is recommended to excavate the material and to carry it somewhere else to build a biopile or a biocell. The difference between these two concepts resides in the fact that in the biopile the contaminated material is piled on the ground surface whereas for the biocell it is necesssary to perform an excavation in a clean site to deposit the material for further treatment. It will be necessary in both cases to place liners to confine the contaminated material and to prevent leachates from seeping toward the clean soil during treatment (Figure 12.3).

When contamination has reached the water table, the bioreactor can be built through the bore-holes. These are always drilled in even numbers; half are used for extraction and half for injection. Depth and location of wells is determined from geohydrological characterization of the site, from which the configuration of the contamination plume and the direction of the underground flow can be determined. The number of wells can be chosen once the radius of influence of each well has been established in terms of porosity and permeability of geological material, and the flow rate and direction of underground flow. The operating procedure of the wells determines the dimension and control of the bioreactor.

As it occurs in wastewater treatment procedures, it is not possible for bioremediation to work under sterile conditions. The diversity of the microbial population is set in terms of the kinetic characteristics of each of the species involved; it usually happens that native microorganisms that have developed a biodegradative capability in the same microenvironment where contaminants are present become the most widespread.

12.7.3 *Oxygen Supply*

Although the degradation of contaminant compounds under anaerobic conditions has been reported elsewhere, aerobic metabolism is the preferred choice to apply bioremediation techniques because reactions occur at a faster rate and complete mineralization is achieved.

When compounds used as a source of carbon lack oxygen in their molecule, such as in the case of hydrocarbons, oxygen demands for their biodegradation are higher than those required by oxidized compounds such as sugars.

Oxygen supply is one of the major engineering challenges for bioremediation applications in soils because a solid medium is encountered; in aquifers, it becomes more difficult to solubilize molecular oxygen as the depth of the subsoil increases. This has promoted the design of *ad hoc* aeration equipment and the search for alternatives to supply oxygen through highly oxidized inorganic compounds not related to alternate energy-yielding pathways so that full mineralization is achieved. As examples of these oxidized compounds mention can be made of peroxides.

12.7.4 *Mass Transfer*

The concept of mass transfer in bioremediation basically refers to the homogeneity of the system – i.e. that in all of the points inside the bioreactor the same microenvironmental conditions exist to promote microbial activity. This includes nutrient concentration, humidity, pH, and concentration of oxygen available.

Bioremediation becomes more difficult when the geological material is basically a clay because its low permeability prevents mass transfer in the system. This is important when contamination has reached the water table and an *in-situ* treatment is available. For superficial soils this problem can be overcome if sand or agroindustrial residues are added to increase the permeability.

In contaminated aquifers, the most popular auxiliary technique is the pump-treat-injection that involves the extraction of groundwater, its treatment at the surface, and its subsequent recharging into the aquifer. For bioremediation purposes treatment is carried out in a bioreactor where the degradative microorganisms may be confined; alternatively, the organisms can be recycled and leave the reactor at the surface to become reactivated (Figure 12.4). Although this technique is widely used, there are certain aspects that determine its successful aplication; for instance, contaminants can be heavily adsorbed by geological material or may be present in low-permeability zones thus restraining the mass transfer. In other cases, it becomes difficult to reach the clean-up levels required because the low concentrations of contaminants that microorganisms use as a substrate are not sufficient to support their microbial activity and they start to die. If this occurs, treatment becomes very costly because of the power requirements demanded by pumping (Kavanaugh, 1995).

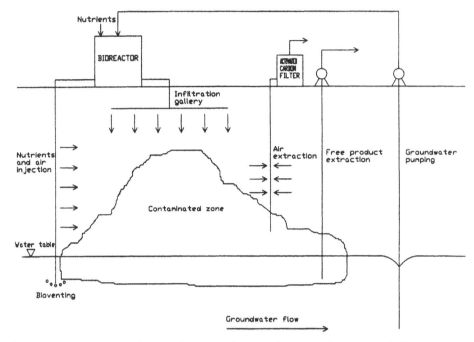

Figure 12.4 Composite diagram for *in-situ* bioremediation techniques, including nutrient percolation, vapour and air extraction, bioventing, pump-treat-injection and free product recovery

If the previous concepts are taken into account, it will be possible to establish that in bioremediation it is not sufficient to work with ideal cases in the laboratory; it is important to maintain the conditions that prevail at the site and to use suitable experimental models so that the results of the study are representative of the scope intended for the field. This is the true objective of performing biotreatability tests before scaling-up is carried out.

12.8   Bioremediation Monitoring in the Field

Monitoring of a bioremediation process is essential to determine two fundamental aspects: the degree of contaminant removal and the catabolic microenvironment. For the former it suffices to determine the residual contaminant concentration by any of the available analytical methods. The second aspect demands the measurement of several parameters such as microbial count of degradative microorganisms, concentration of residual nutrients, pH and humidity (for superficial geological materials), among others.

If the microenvironmental conditions where biodegradation is taking place are periodically determined, it will be possible to maintain each of the parameters within the levels in which the highest metabolic activity can be reached.

12.9   Bioremediation as a Clean Technology

The global market of bioremediation is becoming increasingly wider and more successful because it is regarded as a clean technology. This is mainly due to the fact that contaminants can be transformed into environmentally harmless compounds and some can be fully mineralized. It is also important because microorganisms die when there are no more pollutants to use as substrate.

Something that favours the image of bioremediation is that the soil, once the treatment is over and the degree of pollution very small, can be used for growing plants with the purpose of reintegrating it to its original biological functions.

On the other hand, it is a well known fact that many of the technological advances have been accompanied by environmental deterioration; if bioremediation is carelessly handled, it can lead to failures and to even worse environmental disasters. To keep the concept of bioremediation as a clean technology it will be necessary to perform a rigorous analysis based on ethical environmental concepts and on principles of sustainability that in many cases oppose the economic interests of commercial enterprises.

The increasing number of bioremediation companies on a world-wide scale is mainly due to the fact that this technology is economically feasible (Caplan, 1993). These corporations offer not only environmental services but also related consumables that are basically microbial products, nutrients and commonly biodegradable surfactants. If these products are applied only as recommended, in the minimum necessary amounts and under treatment control, the technology will be indeed successful; otherwise, the surroundings of the site being treated can be adversely affected.

Some of the risks involved in a careless application of bioremediation can be described as follows:

- The uncontrolled addition of surfactants to aquifers can help the dispersion of contaminants rather than their full degradation.

- The excessive use of inorganic compounds used as nitrogen and phosphate sources that can be transported to lakes or lagoons favours the growth of undesirable species; an example is provided by eutrophication.

- Where a native microbial population with degradative capabilities exists, it is better to stimulate its activity *in situ* rather than applying exogenous microorganisms that will eventually die from competition in the natural environment.

- An additional aspect is the increasing interest to apply genetically engineered microorganisms (GEMs) to expedite bioremediation. It is convenient in this case to mention that all the mechanisms that govern a natural environment such as an aquifer are not yet fully understood; furthermore, many factors are involved in the stability of the genetic information within the cells (Harvey, 1993).

12.10 Management Technology Needs

A large number of bioremediation technologies are now being developed and successfully implemented in countries that share certain environmental factors. However, when these technologies are transferred to countries with different environmental characteristics, the results are far from successful (Saval, 1995). It shall be considered in this respect that successful application of any type of technology depends on the need to perform studies for its implementation and innovation that could even result in further developments.

In the case of bioremediation technologies it should be understood that every soil has different characteristics and that no general rule exists for the microorganisms to readily adapt to any habitat. The soils in various parts of the world has distinctive physical, chemical and biological characteristics that make them different from each other.

Literature on new bioremediation technologies is being published every day; competition among big companies who have realized that bioremediation is a profitable money-making opportunity has also become evident. This competitiveness has been focused on the generation of increasingly efficient, but at the same time more sophisticated and expensive, technologies. The secret of a good bioremediation technology is to suit the know-how to every particular problem.

References

AGUILAR, S.A., 1995, Retos y oportunidades de la ciencia del suelo al inicio del siglo XXI (Challenges and opportunities in soil science in the down of the 21st century), *Terra* **13**(1), 3–16.

ALPER, J., 1993, Biotreatment firms rush to marketplace. *Bio/Technology* **11**, 973–975.

AUTRY, A.R. and ELLIS, G.M., 1992. Bioremediation: an effective remedial alternative for petroleum hydrocarbon-contaminated soil. *Environ. Progr.* **11**(4), 318–323.

BLACKBURN, J.W. and HAFKER, W.R., 1993, The impact of biochemistry, bioavailability and bioactivity on the selection of bioremediation techniques. *Trends in Biotechnol.*, **11**, 328–333.

CAPLAN, J.A., 1993, The worldwide bioremediation industry: prospects for profit. *Trends Biotechnol.*, **11**, 320–323.

FINNERTY, W.R., 1994, Biosurfactants in environmental biotechnology. *Curr. Opin Biotechnol.* **5**, 291–295.

GIBSON, D.T. and SUBRAMANIAN, V., 1984, Microbial degradation of aromatic hydrocarbons, in Gibson D.T. (ed.) *Microbial Degradation of Organic Compounds*. Microbiology Series **13**, 181–252.

HARVEY, R.W., 1993, Fate and transport of bacteria injected into aquifers. *Curr. Opin Biotechnol.* **4**, 312–317.

HEITZER, A. and SAYLER, G.S., 1993, Monitoring efficacy of bioremediation. *Trends Biotechnol.* **11**, 334–343.

KAVANAUGH, M.C., 1995, Remediation of contaminated groundwater: A technical and public policy dilemma, in *Proceedings of Forum in Mexico on Ground Water Remediation, December 5–6, Mexico City, Mexico*, 1–5.

MACKAY, D.M. and CHERRY, J.A., 1989, Groundwater contamination: pump-and-treat remediation. *Environ. Sci. Technol.* **23**(6), 630–636.

MACKAY, D.M., ROBERTS, P.V. and CHERRY, J.A., 1985, Transport of organic contaminants in groundwater. *Environ. Sci. Technol.* **19**(5), 384–392.

RISER-ROBERTS, E., 1992, *Bioremediation of Petroleum Contaminated Sites*. Boca Raton, FL: C.K. Smoley, A9–A20.

ROGERS, J.A., TEDALDI, D.J. and KAVANAUGH, M.C., 1993, A screening protocol for bioremediation of contaminated soil. *Environ. Progress* **12**(2), 146–156.

SAVAL, S., 1995, Remediacion y Restauracion (Remediation and Restoration), in Spanish, in *PEMEX: Ambiente y Energia, Los Retos del Futuro*. UNAM-Petroleos Mexicanos, Mexico, 151–189.

13

Increasing Bioavailability of Recalcitrant Molecules in Contaminated Soils

MARIANO GUTIÉRREZ-ROJAS

13.1 Introduction

Soil-persistent compounds are identified as pollutants occurring as a result of industrial activities including fossil fuel exploitation, combustion and spills. Most common examples of such compounds are polycyclic aromatic hydrocarbons (PAHs) and polychlorinated biphenyls (PCBs), generally referred as non-ionic organic contaminants (NOCs). Since some of the NOCs are suspected to be hazardous or even human carcinogens (Cerniglia, 1993), remediation studies and new technology development are now invoked as urgent needs. An interesting way of eliminating NOCs is bioremediation.

Bioremediation can be defined as the manipulation of living systems to bring about desired chemical and physical changes in a confined and regulated environment (Cacciatore and McNeil, 1995). Bioremediation uses microorganisms (bacteria, yeast or fungi) or microbial processes to detoxify and degrade environmental contaminants rather than the conventional approach of disposal (Baker and Herson, 1994). As bioremediation is strongly limited by biodegradation, a serious distinction between the concepts of biotransformation and complete biodegradation or mineralization is required. In this chapter, the term 'biotransformation' means any transformation of the structure of a compound by living organisms or enzymes, while 'biodegradation' involves complete breakdown or mineralization of molecules to carbon dioxide and water. For example, Cerniglia et al. (1994) found that the filamentous marine fungus *Cunninghamella elegans* biotransforms the PAH benz[a]anthracene to *trans*-dihydrodiols and related compounds, which accumulate, but no mineralization was observed. Therefore, while biodegradation assures the elimination of contaminants, biotransformation does not.

Several factors can limit the rate of biodegradation of soil contaminants, including temperature, pH, oxygen content and availability, soil nutrient content and availability, soil moisture content and the physical properties of contaminants. In addition, NOCs have low aqueous solubilities, low dissolution rates and they are, as soil contaminants, strongly bound to or adsorbed onto solids. The biodegradation of such compounds in the natural environment may be restricted, mainly because: (i) the native population of microorganisms is absent or extremely poor and bioreaction limits the process, (ii) the compounds to be degraded are not available to microorganisms and mass transfer becomes limiting, and (iii) compounds or their metabolic intermediates are toxic to the native microorganisms and biodegradation is inhibited. The lack of suitable microorganisms can be overcome by exogenous inoculation, but distinction between availability due to mass transfer limitation, and toxicity has a become matter of discussion (Huesemann, 1997), particularly when residual fraction remains undegraded and accumulates (Pollard *et al.*, 1994) in historically contaminated (i.e. aged) soils.

Bioavailability can be defined as the availability of chemicals to potentially degradative microorganisms. If pollutants become unavailable it is because the rate of mass transfer is zero – as in the case of bound residues in aged soils. Therefore, bioavailability cannot be regarded as a property of one chemical molecule by itself but a property of the complex molecule–environment (e.g. soil moisture, salinity, aqueous phase, pH, etc.).

Recent studies of PAH biodegradation, ranging from the structurally most simple molecules (Guerin and Boyd, 1992; Liu *et al.*, 1995; Volkering *et al.*, 1995; Guha and Jaffé, 1996), to the more complex (Tiehm, 1994; Thibault *et al.*, 1996), agreed that bioavailability was the main constraint to overcome and a relevant question to be answered before planning any bioremediation strategy.

The aims of this chapter are to elucidate the importance of bioavailability compared with other factors also involved in soil bioremediation and to discuss constraints and alternatives to increase bioavailability of recalcitrant compounds found in contaminated soils.

13.2   Soil Bioremediation: An Emerging Technology

Bioremediation of contaminated soils has been used as a safe, reliable, cost-effective and environmentally friendly method for degrading various NOCs, from oil (Bragg *et al.*, 1994, Venosa *et al.*, 1996), diesel and gasoline (Bulman and Newland, 1993), PCBs (Hickey *et al.*, 1993) to dioxin-related compounds (Halden and Dwyer, 1997). In all cases, a common final result prevails: a substantial portion of the pollutant remains unaltered or only partially altered in the bioremediated soil, and target NOCs build up. When incomplete biodegradation is found, it results in accumulation. The reasons why certain pollutants should accumulate in the site are not quite clear. Huesemann, (1997) suggests that incomplete NOC biodegradation is caused not only by bioavailability limitations but also by the inherent resistance of certain compounds to break down in the micro-environment, which may have toxic effects on microbial populations. The relevance of the differences in these two basic concepts is in bioremediation efforts: if biodegradation is incomplete due to bioavailability constraints, one can limit remediation effort because accumulated compounds should be trapped into soil matrix in a non-bioavailable and thus safe form; on the other hand, if some toxins accumulate much more effort must be made. Mistakes underestimating toxicity must be avoided. In other words, adequate attention must be paid to the toxicity of incompletely biodegraded as well as to the biotransformated, still bioavailable, by-products. To understand accumulation, a reasonable explanation is suggested in an interesting critical review (Pignatello and Xing, 1996): sorption (and especially desorption) in natural particles can be extremely slow, which could partially account for the observed difficulties of accumulation. True sorption equilibrium is reached in months or longer, while in the laboratory the time scale is reduced to weeks or even hours. To avoid misunderstanding and erroneous conclusions, two different experimental domains must be distinguished: (i) fast sorption, and (ii) slow sorption. Fast sorption lacks long-term kinetic studies and results are not necessarily conclusive. In the slow sorption domain, great effort is needed to define the origin and constraints of slow desorption for remediation purposes. The list of constraints to study includes (Pollard *et al.*, 1994) composition and nature of the contaminant, temperature-climatic considerations, bioavailability and multiphase partitioning (non-aqueous, aqueous and solid), presence of toxic compounds (the contaminant itself, salinity, heavy metals) and soil texture and structure.

Unfortunately, the vast majority of published work describes experimental techniques in which recalcitrant molecules were added to samples (spiked samples) of soil, sediment (Landrum and Faust, 1994; Harkey *et al.*, 1994) or any other artificial medium, falling in the fast sorption domain. Very few studies describe the biodegradation of such compounds in soils that have been contaminated for years (Frisbie and Nies, 1997).

In conclusion, despite the potential, bioremediation is considered an emerging technology because final effects (e.g. accumulation) are not well understood and are unpredictable. It is possible to surmise the fate of both the biotransformed molecules and exogenous microorganisms, but final consequences in the bioremediated site are uncertain; strict monitoring must be assessed over the next decades and much more predictive research must be done. Finally, the concepts of bioavailability and soil bioremediation are strongly related. Moreover, it is possible to state that soil bioremediation proceeds if bioavailability has been successfully overcome. To assure that bioavailability is not a limitation it is necessary to identify the influencing factors that should be eventually improved.

13.3 Bioavailability Constraints

A number of constraints that could influence bioavailability must be identified before planing a bioremediation strategy. These include (i) the nature of microorganisms, (ii) the composition and properties of pollutants and (iii) the nature of soils.

The influence of the nature of the microorganisms used was studied by Guerin and Boyd (1992). They studied the biodegradation of naphthalene, a minor component of refined petroleum, considered as a model because its moderate solubility ($31.7 \, \mathrm{mg \, l^{-1}}$) and susceptibility to degradation by a large number of microorganisms. The authors examined the rates and extents of degradation in soil-free and soil-containing systems in a comparison of two bacterial species: *Pseudomona putida* ATCC 17484 and a Gram-negative soil isolate, NP-Alk. In the first case, both the rates and extents of naphthalene mineralization exceeded the predicted values, giving enhanced rates of naphthalene desorption from soil (i.e. naphthalene was always available). In the NP-Alk case sorption limited both the rate and extent of naphthalene mineralization (i.e. naphthalene was relatively unavailable). They concluded that there are important organism-specific properties that make it difficult to generalize the concept of bioavailability of soil-sorbed substrates.

Deschênes *et al.* (1996) demonstrated that the biodegradation of the three-ring PAH found in an aged soil were more readily mineralized than the four-ring PAH series. The authors observed no biodegradation at all for PAH having more than four rings. These results confirmed a general rule: the higher the number of fused benzene rings, the more resistant the contaminant to breakdown. Jonge *et al.* (1997) studied the effect of fuel oil composition on bioavailability in a lysimeter and in laboratory scale. They found that two different mechanisms, depending on the *n*-alkane (*n*-C16 to *n*-C20) ratio, controlled the bioavailability. At concentrations higher than 4.0 g kg^{-1}, bioavailability was controlled by solubilization from the non-aqueous liquid phase to the aqueous soil water phase but below this concentration desorption and diffusion became rate limiting. An increase in biodegradation rates with decreasing carbon number was also observed. Finally, they proposed that monitoring *n*-alkane ratios during the course of treatment may be important in identifying bioavailability restrictions in aged soils.

Another interesting approach to elucidate the effect of the nature of contaminants to bioavailability was reported by Sugiura *et al.* (1997). These authors studied the biodegradation by different bacteria of four crude oil samples from which the readily biodegradable fraction had been removed. They found that the same compounds in different crude oil samples were degraded to different extents by the same microorganism. One possible explanation for this was the bioavailability of the monitored alkanes, which were different in different crude oil samples.

A third bioavailability constraint is soil composition. Because it regulates (i) the maximum attainable capacity (partition coefficients) of pollutants in the liquid and solid phase, (ii) the kinetic and desorption rate (Pignatello and Xing, 1996) of chemicals and (iii) the aggregation level, grain size and the porosity of particles in which contaminants are trapped. Soil texture also influences the water regime (e.g. infiltration), gas exchange (producing or avoiding anaerobic cores) (Pollard *et al.*, 1994), local temperature (Chung *et al.*, 1993), effective diffusion (which depends on carbon content of matrix soil) and micropore tortuosity (Huesemann, 1997).

A number of mathematical modelling studies (Chung *et al.*, 1993; Guha and Jaffé, 1996; Huesemann, 1997) followed bioavailability constraints such as the nature and extent of contaminants, and the nature of soils, simultaneously. In models, a set of hypothetical considerations must be made and validated. The final objectives of modelling are to predict transport behaviour of NOCs and discriminate between transport and reaction limitations, an ultimate condition of understanding bioavailability as a transport phenomenon. Bioavailability has been modelled in the traditional chemical reaction engineering way – i.e. through energy and mass balances coupled to reactions in which the complex soil–contaminant–microorganism is assumed to be the catalytic system. For example, a kind of Thiele module has been developed by Chung *et al.* (1993) involving diffusion coefficient, biodegradation rate coefficient, soil aggregate radius and adsorption capacity, which may serve as criteria to assess whether intraparticle diffusion resistance can be ignored. Chemical engineers have devised an interesting generic mathematical concept named bioavailability number (Bn), which resembles the inverse of the dimensionless Damköller number (Bosma *et al.*, 1997). Calculation of Bn can give an idea of the local importance of mass transfer relative to the intrinsic activity of the microorganisms; for example, Bn expresses control by microbial degradation at values greater than, and control by mass transfer at values less than, unity. In general, mathematical models fit well to experimental results when a single contaminant is considered (for example phenanthrene) in an aqueous clean solution (sterile medium, pure inoculum), where such a molecule is always bioavailable (Guha and Jaffé, 1996). However, models fail when complex mixtures of contaminants or soil are present.

In conclusion, the nature of the microorganisms could influence the final results; the bioavailability is not solely determined by the composition and chemical structure of pollutants but other factors such as soil matrix. Until recently, published reports on degradation kinetics of several simultaneously present NOCs, as those win aged soils, have not been available and further research in this direction is expected. Increasing local bioavailability is a serious challenge to overcome in future research.

13.4 General Strategies to Increase Bioavailability

Three general strategies to increase bioavailability have been developed: (i) use of biosurfactants, (ii) addition of organic solvents and (iii) addition of synthetic surfactants. Owing to its importance and potential of application, the addition of synthetic surfactants is separately discussed in a following section.

Plants, animals and microorganisms can naturally produce surfactants; those produced by microorganisms are known as biosurfactants. The principal property of biosurfactants is their capacity to reduce the surface tension of liquid media. Zhang and Miller (1992) postulated that if surface tension is reduced it results in an increase of aqueous dispersion of NOCs, consequently solubility and rates of dissolution are also increased. Once dispersion is increased, bioavailability and likely biodegradation should be enhanced.

Biosurfactants are exopolymers (e.g. polysaccharides, gums) mainly produced by bacteria under growth-limiting conditions (e.g. high carbon to nitrogen ratios and iron limitation) (Guerra-Santos *et al.*, 1984) or when poorly soluble substrates such as *n*-alkanes (Hommel, 1990) are present. Among the most studied biosurfactants are the rhamnolipids produced by *Pseudomonas aeruginosa*. The use of biosurfactants is attractive because they are natural products and therefore biodegradable (Zhang and Miller, 1992). Deschênes *et al.* (1996) showed that purified rhamnolipid produced by *P. aeruginosa* UG2 was readily biodegraded by native PAH degraders found in aged creosote-contaminated soil. These results suggest that two different, and not necessarily coupled, microbial activities must be distinguished: (i) biosurfactants produced under limited conditions by bacteria and (ii) biodegradation of contaminants as a result of increased bioavailability. As microbial growth on hydrocarbons has been associated with the production of biosurfactants, Déziel *et al.* (1996) isolated bacteria with two of the desired characteristics: PAH-metabolizing bacteria capable of producing biosurfactants. They found that in the presence of naphthalene the bacteria produced biosurfactants which promoted the solubility of naphthalene. The authors suggest that if microorganisms promote the solubility of their own substrate, biosurfactants are naturally produced as a part of their strategy for growing on such substrates where bioavailability is essentially reduced. Apparently, the potential advantage of using biosurfactants requires presence of the biosurfactant producer rather than addition of the isolated purified biosurfactant.

Addition of organic solvents changes the polarity of the soil–water environment, influencing partition and consequently bioavailability of non-ionic pollutants. Furthermore, appropriate organic solvents disperse NOCs up to molecular level, promoting a thin layer interface to which NOC degraders could adhere. Despite the potential of this bioremediaton principle few research groups have paid attention to the effect of solvent addition. Efroymson and Alexander (1991) conducted experiments in liquid cultures with an *Arthrobacter* strain and demonstrated that the extent of biodegradation of naphthalene dissolved in heptamethylnonane was increased by addition of the solvent. Biodegradation was proportional to the volume of solvent and attributed to bacteria attached to the solvent–water interface. The authors concluded that adherence of cells to a solvent–water interface was a prerequisite for naphthalene utilization. Moreover, when a surfactant (Triton X-100) was added to prevent adherence of cells to the interface, biodegradation was also prevented. More recently, in an extensive work, Jiménez and Bartha (1996) used different strategies to increase bioavailability and likely biodegradation of pyrene. They studied the effects of non-ionic surfactants, hydrophilic (polyethylene glycol) and hydrophobic (heptamethylnonane, decalin, phenyldecane and diphenylmethane) solvents, which were inhibitory in all cases. They also experimented with the addition of paraffin oil, squalene, squalane, tridecylcyclohexane and tricosene: all increased pyrene biodegradation without being used themselves. Their explanation of enhanced biodegradation agrees with that proposed by Efroymson and Alexander (1991) – that solvents promote adherence of organisms to the interface in the neighbourhood of dispersed NOCs.

In conclusion, the production of biosurfactants can be considered as part of the metabolism of indigenous bacteria in contaminated sites. Microorganisms which produce biosurfactants could create a favourable local environment to enhance bioavailability of the NOCs attached to soil. Because this phenomenon does not assure biodegradation, future research looking for microorganisms and operation conditions that couple both the increased bioavailability and biodegradation is expected. With respect to the addition of solvents to increase bioavailability, the promising results discussed above now constitute a serious prospect for application and appear to justify further studies.

13.5 Addition of Synthetic Surfactants

The uses of synthetic surfactants to promote bioavailability of contaminant molecules attached to soil have recently gained attention, in particular, when insoluble persistent contaminants such as PAHs and PCBs in soils and sediments are present. Volkering *et al.* (1995) investigated the effects of four non-ionic surfactants on the bioavailability and rates of biodegradation of crystalline naphthalene and phenanthrene in agitated flasks. They found that the presence of synthetic surfactants increased both the apparent solubility and the maximal rates of dissolution. The rates of biodegradation of naphthalene and phenanthrene were also enhanced by the addition of surfactants in the dissolution-limited growth phase, indicating that the dissolution rates were higher than in the absence of surfactants. These authors also studied the toxicity of the surfactants, and observed no toxic effects due to the presence of up to 10 g l^{-1} of surfactants. Similar results showed that mineralization of naphthalene (Liu *et al.*, 1995) and phenanthrene (Guha and Jaffé, 1996) was unaffected by the addition of non-ionic surfactants even above their critical micelle concentration (CMC). When solubilized by micelles of surfactants in aqueous phase, naphthalene or phenanthrene became bioavailable and degradable by microorganisms. However, Laha and Luthy (1991) found that at concentrations above the CMC the mineralization of phenanthrene in soil–water systems was inhibited. Similar results were reported with *Mycobacterium* sp. (Jiménez and Bartha, 1996), in which the addition of non-ionic Triton X-100 below the CMC increased pyrene biodegradation but above the CMC severely inhibited mineralization. Furthermore, it was suggested (Aronstein *et al.*, 1991) that non-ionic surfactants at low concentrations may promote the biodegradation of adsorbed aromatic compounds in polluted soils, even when surfactant-induced desorption was not appreciable. Thibault *et al.* (1996) conducted experiments with pyrene in spiked and soil-contaminated samples. They tested four synthetic surfactants, under saturated (soil slurries) and unsaturated conditions. The effectiveness of the surfactants in pyrene desorption increased with increasing concentration. Witconol SN70, a non-ionic ethoxylated alcohol, was the most effective surfactant for pyrene mineralization at unsaturated conditions but inhibited the action of microorganisms at saturated conditions, suggesting that excess of water plays an important role. Researchers need to take account of this in bioavailability and likely biodegradation studies.

Deschênes *et al.* (1996) demonstrated that the addition of sodium dodecyl sulphate (SDS) to creosote-contaminated soil increased the solubility of PAHs but that PAH biodegradation was inhibited, in agreement with the work of Tiehm (1994). Tiehm studied

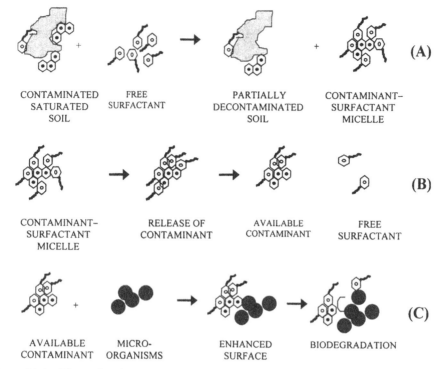

Figure 13.1 Effect of surfactants on bioavailability. (A) Desorption of attached contaminant and micelle formation; (B) release of contaminants trapped into the micelles to the aqueous phase; (C) promotion of enhanced surfaces to facilitate biodegradation

biodegradation in blends of PAHs by pure and mixed cultures in the presence of synthetic non-ionic surfactants and SDS; results indicated that surfactants increased the solubilization of SDS but that its degradation was inhibited. Non-ionic surfactants increased bioavailability and enhanced biodegradation of fluorene, phenanthrene, anthracene, fluoranthene, and pyrene, but differential toxicity was observed. Tiehm concluded that toxicity correlates with structural form: decreasing with increasing length of the surfactant.

To account for the inhibition of SDS degradation in the two previous cases, the same explanation was given: SDS was a preferential growth substrate in liquid cultures (Tiehm, 1994) as well as in microcosm (Deschênes *et al.*, 1996) where 75 % of initial SDS was biodegraded in 15 days.

In an attempt to explain the role played by surfactants in bioavailability a schematic representation is shown in Figure 13.1. Hydrophobic (cyclic ring) and hydrophilic (aliphatic side-chain) moieties illustrate the two functional components of surfactants. As general rule, the presence of a surfactant may help to enhance at least one of the following mechanisms. (i) Desorbing contaminants from soil to the aqueous phase (Figure 13.1a) by releasing a complex contaminant-surfactant, which constitutes a pseudo-phase called micelles, plus partially decontaminated soil. (ii) Increasing the concentration of the hydrophobic compound in the aqueous phase by solubilization into micelles (Figure 13.1a). At a CMC of the surfactant, colloidal aggregates are formed, providing increased solubilization or emulsification of the hydrophobic compound. (iii) Enhancing surfaces to facilitate transport of hydrophobic compounds from the micelle to the aqueous phase, releasing contaminant in available form plus free surfactant (Figure 13.1b). (iv) Enhancing adherence between cells and substrate. Free surfactant could be assimilated and available contaminant might be attacked or removed by microorganisms (Figure 13.1c).

In conclusion, the effectiveness of a synthetic surfactant in increasing bioavailability of soil-bound molecules depends on the nature of the microorganisms and contaminants, the type of surfactant and soil composition. Experimental results on the effects of surfactant addition on biodegradation of NOCs are not consistent and sometimes contradictory (Liu *et al.*, 1995). Apparently, the observed positive effect of non-ionic surfactants in aqueous solutions is absent in the presence of contaminated soil. There is a lack of general explanations on the nature of the physical, physiological or biochemical phenomena involved. Although surfactants have been studied in complex water–soil systems, the mechanisms and effects are not well understood and more research in this direction must be done.

13.6   Increasing Bioavailability: General Recommendations

An appropriate study of surfactant addition to increase bioavailability should include the following separate studies:

- Surfactant to soil sorption kinetics must be assessed. There is a remarkable difference between studies based on clean and contaminated (e.g. aged) soil samples. Figure 13.2 illustrates the comparison: in a clean soil (Figure 13.2a), a higher consumption of synthetic surfactant is required than in a contaminated soil (Figure 13.2b). If some kind of binding force between surfactant and soil matrix (saturated and unsaturated) is observed, then the nature and extent of such bound will determine the role and effectiveness of the surfactant.

- Efficiency of NOCs extraction with and without surfactants must be measured. The presence of the surfactant should not interfere with the NOC extraction methods.

- Presence of the surfactant should not disperse soil samples. Dispersion of soil aggregates will result in an altered form quite different to that found on site.

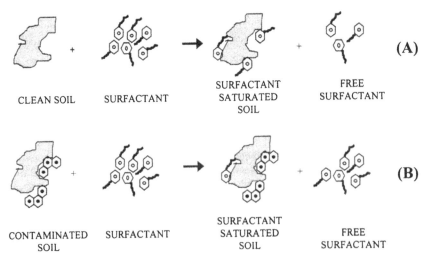

Figure 13.2   Saturation of clean (A) and contaminated (B) soil by a synthetic surfactant

- Partition coefficient and CMC must be evaluated. Use of culture medium, if any, instead of pure water is also suggested. An effective surfactant will produce higher partition and lower CMC values.

- Intrinsic toxicity (e.g. altering cell membrane integrity) of the surfactant must be studied. For example the assimilation of conventional carbon sources such as glucose with and without surfactant can give an idea of the extent of intrinsic toxicity.

- Intrinsic biodegradability of the surfactant must be studied. A careful study regarding the ability of selected microorganisms as NOC degraders to either eliminate or biotransform surfactants is suggested. In some cases, surfactants could be used as preferential source of carbon, thus a compromise between total biodegradability, which avoids surfactant accumulation, and preferential source of carbon must be found.

13.7 Conclusions

1 To achieve an appropriate soil bioremediation approach, a formal distinction between the concept of biotransformation and complete biodegradation must be envisaged.

2 Regarding the role of bioavailability of contaminants in aged soils, spiked samples studies do not allow scale-up of conclusions to the field level.

3 Low bioavailability of hydrophobic compounds bound to soil matrix is now recognized as a major bioremediation constraint to overcome. In most of the cases, mass transfer limitation prevented the success of biodegradation.

4 Mathematical conceptualization of bioavailability allows estimation of such significant factors as to what extent biodegradation is limited by inherent microbial activity in comparison to mass transfer constraints e.g. desorption, transport throughout the solid matrix and dissolution rates.

5 Selection of an appropriate strategy to increase bioavailability strongly depends on concentration of contaminants in soils. Two extreme cases are: (i) low NOC concentration, which probably can be solved by using biosurfactants, and (ii) high NOC concentration by the addition of organic solvents or related additives. However, both alternatives seem to be far from practical.

6 Among the different strategies to increase bioavailability, surfactant addition seems to be the most promising and reliable.

7 Surfactant addition to increase bioavailability involves a complex experimental task. In order to minimize risk and blurred conclusions, a number of parallel studies must be carried out.

Acknowledgements

The Mexican Council of Science and Technology (CONACyT) and the Mexican Institute of Petroleum (IMP FIES 96F-48-VI) provided financial support for this work.

References

ARONSTEIN, B.N., CALVILLO, Y.M. and ALEXANDER, M., 1991, Effect of surfactants at low concentrations on the desorption and biodegradation of sorbed aromatic compounds in soil, *Environ. Sci. Technol.*, **25**, 1728–1731.

BAKER, K.H. and HERSON, D.S., 1994, Introduction and overview of bioremediation, in Baker, K.H., and Herson, D.S. (eds) *Bioremediation*, New York: McGraw Hill, pp.1–7.

BOSMA, T.N.P., MIDDELDORP, P.J.M., SCHRAA, G. and ZEHNDER, A.J.B., 1997, Mass transfer limitation of biotransformation: quantifying bioavailability, *Environ Sc. Technol.*, **31**, 248–252.

BRAGG, J.R., PRINCE, R.C., HARNER, E.J. and ATLAS, R.M., 1994, Effectiveness of bioremediation for the Exxon Valdez oil spill, *Nature*, **368**, 413–418.

BULMAN, T.L. and NEWLAND, M., 1993, In situ bioventing of diesel fuel spill, *Hydrol. Sci.* **8**, 297–308.

CACCIATORE, D.A. and MCNEIL, M.A., 1995, Principles of soil bioremediation, *Biocycle*, **36**, 61–64.

CERNIGLIA, C.E., 1993, Biodegradation of polycyclic aromatic hydrocarbons, *Biodegradation*, **3**, 351–358.

CERNIGLIA, C.E., GIBSON, D.T. and DODGE, R.H., 1994, Metabolism of Benz[a] anthracene by the filamentous fungus *Cunninghamella elegans*, *Appl. Environ. Microbiol.*, **60**, 3931–3938.

CHUNG, G.Y., MCCOY, B.J. and SCOW, K.M., 1993, Criteria to assess when biodegradation is kinetically limited by intraparticle diffusion and sorption, *Biotechnol. Bioeng.*, **41**, 625–632.

DESCHÊNES, L., LAFRANCE, P., VILLENEUVE, J.-P. and SAMSON, R., 1996, Adding sodium dodecyl sulphate and *Pseudomonas aeruginosa* UG2 biosurfactants inhibits polycyclic aromatic hydrocarbon biodegradation in a weathered creosote-contaminated soil, *Appl. Microbiol. Biotechnol.*, **46**, 638–646.

DÉZIEL, E., PAQUETTE, G., VILLEMUR, R., LÉPINE, F. and BISAILLON, J.-G., 1996, Biosurfactant production by a soil *Pseudomonas* strain growing on polycyclic aromatic hydrocarbons, *Appl. Environ. Microbiol.*, **62**, 1908–1912.

EFROYMSON, R. and ALEXANDER, M., 1991, Biodegradation by an *Arthrobacter* species of hydrocarbons partitioned into an organic solvent, *Appl. Environ. Microbiol.*, **57**, 1441–1447.

FRISBIE, A.J. and NIES, L., 1997, Aerobic and anaerobic biodegradation of aged pentachlorophenol by indigenous micro-organisms, *Bioremediation J.*, **1**, 65–75.

GUERIN, W.F. and BOYD, S.A., 1992, Differential bioavailability of soil-sorbed naphthalene to two bacterial species, *Appl. Environ. Microbiol.*, **58**, 1142–1152.

GUERRA-SANTOS, L.H., KÄPPELI, O. and FIECHTER, A., 1984, *Pseudomonas aeruginosa* biosurfactant production in continuous culture with glucose as carbon source, *Appl. Environ. Microbiol*, **48**, 301–305.

GUHA, S. and JAFFÉ, P.R., 1996, Bioavailability of hydrophobic compounds partitioned into the micellar phase of nonionic surfactants, *Environ. Sci. Technol.*, **30**, 1382–1391.

HALDEN, R.U. and DWYER, D.F., 1997, Biodegradation of dioxin-related compounds: a review, *Bioremediation J.*, **1**, 11–25.

HARKEY, G.A., LYDY, M.J., KUKKONEN, J. and LANDRUM, P.F., 1994, Feeding selectivity and assimilation of PAH and PCB in *Diporeia* spp, *Environ. Toxicol. Chem.*, **13**, 1445–1455.

HICKEY, W.J., SEARLES, D.B. and FOCHT, D.D., 1993, Enhanced mineralization of polychlorinated biphenyls in soil inoculated with chlorobenzoate degrading bacteria, *Appl. Environ. Microbiol.*, **59**, 1194–1200.

HOMMEL, R.K., 1990, Formation and physiological role of biosurfactants produced by hydrocarbon-utilising microorganisms, *Biodegradation*, **1**, 107–119.

HUESEMANN, M.H., 1997, Incomplete hydrocarbon biodegradation in contaminated soils: limitations in bioavailability or inherent recalcitrance?, *Bioremediation J.*, **1**, 27–39.

JIMÉNEZ, I.Y. and BARTHA, R., 1996, Solvent-augmented mineralization of pyrene by *Mycobacterium* sp., *Appl. Environ. Microbiol.*, **62**, 2311–2316.

JONGE DE, H., FREIJER, J.I., VERSTRATEN, J.M. and WESTERVELD, J., 1997, Relation between bioavailability and fuel oil hydrocarbon composition in contaminated soils, *Environ. Sci. Technol.*, **31**, 771–775.

LAHA, S. and LUTHY, R.G., 1991, Inhibition of phenanthrene mineralization by nonionic surfactants in soil-water systems, *Environ. Sci. Technol.*, **25**, 1920–1930.

LANDRUM, P.F. and FAUST, W.R., 1994, The role of sediment composition on the bioavailability of laboratory-dosed sediment-associated organic contaminants to the amphipod, *Diporeia* (spp.), *Chem. Spec. Bioavailability*, **6**, 85–93.

LIU, Z., JACOBSON, A.M. and LUTHY, R.G., 1995, Biodegradation of naphthalene in aqueous nonionic surfactant systems, *Appl. Environ. Microbiol.*, **61**, 145–151.

PIGNATELLO, J.J. and XING, B., 1996, Mechanisms of slow sorption of organic chemicals to natural particles, *Crit. Rev. Environ. Sci. Technol.*, **30**, 1–11.

POLLARD, S.J.T., HRUDEY, S.E. and FEDORAK, P.M., 1994, Bioremediation of petroleum- and creosote-contaminated soils: a review of constraints, *Waste Man. Res.* **12**, 173–194.

SUGIURA, K., ISHIHARA, M., SHIMAUCHI, T. and HARAYAMA, S., 1997, Physicochemical properties and biodegradability of crude oil, *Environ. Sci. Technol.*, **31**, 45–51.

THIBAULT, S.L., ANDERSON, M. and FRANKENBERGER, W.T., 1996, Influence of surfactants on pyrene desorption and degradation in soils, *Appl. Environ. Microbiol.*, **62**, 283–287.

TIEHM, A., 1994, Degradation of polycyclic aromatic hydrocarbons in the presence of synthetic surfactants, *Appl. Environ. Microbiol.*, **60**, 258–263.

VENOSA, A.D., SUIDAN, M.T., WREN, B.A., STROHMEIER, K.L., HAINES, J.R., EBERHART, B.L. *et al.*, 1996, Bioremediation of an experimental oil spill on the shore line of Delaware Bay, *Environ. Sci. Technol.*, 30, 1764–1775.

VOLKERING, F., BREURE, A.M., ANDEL, J.G. and RULKENS, W.H., 1995, Influence of nonionic surfactants on bioavailability and biodegradation of polycyclic aromatic hydrocarbons, *Appl. Environ. Microbiol.*, **61**, 1699–1705.

ZHANG, Y. and MILLER, R.M., 1992, Enhanced octadecane dispersion and biodegradation by a *Pseudomonas* rhamnolipid surfactant (biosurfactant), *Appl. Env. Microbiol.*, **58**, 3276–3282.

14

Bioremediation of Contaminated Soils

MARÍA TREJO AND RODOLFO QUINTERO

14.1 Introduction

Bioremediation is defined as the use of biological treatment systems to destroy, or reduce the concentrations of, hazardous wastes from contaminated sites. Such systems have potentially numerous applications, including clean-up of ground water, soils, lagoons, sludge and process waste streams. Bioremediation has been used on very-large-scale applications such as the shoreline-clean-up efforts in Alaska resulting from the Exxon Valdez oil spill in 1989 (Caplan, 1993). At the same time bioremediation is still considered an innovative technology which has been used in only a limited number of cases, most of them in USA. Several advantages of bioremediation systems have been identified (Levin and Gealt, 1993). It:

- can be done on site
- keeps site disruption to a minimum
- eliminates transportation costs and liabilities
- eliminates long-term liability
- uses biological systems, often less expensive
- can be coupled with other treatment techniques into a treatment train.

Some authors consider that bioremediation is still unproven and that there are some disadvantages and/or limitations when it is applied to specific problems, such as:

- some chemical compounds are not biodegradable
- extensive monitoring is required
- each site has specific requirements
- potential production of toxic unknown subproducts
- strong scientific support is needed.

This chapter describes in detail the main aspects and components of bioremediation systems.

14.2 Current Market for Bioremediation

Bioremediation comprises today only a small fraction of the very large market for hazardous waste treatment, but it is one of the fastest growing sectors in environmental management. In the USA particularly the bioremediation industry is still limited, but progresses toward commercialization at a faster rate that in other countries (Jespersen *et al.*, 1993). The key reasons for this situation are:

- Most bioremediation research and development have been carried out in the USA.

- The scope and enforcement of USA environment law exceeds that of other nations.

- Public acceptance of bioremediation in the USA is high due to the publicity generated by the use of bioremediation for clean-up of large oil spills.

There are several estimations about the world market for bioremediation, indicating that the most important sector of the bioremediation market is petroleum-contaminated soils and groundwater, resulting from leaking underground storage tanks (USTs). In Europe, within the next 10 years soil clean-up costs alone are estimated to exceed US$30 billion; if 5 % of this soil is cleaned using biotreatments then 1.5 billion dollars will be spent by bioremediation methods. In the USA there are around 750 000 existing UST facilities, over 50 % of which are used for storage of petroleum hydrocarbons, which are biodegradable. Conservative estimates indicate that one out of three of these tanks is leaking and if 10 % of these sites undergo biological clean-up the total revenue would be US$4 billion. Other future markets for biological treatment include process waste pretreatment, industrial lagoons, municipal landfill leachates and general chemical spills. The potential world-wide revenues for bioremediation systems are US$11.5 billion over the next decade.

In practice the situation is different. In a review of the Superfund Program Remedial Actions it was found that nearly half of the remedial treatment for source control (primarily soils) involves technologies that were not available when the Superfund Program was authorized in 1986. The study, which was performed in 1995, evaluated more than 300 innovative technologies projects for soil and groundwater, the most frequent being solidification/solubilization (29 %), followed by off-site incineration (15 %) and on-site incineration (11 %). Bioremediation (*ex situ* and *in situ*) accounts for only 10 %, but among innovative technologies was second only to soil vapour extraction (19 %) (Fiedler and Quander, 1996).

14.3 Bioremediation Systems

In order to carry on bioremediation, several biological systems and techniques have been developed in the last few years (Table 14.1). The diversity of bioremediation technologies is an indication of the complexity that one encounters when trying to clean soils contaminated with some hazardous organic matter.

The selection of a bioremediation system is not a simple task because there are many choices and it is solved on a case-by-case basis. We will not describe in detail any single bioremediation system; instead we will discuss their main general components and characteristics.

The successful implementation of bioremediation techniques will involve a multidisciplinary approach requiring knowledge from individuals with expertise in chemistry, microbiology, geology, environmental engineering, chemical engineering, soil science,

Table 14.1 Bioremediation treatment technologies

Bioaugmentation	Addition of bacterial cultures to a contaminated medium; frequently used in bioreactors and *ex situ* systems
Biofilters	Use of microbial stripping columns to treat air emissions
Biostimulation	Stimulation of indigenous microbial populations in soils and/or ground water; may be done *in situ* or *ex situ*
Bioreactors	Biodegradation in a container or reactor; may be used to treat liquids or slurries
Bioventing	Method of treating contaminated soils by drawing oxygen through the soil to stimulate microbial growth and activity
Composting	Aerobic, thermophilic treatment process in which contaminated material is mixed with a bulking agent; can be done using static piles, aerated piles, or continuously fed reactors
Land farming	Solid-phase treatment system for contaminated soils; may be done *in situ* or in a constructed soil treatment cell

Table 14.2 Classification of microorganisms

Group	*Carbon source*	*Energy source*
Photoautotrophs	Carbon dioxide	Light
Photoheterotrophs	Organic carbon	Light
Chemoautotrophs	Carbon dioxide	Inorganic chemical (e.g. nitrate)
Chemoheterotrophs	Organic carbon	Light

Source: Baker and Herson (1994)

etc. In order to solve an environmental contamination problem using bioremediation, it is necessary to study three important aspects: microbial systems, type of contaminant and geological and chemical conditions at the contaminated site.

14.3.1 *Microbial Systems*

The key players in bioremediation are microscopic organisms that live virtually everywhere. Microorganisms are the catalyst generator, and the enzymes are the catalysts. These enzymes are involved in degradative (catabolic) reactions to provide energy and material for synthesis of additional microbial cells. Microbial transformation of organic contaminants usually occurs because the organisms can use the contaminant for their own growth as a source of carbon and energy. The mechanisms by which microorganisms obtain energy are the basis for their major classification (Baker and Herson, 1994); see Table 14.2.

The chemoheterotrophic microorganisms are mainly responsible for the degradation of organic contaminants in the environment. Chemoheterotrophs are a large group of organisms containing numerous genera and species.

Microorganisms gain energy by catalysing energy-producing chemical reactions that involve breaking chemical bonds and transferring electrons away from the contaminant through oxidation–reduction reactions. The organic contaminant is oxidized (i.e. it loses electrons); correspondingly, the chemical that gains the electrons is reduced. The contaminant is the electron donor, while the electron recipient is the electron acceptor. The energy released in these reactions is conserved in the form of the high-energy-phosphate bond of ATP for use in fuelling biosynthetic reactions. In general, the biochemical process can be divided in two groups: fermentation and respiration. These can be distinguished by the nature of the redox reaction on the basis of the terminal electron acceptor (Commission on Engineering and Technical Systems, National Research Council, 1993).

Fermentation

In fermentation, the microorganisms use the same organic compound as electron donor and electron acceptor. This process does not lead to complete oxidation of the entire original substrate to carbon dioxide. Fermentative metabolism is characterized by production of a mixture of end products with different levels of oxidation, such as organic acids, alcohols, hydrogen and carbon dioxide.

Aerobic respiration

This is the process of destroying organic compounds with the aid of molecular oxygen as the electron acceptor. In these conditions, microorganisms use oxygen to oxidize part of the carbon in the contaminant to carbon dioxide; the rest of the carbon is used to produce new cell mass.

Anaerobic respiration

Anaerobic respiration occurs only in the absence of molecular oxygen. In these conditions, nitrate (NO_3^-) sulphate (SO_4^{2-}), iron (Fe^{3+}), manganese (Mn^{4+}) and even carbon dioxide serve as electron acceptor to degrade contaminants. In addition to new cell mass, the by-products of anaerobic respiration may include nitrogen gas, hydrogen sulphide, reduced forms of metals, and methane, depending on the electron acceptor.

Reductive dehalogenation

Reductive dehalogenation is a variation of the microbial metabolism, which is potentially important in the detoxification of halogenated organic contaminants. Microorganisms catalyse a reaction in which a halogen atom (such as chlorine) of the contaminant molecule is replaced by a hydrogen atom. The reaction adds two electrons to the contaminant, thus reducing it. In most cases, reductive dehalogenation generates no energy but is an incidental reaction that may benefit the cell by eliminating a toxic compound. Most dehalogenated products tend to be less toxic and more susceptible to further microbial decay than the parent compounds.

Cometabolism

In some cases, microorganisms can transform contaminants even though the conversion reaction yields no benefit to the cell. The term for such non-beneficial biotransformations is secondary utilization or cometabolism. Cometabolism is defined as degradation of a compound only in the presence of other organic material that serves as the primary energy source. Cometabolism is the predominant mechanism for the transformation of many substrates, including some polynuclear aromatic hydrocarbons, halogenated aliphatic and aromatic hydrocarbons and pesticides. The organisms that carry out cometabolic reactions include species of *Pseudomonas, Acinethobacter, Norcadia, Bacillus, Mycococcus, Methylosinus*, and *Arthrobacter* among bacteria and *Penicillium* and *Rhizoctonia* among the fungi (Alexander, 1994).

14.3.2 *Bioremediation Organisms*

Microorganisms carry out biodegradation in many different environments. Of particular relevance for pollutants are sewage-treatment systems, soils, underground sites for disposal of chemicals wastes, groundwaters, etc. Natural communities of microorganisms in these various habitats have an amazing physiological versatility; they are able to metabolize and often mineralize an enormous number of organic molecules. Certain communities of bacteria and fungi metabolize a multitude of synthetic chemicals; the number of molecules that can be degraded is not known but thousands are known to be destroyed as a result of microbial activity in one environment or another. Hazardous compounds result in the selection of a mixed microbial population with improved abilities to tolerate and extract energy from these contaminants. Among the bioremediation microorganisms that frequently are identified as active members of microbial consortiums are: *Acinethobacter, Actinobacter, Alcaligenes, Arthrobacter, Bacillus, Berijerinckia, Flavobacterium, Methylosinus, Mycobacterium, Mycococcus, Nitrosomonas, Nocardia, Penicillium, Phanerochaete, Pseudomonas, Rhizoctonia, Serratia, Trametes and Xanthobacter* (Cookson, 1995).

Microbial consortiums

In nature there is a diversity of types of microorganisms and energy sources. This diversity makes it possible to break down a large number of different organic chemicals. This capacity is used to fill the needs of the microbial population when degrading complex organic compounds or when a site is contaminated with a mixture of organic materials. Microorganisms individually cannot mineralize most hazardous compounds. Complete mineralization results in a sequential degradation by a consortium of microorganisms and involves synergism and cometabolism actions. As initial transformations often depend on the efficiency of reactions that remove intermediates, the complete degradation of a contaminant is related to the nature of the microbial consortium.

Adaptation

Often, microorganisms do not degrade contaminants upon initial exposure but may develop the capability to degrade the contaminant after prolonged exposure. Adaptation is important because it is a process to ensure the existence of microorganisms that can use xenobiotics which were recently created and introduced in the environment. Adaptation may be defined as the length of time between the entry of a chemical into an environment and evidence of its detectable loss. Adaptation occurs not only within single communities but also among distinct microbial communities that may evolve a cooperative relationship in the destruction of toxic compounds. It is generally desirable to enhance the microbial activity of indigenous organisms, rather that using exogenous organisms, because they have already acclimatized to the waste material. Only if the environment is sterile, or the present microbial population does not degrade the contaminant, are exogenous microorganisms considered.

Biological process requirements

Microorganisms carry out biodegradation in many different environments. To deal with the assortment of systems of biodegradation, it is necessary to satisfy several conditions for degradation to take place in an environment. A summary of biological process requirements are: (a) the presence of organisms with the capability to degrade the target compound or compounds; (b) the chemical substrate must be accessible to the organism and capable of being used as energy and carbon source; (c) the presence of an inducer to cause synthesis of specific enzymes for the target compounds; (d) the presence of an appropriate electron acceptor–donor system; (e) environmental conditions that are adequate for enzymatically catalysed reactions (moisture and pH); (f) nutrients necessary to support the microbial growth and enzyme production (nitrogen and phosphorus are essential); (g) a temperature range that supports microbial activity and catalysed reactions; (h) absence of toxic substances; (i) presence of microorganisms to degrade metabolic products; (j) presence of microorganisms to prevent transit build up of toxic intermediates; (k) environmental conditions that minimize competitive organisms with those conducting the desired reactions. It is important to adjust these biological requirements as well as the environmental conditions in order to have a successful bioremediation (Baker, 1994).

Microbial nutrition

Microorganisms require food, water, and a suitable environment in which to live, grow and multiply. Food for microbes must provide a source of carbon for synthesis of essential biochemical and cellular components. The carbon source may consist of organic carbon of many varieties (including hydrocarbons), or inorganic carbonate or carbon dioxide. Bacteria used in bioremediation are the hydrocarbon-degrading heterotrophs and those which can use inorganic carbon and salts (King *et al.*, 1992).

Microorganisms involved in the degradation of xenobiotics in natural systems are dependent on fixed forms of nitrogen (nitrate, nitrite, ammonia or organic nitrogen) to meet their requirements. These forms of nitrogen are frequently the limiting factor for microbial populations in soil.

Phosphorus is essential to microbial cells for the synthesis of ATP, nucleic acids and cell membranes. Because of the low water solubility of phosphates, phosphorus is limiting to bacterial growth in soil environments. In general it is suitable to introduce in culture media in optimal carbon–nitrogen and carbon–phosphorus ratios.

14.3.3 *Type of Contaminant*

The variety of materials and processes used in industrial activities causes different types of contamination in soils. As a rule, the contaminants are not found individually but in simple or complex mixtures. The mixtures may be associated with the release, storage or transport of many chemicals in surface or groundwater, waste-treatment systems, soils or sediments. The number of chemicals found to date is enormous, and the types of mixtures are similarly countless. Sites are frequently contaminated with complex mixtures of organic compounds such as creosote, petroleum or combinations of industrial solvents. Concentrations of individual contaminants may vary significantly within the site, with some areas having extremely low concentrations and others having concentrations over a millionfold higher. Simple hydrocarbons degrade faster than complex hydrocarbons (Norris, 1994).

Organic contaminants

Most of the organic chemicals that are found in the soil system are a result of human activities. These include agricultural amendments (such as fertilizers, sewage, sludge, pesticides), atmospheric deposition (such as automobiles, power plants, municipal incinerators, paint manufacturing industries, etc.), oil-field brines (hydrocarbons from petroleum activities, spills, etc.), inert-fill materials (including foundry sand waste, municipal incinerator ash, power plant ash and manufacturing and processing aggregates), septic systems discharge a large quantity of waste water into soil. Benzene, toluene, xylene, ethyl benzene and chlorinated organic chemicals are often detected in organic solvents used for commercial and household products (Bandyopadhyay *et al.*, 1994).

Inorganic contaminants

- Metals, such as copper, nickel, zinc and cadmium, are often detected in contaminated soil in varying concentrations. The toxic effects are observed at higher concentrations, resulting in the inhibition of metabolic activities of the microorganisms. Microorganisms can not destroy metals but they can alter the microbial reactivity and mobility.
- Cyanide contamination results mainly from disposal of plating bath wastes and plating shop waste. In acidic environment cyanides release toxic hydrogen cyanide. The cyanide ion is a non-specific enzyme inhibitor but exerts its powerful toxic effects by inhibiting the enzyme cytochrome oxidase, and thus preventing the uptake of oxygen by living cells.
- Sulphate contamination mainly results from various human activities with sulphur and its compounds, ranging from acid rain to the disposal of sulphur wastes.

Physicochemical properties

Determination of the chemical properties of the contaminant is necessary to assess its fate and transport potential, as well as to evaluate its potential for bioremediation. These properties include density, adsorption coefficient for soils, solubility in water, solubility in various solvents, volatilization from soil, volatilization from water, etc. (Troy, 1994).

- *Electron acceptors.* All biological reactions that yield energy are redox reactions. Thus adequate electron acceptors are important to control for bioremediation. Type of electron acceptor establishes the metabolism and therefore the specific degradation reactions. From thermodynamics considerations, the electron acceptors are preferred in a definite order: oxygen, nitrate, sulphate, carbon dioxide, organic chemical (Fletcher, 1994).

- *pH.* For most microbiological activities pH will need to be kept in the range 6–8. The pH affects the microorganisms' ability to conduct cellular functions, cell membrane transport and equilibrium of catalysed reactions.

- *Temperature* has a marked influence on the rate of bioremediation because bioremediation temperature affects the rate of degradation. There are specific ranges in which the microbial metabolic pathways and enzymes can operate. Each microorganism has a minimum temperature below which growth no longer occurs. Some organisms have the ability to adapt to temperature changes. In general, microorganisms commonly found effective in bioremediation perform well in the temperature range of 10–40°C (Cookson, 1995).

- *Moisture* is an essential parameter of bioremediation. Moisture content of soils affects the availability of contaminants, the transfer of gases, the effective toxicity level of contaminants, the movement and growth stage of microorganisms and species distribution. During bioremediation, if the water content is too high, it will be difficult for transfer oxygen into the soil, and can be a factor that limits growth efficiency. In general, biodegradation of contaminants in soil systems is optimal at a soil moisture content of 10–20 %. This translates to a value of about 40–75 % of field capacity (King *et al.*, 1992).

Bioavailability of contaminant

The poor availability or the total lack of bioavailability of a compound may be a major determinant of persistence. Salkinoja-Salonen *et al.* (1990) summarize the reasons for the persistence of an organic chemical in soil: (a) incompatibility of environmental conditions with microbial growth; (b) insufficient nutrients for degraders to grow; (c) easily degradable carbon is available and causes catabolic repression of degrading enzymes; (d) chemical compound unavailable for the microorganisms; (e) biophysical factors are unfavourable for the degrading enzymes.

In general, adsorption of organics to soil particles reduces the availability of these materials to microorganisms and hence reduces their biodegradability. For instance, microbial degradation of water-insoluble compounds such as petroleum hydrocarbons require the microbial production of an emulsifying agent to increase the solubility and hence bioavailability of these compounds (Cooper, 1986; Berg *et al.*, 1990). Sanseverino *et al.* (1994) reported that surfactant enhanced solubility and biodegradation of polynuclear aromatic hydrocarbons in coke waste.

In Figure 14.1 we present a gas chromatogram obtained in our laboratory when a naturally selected consortium isolated from a petroleum-contaminated soil was incubated with refined oil in liquid medium. After 3 weeks, under controlled conditions, many of the components of the oil were degraded, showing that bacteria uses hydrocarbons according to their complexity and structure.

14.3.4 *Site Characterization*

Information about site characteristics should be examined for evaluating the viability of bioremediation technology. The factors, which should be studied, include the chemical

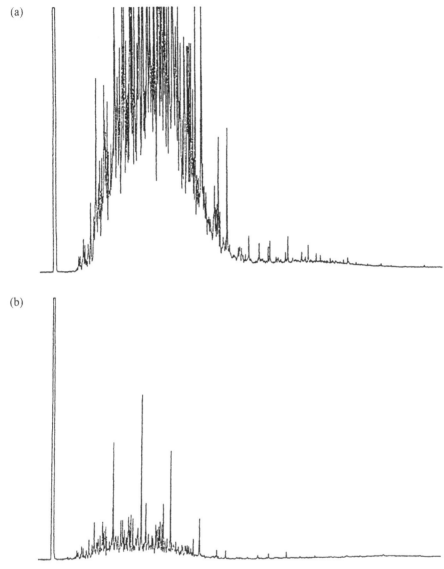

Figure 14.1 Gas chromatogram of refined oil in liquid medium incubated with a mixture of soil microorganisms. The solution was supplemented with nitrogen and phosphorus (a) at 0 h; (b) after 3 weeks

characteristics of contaminants as well as the hydrogeological characteristics of the site. A summary of these factors is presented in Table 14.3 (Wagner *et al.*, 1986).

Geological considerations should include stratigraphic effects such as horizontal extent of the aquifer and heterogeneity of the soil. Hydrogeological data include porosity, permeability, and groundwater velocity, direction and recharge/discharge (Rogers *et al.*, 1993). In addition, hydraulic connection between aquifers, potential recharge/discharge areas, and water table fluctuations must be considered. It is important to initially identify the contaminants present and their concentrations, because the microbial systems capable of biotransformation and rates are compound specific.

Table 14.3 Favourable and unfavourable chemical and hydrogeological site conditions for implementation of *in-situ* bioremediation

	Favourable factors	*Unfavourable factors*
Chemical characteristics	• Small number of organic contaminants • Non-toxic concentrations • Diverse microbial populations • Suitable electron acceptor condition • pH 6–8	• Numerous contaminants • Complex mixture of inorganic and organic compounds • Toxic concentrations • Sparse microbial activity • Absence of appropriate electron acceptors • pH extremes
Hydrogeological characteristics	• Granular porous media • High permeability $(K > 10^{-4}\ \mathrm{cm\ s^{-1}})$ • Uniform mineralogy • Homogeneous medium • Saturated medium	• Fractured rock • Low permeability $(K < 10^{-4}\ \mathrm{cm\ s^{-1}})$ • Complex mineralogy • Heterogeneous medium • Unsaturated–saturated conditions

14.4 Concluding Remarks

In bioremediation, xenobiotics comprise naturally occurring and synthetic compounds that, by nature, only a small part of which can be biodegraded. Under the proper environmental conditions, most xenobiotics are biodegraded by microorganisms already present in the environment. The persistence of xenobiotic compounds is a consequence of the difficulties of the microbiota to reach them. In addition to the bioavailability, a number of environmental factors can affect both the rate and decrease of toxicity when a contaminant is biodegraded. The ability to manipulate all of these parameters will improve current bioremediation efforts by increasing their effectivity while decreasing cost of the treatment. In order to overcome these limitations, researchers must try to develop more competitive bioremediation technologies. The areas of research include the development of methods for increasing the bioavailability of persistent materials; development of better qualitative and quantitative methods for sampling and analysis of the contamination; improved and standardized techniques for assessment of biodegradation methods and microbial systems; standarized methods to determine levels of toxicity and monitoring toxic compounds during and at the end of the treatment.

Bioremediation technologies offer a cost-efective, permanent solution to clean-up of soils contaminated with xenobiotics compounds. It is a new and exciting field and its multidisciplinary nature is a challenge for those interested in the remediation of contaminated soils.

References

ALEXANDER, M., 1994, *Biodegradation and Bioremediation*, San Diego, California: Academic Press Inc.

BAKER, K.H., 1994, Bioremediation of surface and subsurface soils, in Baker, K.H. and Herson, D.S. (eds) *Bioremediation*, New York: McGraw Hill Inc, pp.203–59.

BAKER, K.H., and HERSON, D.S., 1994, Microbiology and degradation, in Baker, K.H. and Herson, D.S. (eds) *Bioremediation*, New York: McGraw Hill Inc, pp.9–60.

BANDYOPADHYAY, S., BHATTACHARYA, S. and MAJUMBAR, P., 1994, Engineering aspects of bioremediation, in Wise, D. and Trantolo, D. (eds) *Remediation of Hazardous Waste Contaminated Soil*, New York: Marcel Dekker Inc, pp.55–76.

BERG, G.A., SEECH, G., LEE, H. and TREVORS, J.T., 1990, Indentification and characterization of a soil bacterium with extracellular emulsifing activity, *J. Environ Sci. Health*, **A25**, 753–764.

CAPLAN, J.A., 1993, The worldwide bioremediation industry: prospects for profit, *Trends in Biotechnology*, **11**, 320–323.

COMMISSION on ENGINEERING and TECHNICAL SYSTEMS, NATIONAL RESEARCH COUNCIL, 1993, *In situ Bioremediation*, Washington, DC: National Academic Press.

COOKSON, J.T. JR, 1995, *Bioremediation Engineering. Design and Application.* New York: McGraw Hill Inc.

COOPER, D.G., 1986, Biosurfactants, *Microbiol. Sci.*, **3**, 145–149.

FIEDLER, L. and QUANDER, J., 1996, Over 300 innovative treatment technologies selected at Superfund sites, new database provides search capability, *EM*, **October**, 21–25.

FLETCHER, R.D., 1994, Practical considerations during bioremediation, in Wise, D. and Trantolo, D. (eds) *Remediation of Hazardous Waste Contaminated Soil.*, New York: Marcel Dekker Inc, pp.39–54.

JESPERSEN, C., JERGER, C. and EXNER, J., 1993, Bioremediation tackles hazwaste, *Chemical Engineering*, **100**, 116–122.

KING, R.B., LONG, G.M. and SHELDON, J.K., 1992, *Practical Environmental Bioremediation*, Boca Raton, FL: Lewis Publishers.

KOVALICK, W.W. and KINGSCOTT, J.W., 1993, Hazardous waste site remediation in the United States: information sources and selection trends for innovative technologies, *UNEP Industry and Environment*, **16**, 10–14.

LEVIN, M.A. and GEALT, M.A., 1993, Overview of biotreatment practices and promises, in Levin, M.A. and Gealt, M.A. (eds), *Biotreatment of Industrial and Hazardous Wastes*, New York: McGraw Hill Inc, pp.1–19.

NORRIS, R.D., 1994, In-situ bioremediation of soils and ground water contaminated with petroleum hydrocarbons, in Kerr, R.S. (ed.) *Handbook of Bioremediation*, Boca Raton, FL: CRC Press Inc, pp.17–34.

ROGERS, J.A., TEDALDI, D.J. and KAVANAUGH, M.C., 1993, A screening protocol for bioremediation of contaminated soil, *Environmental Progress*, **12**(2), 146–56.

SALKINOJA-SALONEN, M., MIDDELORP, P., BRIGLIA, M., VALO, R., HŠGGBLOM, M. and MCBAIN, A., 1990, Cleanup of old industrial sites, in Kamely, D., Chakrabarty, A. and Omenn, G.S. (eds) *Advances in Applied Biotecnology Series. Biotechnology and Biodegradation* Vol. 4. Houston, Texas: Portfolio Publishing Co. and Gulf Publishing Co, pp.347–367.

SANSEVERINO, J., GRAVES, D.A., LEAVITT, M.E., GUPTA, S.K. and LUTHY, R.G., 1994, Surfactant-enhanced bioremediation of polinuclear aromatic hydrocarbons, in Wise, D. and Trantolo, D. (eds) *Remediation of Hazardous Waste Contaminated Soils*, New York: Marcel Dekker Inc, pp.97–124.

TROY, M.A., 1994, Bioingineering of soils and ground waters, in Baker, K.H., Herson, D.S. (eds) *Bioremediation*, New York: McGraw Hill Inc, pp.173–220.

WAGNER, K., BAYER, K., CLAFF, R., EVANS, M., HENRY, S., HODGE, V. *et al.*, 1986, *Remediation Action Technology for Waste Disposal Sites*, 2nd edition, Park Ridge, New Jersey: Noyes Data Corporation.

15

Environmental Oil Biocatalysis

RAFAEL VAZQUEZ-DUHALT

15.1 Introduction

Oil is the main world energy source and a very important raw material. In 1998 more than 3500 million tonnes were extracted and consumed world-wide. The use of fossil fuels for energy and raw materials during the last century has been the origin of some widespread environmental pollution. Among these pollutants are the polycyclic aromatic hydrocarbons (PAHs), which are considered to be a potential health risk because of their possible carcinogenic and mutagenic activities.

The capacity of microorganisms to degrade hydrocarbons has been well known for 50 years (Zobell, 1946). A large number of bacteria (Walter *et al.*, 1975; Austin *et al.*, 1977), yeasts and fungi (Davies and Westlake, 1979; Cerniglia *et al.*, 1980) and algae (Champagnat *et al.*, 1963; Al Hasan *et al.*, 1994) are able to metabolize hydrocarbons. Many studies have been carried out during the last 40 years. Biodegradation of oil and its derivatives takes place in all kinds of natural ecosystems, such as marine (Atlas, 1981; Fedorak and Westlake, 1981), lake (Ward and Brock, 1976; Horowitz and Atlas, 1977), sediment (Hambrick *et al.*, 1980), and soil (Jobson *et al.*, 1974; Sexstone and Atlas, 1977; Westlake *et al.*, 1978; Ismailov, 1983) environments. In all these systems there is a large diversity of microorganisms able to metabolize various oil components, establishing in these compounds the source of energy and carbon to support their growth. In most cases, hydrocarbon biodegradation is an aerobic process; however, oil biodegradation is also possible under anaerobic conditions (Chouteau *et al.*, 1962; Nazina and Rozanova, 1978; Nel *et al.*, 1984). There are also thermophilic bacteria and fungi able to degrade hydrocarbons at temperatures from 55° to 70°C (Fergus, 1966; Mateles *et al.*, 1967; Markel *et al.*, 1978). Similarly, solid (Lonsane *et al.*, 1977) and gaseous (Patel *et al.*, 1980a,b; Hou *et al.* 1981) hydrocarbons can be degraded by different species of microorganisms.

World oil production exceeds 3000 million tonnes per year (Petroleos Mexicanos, 1997). Estimations show that 0.9 % of oil is spread into the environment during the processes of extraction, refining, transport, and use. Thus, more than 27 million tonnes are spread into the sea, lakes, soils, and the atmosphere every year. More than 8 million tonnes of oil derivatives reach the oceans every year. The main input is from non-oil transportation vessels and river runoff. Marine transportation and river contributions represent 39 % of total oil input into the oceans. There are no rigorous estimations for terrestrial oil contamination. However, it is well known that there are several very large contaminated sites world-wide. It could be estimated that terrestrial oil pollution represents twice that of marine pollution.

This chapter is a review of the biocatalytic processes involved during oil biodegradation. Aromatic constituents of oil seem to be a public health risk, due to their potential mutagenic and carcinogenic activities. For this reason, this work is mainly focused on the enzymatic reactions involved in the degradative pathways of aromatic compounds and the genetics of the implicated enzymes.

15.2 Pathways in Hydrocarbon Degradation

Enormous research work has been carried out on the mechanisms of hydrocarbon degradation by microbial cells (McKena and Kallio, 1965; Van der Linden and Thijesse, 1965; Watkinson, 1977). The pathway depends on the physicochemical conditions of the medium, the nature of hydrocarbon substrate, and the microbial strain.

15.2.1 Aliphatic Hydrocarbons

Several forms of *n*-alkane attack have been reported, but the most common involve an oxidation of the terminal methyl group to form a primary alcohol. This oxidation is carried out using molecular oxygen. The *n*-alkane oxidation is started, in most cases, by a mono-oxygenase system, which oxidizes the alkane molecule to form a primary alcohol. Two enzymatic systems have been elucidated (Figure 15.1). The first involves cytochrome P-450, an iron protein and a flavoprotein. This system, which is sensitive to carbon monoxide, has been found in *Corynebacterium* sp. and in some *Candida* strains. Cytochrome P-450 was initially reported as an enzymatic mechanism in mammals for transforming toxic substances. The first isolation and characterization of a cytochrome P-450 system was reported using liver cells. However, it is now well known that this is a superfamily of hemoproteins, which are found in all kinds of organism, bacteria, fungi, animals and plants. The second system is found mainly in bacteria from the *Pseudomonas* group. A flavoprotein and a rubredoxin are components for electron transfer in this catalytic system. In bacteria, the information for the synthesis of this enzymatic system is encoded into plasmids, and can be transferred. In addition, this enzymatic complex is able to perform fatty acid oxidation *in vitro* (Nieder and Shaprio, 1975).

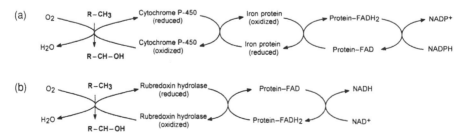

Figure 15.1 Mechanisms for oxidation of *n*-alkanes to primary alcohols. (a) The cytochrome P-450 system; (b) the rubredoxin system

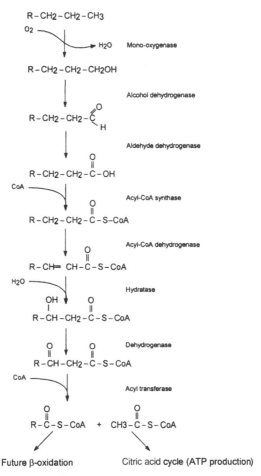

Figure 15.2 β-Oxidation of *n*-alkanes

It has been proposed that the hydroperoxides could be intermediates during the hydroxylation process, but this has not been clearly demonstrated. Some reports have suggested 1-alkene formation during the process. However, not all the organisms able to grow on alkanes are able to grow on the corresponding 1-alkene, which suggests that 1-alkene formation is not a general pathway for primary alcohol formation. Some microorganisms are able to start *n*-alkane oxidation in a non-terminal site, forming ketones. This oxidation may also be catalysed by mono-oxygenases. The ketones are transformed to esters by insertion of oxygen, and then hydroxylated to form one acid and one alcohol.

The degradation of the primary alcohol is carried out by the alcohol–aldehyde dehydrogenase pathway to obtain the corresponding fatty acid. These enzymes usually have a bound NAD, and are induced by the presence of hydrocarbons. Generally, the fatty acid is transformed to the acyl-CoA derivative, and then cleaved to form acetyl CoA. Many microorganisms are also able to make oxidations to produce α-hydroxyl fatty acids, which are decarboxylated later (Figure 15.2).

Terminal alkenes are attacked, in most cases, starting at both ends of the molecule, producing a large variety of products. Internal alkenes are modified by the addition of a water molecule across the double bond. Secondary alcohols or ketones are transformed to esters, and then hydrolysed to form the alcohol and carboxylic acid.

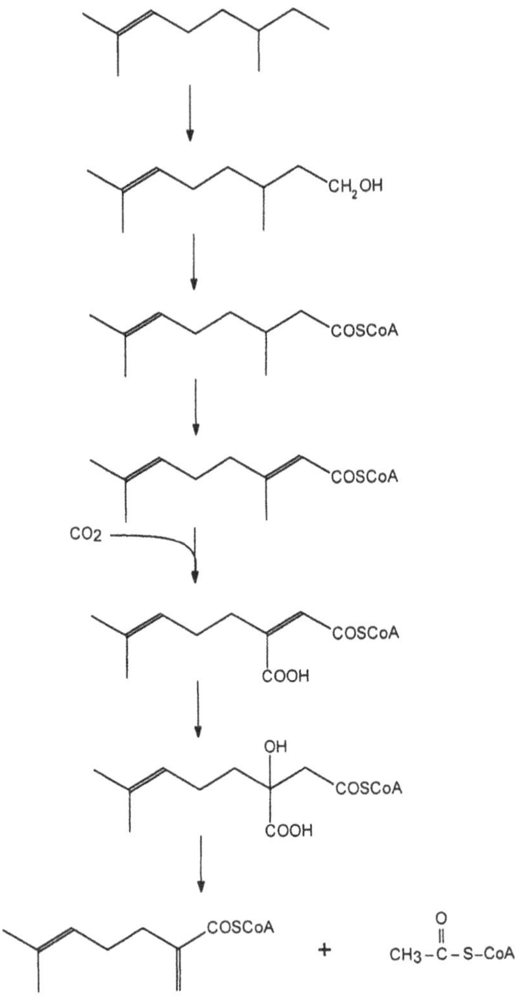

Figure 15.3   Terpenic hydrocarbon oxidation

Few hydrocarbon-degrading microorganisms are able to metabolize branched hydrocarbons. Degradation of isoprenoids by *Pseudomonas citronellolis* has been studied, and the enzymes involved have been partially identified (Seubert, 1960; Cantwell *et al.*, 1978; Fall *et al.*, 1979). This degradation pathway is shown in Figure 15.3.

Cycloalkanes are substrates for few microorganisms (Tonge and Higgins, 1974; Donoghue *et al.*, 1976). Degradation mechanisms have been studied in *Nocardia* sp. and *Pseudomonas* sp. (Stirling *et al.*, 1977; Anderson *et al.*, 1980). The pathway follows the sequence: cyclohexanol, cyclohexanone, ε-caprolactone, 6-hydroxycaprolactone, and adipic acid (Figure 15.4). The cyclohexane hydrolase activity uses molecular oxygen and is specific for NADH.

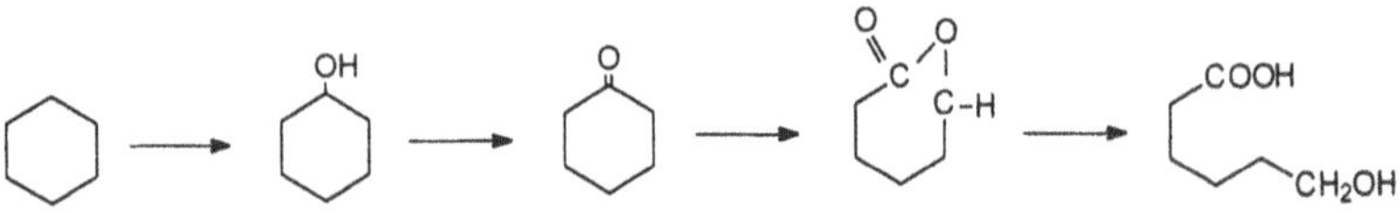

Figure 15.4   Cycloalkane oxidation pathway

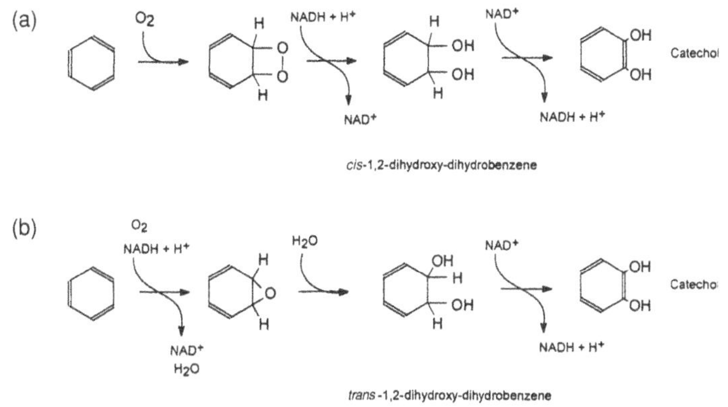

Figure 15.5 Pathways for conversion of benzene to catechol. (a) Dioxygenase mechanism; (b) mono-oxygenase mechanism

15.2.2 *Aromatic Hydrocarbons*

Aromatic hydrocarbon degradation is performed by a large number of microorganisms, and several reports on the mechanisms are available (Dagley *et al.*, 1964; Hayaishi, 1966; Gibson, 1968). Dihydroxylation is necessary to break the aromatic ring. Catechol and protocatechuic acid are substrates for the aromatic ring cleavage during the biodegradation of aromatic compounds. Catechol is also an intermediate in the biodegradation of PAHs, such as anthracene, phenanthrene, pyrene, and benzo[a]pyrene. Dihydroxylated polyaromatic hydrocarbons are substrates for the first aromatic ring breakage. Alkylaromatics are first oxidized on their aliphatic chain to form benzoic acid derivatives, and transformed successively to catechol: oxidation of the aliphatic chain is carried out by β-oxidation. Alkylaromatic compounds, containing an aliphatic chain with an even number of carbons, are degraded by β-oxidations to form phenylacetic acid, which could be transformed to 2,5-dihydroxyphenylacetic acid or to 3,4-dihydroxyphenylacetic acid, depending on the microorganisms involved. In the cases of toluene and isopropylbenzene, dihydroxylation of the aromatic ring could be carried out without oxidation of the aliphatic chain.

Two main mechanisms have been proposed for the incorporation of oxygen before aromatic ring cleavage. The oxygen atom could be incorporated through a mono-oxygenase, such as cytochrome P-450. This mechanism involves an epoxide formation, which is then hydrolysed to obtain *trans*-1,2-dihydroxydihydrobenzene. This product is then transformed to catechol (Figure 15.5). The second mechanism is catalysed by a dioxygenase that incorporates both oxygen atoms from the oxygen molecule, and where peroxide is the intermediate. This peroxide is then transformed to *cis*-1,2-dihydroxydihydrobenzene, and then to catechol (Figure 15.5).

Catechol oxidation is performed with the addition of one oxygen molecule and catalysed by the enzyme catechol dioxygenase. Cleavage of aromatic ring could take place in two different positions (Figure 15.6): *ortho* cleavage, which produces muconic acid by an intradiol dioxygenase (catechol-1,2-dioxygenase), and *meta* cleavage to form hydroxymuconic aldehyde by an extradiol dioxygenase (catechol-2,3-dioxygenase). A similar reaction has been found for protocatechuic acid, where the enzyme involved in the ring cleavage is protocatechuic acid-2,3-dioxygenase. The *ortho* cleavage, between the hydroxyl

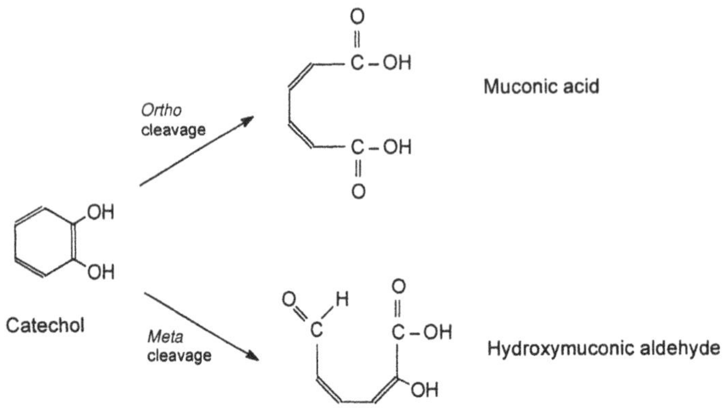

Figure 15.6   Aromatic ring cleavage

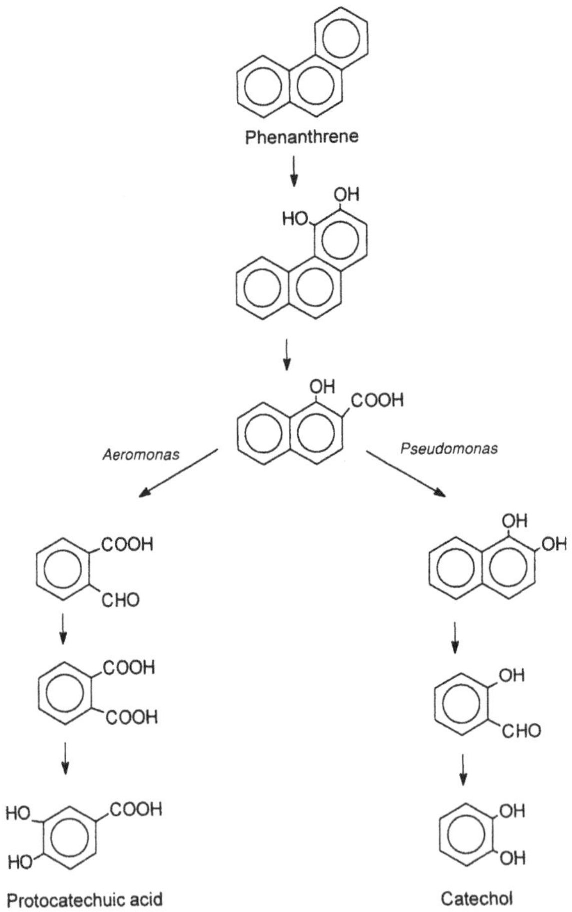

Figure 15.7   Pathway for phenanthrene oxidation

groups, produces β-carboxylated muconic acid. *Ortho* cleavage is also known as endocleavage, while *meta* cleavage is known as exocleavage. Dioxygenase enzymes contain iron, which is at the active site of the molecule. Iron can be extracted by chelating compounds in the presence of reducing agents. Ironless enzyme is inactive, and can often be regenerated in the presence of ferric iron and an oxidizing agent. Polycyclic aromatic hydrocrabon degradation is carried out by a sequence of ring oxidation reactions, ring cleavage reactions and β-oxidations. Figure 15.7 shows phenanthrene degradation by two different microorganisms.

15.3 Genetics of Aromatic Hydrocarbon Biodegradation

15.3.1 *Gene Organization for Aromatic Degradation*

Since 1972 it has been known that in some *Pseudomonas* species the genes containing information for the degradation of certain hydrocarbons are contained in extrachromosomal DNA, called plasmids (Chakrabarty, 1972). Subsequent reports have noted the role of plasmids in degradation of alkanes (Chakrabarty *et al.*, 1973), naphthalene (Dunn and Gunsalus, 1973), toluene, *m*- and *p*-xylenes (Worsey and Williams, 1975; Friello *et al.*, 1976) and other aromatics. Plasmids can be transferred between bacteria *in vivo* or *in vitro* and, with the more recent development of sophisticated techniques, the DNA of any source, prokaryotic or eukaryotic, can be incorporated into bacterial plasmids. With these advances we have the possibility of building bacterial strains in the laboratory able to biodegrade hydrocarbons with new and better characteristics.

One reason for the low biodegradability of some pollutant compounds could be the inability of native microorganisms to metabolize these exogenous compounds, which are not substrates for existing enzymes. However, microbial communities exposed to these pollutants usually adapt and then are able to metabolize new chemical compounds. Little information is available on the adaptation mechanisms of microbial communities. Adaptation is a term used in literature on biodegradation unclearly, because it is used for both microbial communities and pure laboratory strains. A typical adaptation process occurs just after exposure to the pollutant, on which no biodegradation occurs. Nevertheless, after a time, which varies from hours to months, mineralization starts. Different molecular and biochemical processes could be the origin of the adaptation response:

1 Induction of specific enzymes by some members of the microbial communities that increase the degradation capacity of the whole community.

2 Growth of some subpopulations able to take up and metabolize the polluting substrate.

Adaptation could mean selection of mutants that have acquired an altered enzymatic specificity, or new metabolic activities, which were not present in the community before the presence of the pollutant.

Degradation mechanisms for aromatic compounds have been extensively studied, and are good examples of the genetics of organisms able to degrade polluting compounds. The analysis of the different pathways for aromatic biodegradation by bacteria shows that, although the initial degradation may be carried out by different enzymes, these compounds are transformed to a limited number of intermediate metabolites, such as substituted protocatechuates and catechols (Clarke, 1982; Leahy and Colwell, 1990). Dihydroxylated intermediates can be metabolized by two different ways – *meta* cleavage or *ortho* cleavage (Harayama and Rekik, 1989) (Figure 15.7) – but both pathways produce intermediates of central metabolic pathways. This process suggests that microorganisms have developed systems to increase the range of substrates by using peripheral enzymes which transform initial substrates to an intermediate metabolite of the central

Table 15.1 Extradiol dioxygenases, genes, and substrates

Enzyme	Gene	Microorganism
2,3-Dihydroxybiphenyl-1,2-dioxygenase	*bphC*	*Pseudomonas pseudoalcaligenes*
3-Methylcatechol-2,3-dioxygenase	*todE*	*Pseudomonas paucimobilis* Q1
2,3-Dihydroxybiphenyl-1,2-dioxygenase	*cbpC*	*Pseudomonas putida*
1,2-Dihydroxynaphthalene dioxygenase		Not identified
Catechol-2,3-dioxygenase		*Acinetobacter eutrophus*
Protocatechuate-4,5-dioxygenase		*Pseudomonas paucimobilis*

metabolic pathways. Some examples of genes involved in the metabolic cleavage of aromatic compounds are described.

15.3.2 *Genes of Meta Cleavage*

Enzymes (extradioxygenases) for degradation of methylbenzenes such as toluene, xylenes and 1,2,4-methylbenzene are encoded into self-transmissible TOL plasmids containing two or more operons (Harayama and Rekik, 1989; Burlage *et al.*, 1989; Assinder and Williams, 1990; Harayama and Rekik, 1990). Enzymes of the upper metabolic pathway, which oxidize methylbenzene to methylbenzoate, are encoded by an operon (*xylCMABN*) in the TOL plasmid (Harayama *et al.*, 1989). Enzymes of the lower pathway, for the transformation of methylbenzoates to pyruvate, acetaldehyde, and acetate by the methylcatechol pathway, are encoded in the operon *meta* that is composed of 13 genes (Harayama and Rekik, 1990; Franklin *et al.*, 1983).

Enzymes for naphthalene degradation by the salicylicate pathway are encoded by catabolic genes included in two operons (*nah* and *sal* operons) from NAH7 plasmid (Yen and Serdar, 1988). The *sal* operon contain *meta* cleavage genes that are similar to those found in the pWW0 plasmid, and some parts of DNA sequences from both operons are homologous (*xylEGFJ* and *nahHINL*). In toluene degradation, an encoded pathway may be also found in the *tod* operon containing genes related to the *xylF* and *xylJ*. Here, *todF* and *todJ* genes are separated by genes encoding a specific dioxygenase for aromatic ring, and a 3-methyl-2,3-dioxygenase (Horn *et al.*, 1991). Homology has also been found between catechol dioxygenases encoded in the TOL and NAH plasmids and other extradiol dioxygenases. Table 15.1 shows some extradiol dioxygenases, their genes and their substrates (van der Meer *et al.*, 1992).

15.3.3 *Genes of Ortho Cleavage*

Catechol and protocatechuic acid biodegradation can be carried out by the *ortho* cleavage mechanism. The *ortho* cleavage reaction produces a common intermediate, 3-oxoadipate enol-lactone, which is then transformed to succinate and acetyl CoA. Genes for *ortho* cleavage are located in the chromosome (Neidle and Ornston, 1986; Zylstra *et al.*, 1989). Organization of the genes, *cat* encoding catechol degradation and *pca* encoding protocatechuate degradation, varies in the gene order and operon distribution. For example, in some microorganisms, such as *Acinetobacter calcoaceticus*, there are two enzymatic systems for the degradation of 3-oxodipate enol-lactone (encoded by *catEFD* and *pcaDEF*) while in others, such as *Pseudomonas putida*, there is only one enzymatic system encoded apparently by different genes (Doten *et al.*, 1987; Hughes *et al.*, 1988).

Catechol-1,2-dioxygenase, protocatechuate-3,4-dioxygenase and chlorocatechol-1, 2-dioxygenase are the enzyme groups that constitute the intradiol dioxygenase family. These enzymes break the aromatic ring between the two adjacent hydroxyl moieties of catechol or protocatechuate. Catechol-1,2-dioxygenase and protocatechuate-2,3-dioxygenase have been detected in a large number of bacteria, including *P. putida*, *P. aeruginosa*, *P. cepacia*, and *A. calcoaceticus*. The genes encoding catechol-1,2-dioxygenase (*catA*), and two different subunits of the protocatechuate-3,4-dioxygenase (*pcaH* and *pcaG*) are transcribed separately from the other genes of the *ortho* cleavage pathway (Aldrich *et al.*, 1987; Hartnett *et al.*, 1990). According to the three-dimensional structure of protocatechuate-3,4-dioxygenase (Ohlendorf *et al.*, 1988), two histidines and two tyrosines seem to be involved as ligands for catalytic iron, and these amino acid residues are strongly conserved among all the members of intradiol dioxygenases. Homology between catechol-1,2-dioxygenase, protocatechuate-3,4-dioxygenase, and chlorocatechol-1,2-dioxygenase vary between 53.5 % and 63.1 % in amino acid sequence basis. However, these enzymes differ in their range of substrates, and this can be very important for studies on enzyme adaptation to xenobiotic compounds.

15.3.4   Genes of Peripheral Enzymes

Dioxygenases

Enzymes involved in the first oxidation of polycyclic aromatic hydrocarbon show a greater variability than those for the ring cleavage mentioned above. Nevertheless, some genes of these enzymes are evolutionarily related. Multicomponent dioxygenases catalyse the incorporation of two hydroxyl groups, in adjacent positions, into the aromatic ring in an NADH dependent reaction. This enzymatic system is able to hydroxylate benzene, toluene, biphenyl and benzoates (Kurkela *et al.*, 1988; Neidle *et al.*, 1991), and is constituted by three different components: an oxygenase terminal protein – also called iron-sulphur protein and composed of two subunits (a and b) – a ferredoxin, and a ferredoxin reductase. Genes for dioxygenases are clustered with an order different in each microbial system, such as happens with toluene, benzene and naphthalene dioxygenases from *P. putida*, and biphenyl dioxygenase from *P. pseudoalcaligenes* and other *Pseudomonas* strains. Three-component systems, which have not been completely elucidated in their DNA sequence, are naphthalene dioxygenase from NAH7 plasmid (Yen and Serdar, 1988), and biphenyl dioxygenase from *P. paucimobilis* (Taira *et al.*, 1988).

Two-component systems are toluate dioxygenase from *P. putida* and benzoate dioxygenase from *A. calcoaceticus*. These enzymes are encoded by *xylXYZ* (Harayama and Rekik, 1990) and *benABC* (Neidle *et al.*, 1991) genes, respectively. In these cases, the electron transport is carried out by a simple protein. The *n*-terminal structure of this protein is similar to those from ferredoxin, and its *c* terminus shows a similar structure to the NADH-ferredoxin reductase (Neidle *et al.*, 1991). Both a and b subunits from the different terminal dioxygenases have similarity in their amino acid sequence. Genes for dioxygenases are followed by a gene encoding dihydrodiol dehydrogenase in all the multicomponent gene clusters. These enzymes are members of the family of short-chain alcohol dehydrogenases.

Mono-oxygenase

Toluene-4-mono-oxygenase from *P. mendocina* catalyses the oxidation of toluene to form *p*-cresol, and is encoded by five genes, *tmoA–E* (Yen *et al.*, 1991). This is a multicomponent enzyme system that, as other multicomponent dioxygenases, has an electron transfer component similar to those found in other metabolic mono-oxygenases. The ferredoxin component from toluene-4-mono-oxygenase is encoded by the *tmoC* gene, and has 32 % and 28 % homology, in amino acid sequence basis, to ferredoxins from the toluene and naphthalene dioxygenases, respectively (Yen *et al.*, 1991). The other components of this complex are not significantly related to those from other multicomponent dioxygenases. However, the three components of toluene-4-mono-oxygenase are related to those of phenol hydrolase from *Pseudomonas* sp. (Yen *et al.*, 1991). Phenol hydrolase is a multienzymatic system that transforms phenol to catechol, and it is encoded by the genes *dmpK–P*. Both mono-oxygenase systems contain monopeptides structurally related to the methane mono-oxygenase. Hydrolases and multicomponent mono-oxygenases, which take part in various metabolic pathways, have the same conserved domains, such as in the case of salicylate and *p*-hydroxybenzoate hydrolases (Yen and Serdar, 1988).

Xylene mono-oxygenase has two components, *XylM* and *XylA*. *XylA* seems to have an *n*-terminal domain similar to that found in the ferredoxin from chloroplasts, and a *c*-terminal a sequence close to NADP$^+$-ferredoxin reductase (Suzuki *et al.*, 1991). Although xylene mono-oxygenase and 4-toluene sulphonate mono-oxygenase catalyse the same reaction, they have different biochemical properties (Locher *et al.*, 1991).

Regulatory genes

Specific proteins that recognize inducer molecules and interact with promoter–operator regions of operons are known as regulatory proteins. A large number of genes for catabolic regulators have been studied. In the TOL-encoded pathways *XylR* and *XylS* together regulate the complete transcription of *xyl* operon (Spooner *et al.*, 1986): *XylR* responds to the presence of xylene, methylbenzene and benzylalcohols, while *XylS* recognize benzoates and their analogues, and activates transcription of the *meta* cleavage genes. The catalytic operons *nah* and *sal* in the NHA7 plasmid are regulated by only one regulatory protein, *NahR* (Schell, 1985). Salicylate is the effector for *NahR*.

15.4   Mechanisms of Genetic Adaptation

Genetic mechanisms are involved in the evolution of catabolic pathways for hydrocarbons, and in the adaptation processes of microorganisms to metabolize xenobiotic substrates. Adaptation mechanisms can be classified into three different groups:

1   Gene transfer

2   Random mutation

3   Genetic recombination and transposition

These mechanisms could be involved in the evolution of metabolic pathways for oil hydrocarbon degradation. Some of these mechanisms are difficult to prove experimentally, because it is only possible to obtain a final result and make inferences, such as happens with the existence of homologous genes in different microorganisms. The next paragraphs will deal, briefly, with these different adaptation mechanisms.

15.4.1 Gene Transfer

Plasmids in bacteria are certainly a common fact in nature. Plasmids are a huge reserve of genetic information, which is placed in plasmatic vehicles that flow among the microorganisms in nature. There is strong evidence that there was common ancestral gene that was distributed among the microbial population. Thus, these observations support the occurrence of extensive horizontal gene transfer during evolution. Differences between the descendants are obtained, but these divergences do not mean evolutionary distances. Strong DNA homologies have been found in TOL, NAH and SAL plasmids, both inside and outside of catabolic gene clusters (Lehrbach *et al.*, 1983). Microbial interactions could have occurred between different communities by different mechanisms, such as conjugation by plasmid replicons, transduction and transformation (Saye *et al.*, 1990; Lorenz *et al.*, 1991). So far, there are many examples of self-transmissible plasmids that carry degradation genes for aromatic compounds (Chakrabarty *et al.*, 1978; Sayler *et al.*, 1990). Doubtless these plasmids have been very important in gene transfer to other microorganisms.

Adaptation by gene transfer is an important mechanism, as demonstrated in several studies on the experimental evolution of novel metabolic activities. Inability to degrade a compound by natural microorganisms could be eliminated by transferring appropriate genes. This could be accomplished by replacing specific enzymes by non-specific ones, or by providing peripheral enzymes in order to transform substrates into appropriate metabolites for existing catabolic pathways.

15.4.2 Random Mutation

A simple mutation can alter enzyme specificity, as demonstrated in several studies. The change of one amino acid residue has modified the specificity of 2,3-dioxygenase encoded by the TOL *pWW0* plasmid (Ramos *et al.*, 1987). A single amino acid change has also modified the specificity of xylene mono-oxygenase encoded by *xylMA* (Abril *et al.*, 1989). Point mutation can occur continuously and randomly at any time, as a result of mistakes in DNA replication. However, some researchers claim that directed mutation may occur under selective pressure. Even if this statement is still debatable, it seems possible that stress factors, including chemical pollutants, stimulate error-prone DNA replication, accelerating DNA evolution.

15.4.3 Genetic Recombination and Transposition

Four different mechanisms of change could be classified in this part of genetic adaptation to polluting compounds:

- DNA rearrangement
- Gene duplication
- Transposition
- Activation by insertion

Only some examples in each type of change will be described.

DNA rearrangement is supported by the fact that genes encoding *ortho* cleavage in *A. calcoaceticus* and in *P. putida* are different from one to another, suggesting some DNA rearrangement (van der Meer *et al.*, 1991). Another example is the case of catabolic genes encoded by TOL and NAH, for which the operons for *meta* cleavage are identical, while upstream and downstream genes are different (Harayama and Rekik, 1990). Genes for a three-component dioxygenase could have been inserted in an ancestor of operon of *meta* cleavage pathway giving a new toluene degradation pathway (Horn *et al.*, 1991). Genes for other multicomponent dioxygenases differ in their order, such as biphenyl dioxygenase.

Gene duplication has been considered as an important mechanism during evolution. After gene duplication in a microorganism, there is an extra gene that may remain free of selective restrictions, and thus could accumulate mutations. When the mutation results in an inactivation, the gene becomes a silent copy. Reactivation of a silent copy may occur by insertion of some elements, as mentioned below. Comparing various TOL plasmids, it is possible to find that regulatory genes and operons are sometimes in an inverted position, changed, or in a higher number of copies (Assinder and Williams, 1990).

Transposition and element insertion are considered two important ways of obtaining DNA rearrangements, as in the case of the activation and inactivation of silent copies. The catabolic operon TOL is a good example; this operon is part of a large transposable element, *Tn4651*, which is a member of the transposon family type *Tn3*. This transposon was also reported as a part of a larger and more mobil element (Tsuda and Iino, 1988). Another similar element has been found in the NAH7 plasmid (Tsuda and Iino, 1990).

Recently, a substantial debate has taken place on the possibility of using genetically altered microorganisms in bioremediation processes. The main issues discussed involve biosecurity, social, scientific and ethics aspects and the limited competition. It is not clear that genetically engineered organisms are the best solution for open systems. From the information shown here, it is clear that microorganisms are able to obtain new degradative pathways by themselves. In fact microorganisms, which have evolved naturally, tend to be more competitive, because they have survived a natural selection.

In bioremediation processes, or in natural conditions, the degradation of polluting compounds is limited and controlled by chemical and physical factors. Temperature, pH and availability of nutrients and substrates are the main factors on bioremediation, and not so much the presence of appropriate microorganisms. Adaptation and evolution will be better under optimal chemical and physical conditions, because the microbes are able to grow. If in optimal conditions the microbial adaptation is slow, the addition of exogenous microbial populations carrying desirable genes, may speed up the evolution. This directed evolution has been proved in some laboratories, and is potentially very useful for obtaining a better degradation in natural environments.

15.5 Final Remarks

The use of fossil fuels for energy and raw materials during the last century has been the origin of widespread environmental pollution. Petroleum hydrocarbons are a public health risk, because of their potential mutagenic and carcinogenic activities.

The capacity of microorganisms to degrade hydrocarbons has been well known for 50 years. A large number of bacteria, yeasts fungi and algae are able to metabolize hydrocarbons. Individual microorganisms can metabolize only a limited range of hydrocarbons, so mixed microbial populations are needed to degrade complex mixtures of hydrocarbons such as crude oil. Hydrocarbon degradation depends on several factors, such as temperature, concentration and chemical nature of nutrients, phosphorus and nitrogen, iron availability, oil composition and the presence of other organic substances. Thus, each ecosystem and each contaminated site should be considered as a particular case.

Adaptation of microbial population to hydrocarbons can be obtained by interrelated mechanisms: selective enrichment of organisms able to transform the specific substrate, induction or depression of specific enzymes, and genetic changes, which result in new metabolic capabilities. Genetic mechanisms of adaptation can be carried out by gene transfer, random mutation and genetic recombination and transposition.

Oil biodegradation is a biocatalytic process in which several enzymes are involved. The first limiting factor for bioremediation is the presence of the degradative enzymes. If in the polluted site there are the appropriate biocatalytic systems – microorganisms containing genetic information for the production degradative enzymes – the biodegradation processes will be determinated by the environmental conditions.

References

ABRIL, M.A., MICHAN, K.N., TIMMIS, K.N. and RAMOS, J.L., 1989, Regulator and enzyme of the TOL plasmid-encoded upper pathway for degradation of aromatic hydrocarbons and expansion of the substrate range of the pathway, *J. Bacteriol.*, **171**, 6782–6790.

ALDRICH, T.L., FRANTZ, B., GILL, J.F., KILBANE, J.J. and CHAKRABARTY, A.M., 1987, Cloning and complete nucleotide sequence determination of the catB gene encoding cis, cis-muconate lactonizing enzyme, *Gene*, **52**, 185–195.

AL HASAN, H., SORKHOH, N.A., AL BADER, D. and RADWAN, S.S., 1994, Utilization of hydrocarbons by cyanobacteria from microbial mats on oily coast of the Gulf, *Appl. Microbiol. Biotechnol.*, **41**, 615–619.

ANDERSON, M.S., HALL, R.A. and GRIFFIN, M., 1980, Microbial metabolism of alicyclic hydrocarbons: Cyclohexane catabolism by a pure strain *of Pseudomonas* sp., *J. Gen. Microbiol.*, **120**, 89–94.

ASSINDER, S.J. and WILLIAMS, P.A., 1990, The TOL plasmids, determination of the catabolism of toluene and the xylenes, *Adv. Microb. Physiol.*, **31**, 1–69.

ATLAS, R.M., 1981, Microbial degradation of petroleum hydrocarbons: and environmental perspective, *Microbiol. Rev.*, **45**, 180–209.

AUSTIN, B., CALOMIRIS, J.J. and COLWELL, R.R., 1977, Numeral taxonomy and ecology of petroleum-degrading bacteria, *Appl. Environ. Microbiol.*, **34**, 60–68.

BURLAGE, R.S., HOOPER, S.W. and SAYLER, G.S., 1989, The TOL (pWW0) catabolic plasmid, *Appl. Environ. Microbiol.*, **55**, 1323–1328.

CANTWELL, S.G., LAU, E.P., WATTS, D.S. and FALL, R.R., 1978, Biodegradation of acyclic isoprenoids by Pseudomonas species, *J. Bacteriol.*, **135**, 324–333.

CERNIGLIA, C.E., VAN BAALRN, C. and GIBSON, D.T., 1980, Metabolism of naphtalene by cyanobacterium *Oscillatoria* sp., strain J.M.C., *J. Gen Microbiol.*, **116**, 485–494.

CHAKRABARTY, A.M., 1972, Genetic basis of the biodegradation of salicylate in Pseudomonas, *J. Bacteriol.*, **112**, 815–823.

CHAKRABARTY, A.M., CHOU, G. and GUNSLAUS, I.C., 1973, Genetic regulation of octane dissimilation plasmid in Pseudomonas. *Proc. Natl. Acad. Sci. USA*, **70**, 1137–1140.

CHAKRABARTY, A.M., FRIELLO, D.A. and BOPP, L.H., 1978, Transposition of plasmid DNA segments specifying hydrocarbon degradation and their expression in various microorganism. *Proc. Natl. Acad. Sci. USA*, **75**, 3109–3112.

CHAMPAGNAT, A., VERNET, C., LAINÉ, B. and FILOSA, J., 1963, Biosynthesis of protein-vitamin concentrates from petroleum, *Nature*, **197**, 13–14.

CHOUTEAU, J., AZOULAY, E. and SENEZ, J.C., 1962, Anaerobic formation of n-hept-1-ene from n-heptane by resting cells of *Pseudomonas aeruginosa, Nature*, **194**, 576–578.

CLARKE, P.H., 1982, The metabolic versatility of pseudomonads, *Antoine van Leeuwenhoek*, **48**, 105–130.

DAGLEY, S., CHAPMAN, P.J., GIBSON, D.T. and WOODS, J.M., 1964, Degradation of the benzene nucleus by bacteria, *Nature*, **202**, 775–778.

DAVIES, J.S. and WESTLAKE, D.W.S., 1979, Crude oil utilization by fungi, *Can. J. Microbiol.*, **25**, 46–156.

DONOGHUE, N.A., GRIFFIN, M., NORRIS, D.B. and TRUDGILL, P.W., 1976, The microbial metabolism of cyclohexane and related compounds. *Proceedings of the Third International Biodegradation Symposium*, Essex: Applied Science. pp.43–56.

DOTEN, R.C., NGAI K.L., MITCHELL, D.J. and ORNSTON, L.N., 1987, Cloning and genetic organization of the pca gene cluster from *Acinetobacter calcoaceticus, J. Bacteriol.*, **169**, 3168–3174.

DUNN, N.W. and GUNSALUS, I.C., 1973, Transmissible plasmid coding early enzymes of naphthalene oxidation in *Pseudomonas putida, J. Bacteriol.*, **114**, 974–979.

FALL, R.R., BROWN, J.L. and SCHAEFFER, T.L., 1979, Enzyme recruitment allows the biodegradation of recalcitrant branched hydrocarbons by *Pseudomonas citronellolis, Appl. Environ. Microbiol.*, **38**, 715–722.

FEDORAK, P.M. and WESTLAKE, D.W.S., 1981, Microbial degradation of aromatics and saturates in Prudhoe bay crude oil as determined by glass capillary gas chromatography, *Can. J. Microbiol.*, **27**, 432–443.

FERGUS, C., 1966, Paraffin utilization by thermophilic fungi, *Can J. Microbiol.*, **12**, 1067–1068.

FRANKLIN, F.C.H., BAGDASARIAN, M., BAGDASARIAN, M.M. and TIMMIS, K.N., 1983, Localization and functional analysis of transposon mutations in regulatory genes of the TOL catabolic pathway, *J. Bacteriol.*, **154**, 676–685.

FRIELLO, D.A., MYLROIE, J.R., GIBSON, D.T., ROGERS, J.E. and CHAKRABARTY A.M., 1976, XYL a nonconjugative xylene-degradative plasmid in *Pseudomonas* pxy,. *J. Bacteriol.*, **127**, 1217–1224.

GIBSON, D.T., 1968, Microbial degradation of aromatic compounds, *Science*, **161**, 1093–1097.

HAMBRICK III, G.A., DE LAUNE, R.D. and PATRICK, JR. W.H., 1980, Effect of estuarine sediment pH and oxidation-reduction potential on microbial hydrocarbon degradation, *Appl. Environ Microbiol.*, **40**, 365–369.

HARAYAMA, S. and REKIK, M., 1989, Bacterial aromatic ring-cleavage enzymes are classified into two different gene families, *J. Biol. Chem.*, **264**, 15328–15333.

HARAYAMA, S. and REKIK, M., 1990, The meta cleavage operon of TOL degradative plasmid pWW0 comprises 13 genes, *Mol. Gen. Genet.*, **221**, 113–120.

HARAYAMA, S., REKIK, M., WUBBOLTS, M., ROSE, K., LEPPIK, R.A. and TIMMIS, K.N., 1989, Characterization of five genes in the upper-pathway operon of TOL plasmid pWW0 from *Pseudomonas putida* and identification of the gene products, *J. Bacteriol.*, **171**, 5048–5055.

HARTNETT, C., NEIDLE, E.L., NGAI, K.L. and ORNSTON, L.N., 1990, DNA squence of genes encoding *Acinetobacter calcoaceticus* protocatechuate-3, 4-dioxygenase, evidence indicating shuffling of genes and DNA sequences within genes during their evolutionary divergence, *J. Bacteriol.*, **172**, 956–966.

HAYAISHI, O., 1966, Enzymatic studies on the mechanism of double hydroxylation, *Pharmacol. Rev.*, **18**, 71–75.

HORN, J.M., HARAYAMA, S. and TIMMIS, K.N., 1991, DNA sequence determination of the TOL plasmid (pWW0) xylGFJ genes of *Pseudomonas putida*, implications for the evolution of aromatic catabolism, *Mol. Microbiol.*, **5**, 2459–2474.

HOROWITZ, A. and ATLAS, R.M., 1977, Response of microorganisms to an accidental spillage in artic freshwater ecosystem, *Appl. Environ. Microbiol.*, **33**, 1252–1258.

HOU, C.T., PATEL, R.N., LASKIN, A.I., MARCZAK, Y. and BERNABE, E.M., 1981, Microbial oxidations of gaseous hydrocarbons, production of alcohols and methyl ketones from their corresponding n-alkanes by methylotrophic bacteria, *Can. J. Microbiol.*, **27**, 107–115.

HUGHES, E.J., SHAPIRO, J.E., HOUGHTON, J.E. and ORNSTON, L.N., 1988, Cloning and expression of pca genes from *Pseudomonas putida* in *Escherichia coli, J. Gen. Microbiol.*, **134**, 2877–2887.

ISMAILOV, N.M., 1983, Effect of oil pollution on the nitrogen cycle in soil, *Mikrobiologiya*, **52**, 1003–1007.

JOBSON, A.M., McLAUGHLIN, M., COOK, F.D. and WESTLAKE, D.W.S., 1974, Effect of amendments on the microbial utilization of oil applied to soil, *Appl. Microbiol.*, **27**, 166–171.

KURKELA, S., LEHVASLAIHO, H., PALVA, E.T. and TEERI, T.H., 1988, Cloning, nucleotide sequence and characterization of genes encoding naphthalene dioxygenase of *Pseudomonas putida* strain NCIB9816, *Gene,* **73**, 355–362.

LEAHY, J.K. and COLWELL, R.R., 1990, Microbial degradation of hydrocarbons in the environment, *Microbiol. Rev.*, **54**, 305–315.

LEHRBACH, P.R., McGREGOR, Y., WARD, J.M. and BRODA, P., 1983, Molecular relationship between *Pseudomonas* Inc P-9 degradative plasmids TOL, NAH, and SAL, *Plasmid,* **10**, 164–174.

LOCHER, H.H., LEISINGER, T. and COOK, A.M., 1991, 4-Toluene sulfonate methyl-monooxygenase from *Comamonas testosteroni* T-2, purification and some properties of the oxygenase component, *J. Bacteriol.*, **173**, 3741–3748.

LONSANE, B.K., SINGH H.D., NIGAM, J.N. and BARUACH, J.N., 1977, Fermentation studies on solid hydrocarbons utilizing bacterial isolates, *Indian J. Exp. Biol.*, **17**, 1263–1265.

LORENZ, M.G., GERJETS, D., and WACKERNAGEL, W., 1991, Release of transforming plasmid and chromosomal DNA from two cultured soil bacteria, *Arch. Microbiol.*, **156**, 319–327.

MARKEL, G.J., STAPLETON, S.S. and PERRY, J.J., 1978, Isolation and peptidoglycan of gram-negative hydrocarbon-utilizing thermophilic bacteria, *J. Gen Microbiol.*, **109**, 141–148.

MATELES, R.I., BARUACH, J.N. and TANNENBAUM, S.R., 1967, Growth of a thermophilic *bacterium* on hydrocarbons, a new source of single-cell protein, *Science*, **157**, 1322–1323.

McKENA, E.J. and KALLIO, R.E., 1965, The biology of hydrocarbons, *Annu. Rev. Microbiol.*, **19**, 183–208.

NAZINA, T.N. and ROZANOVA, E.P., 1978, Thermophilic sulfate reducing bacteria from oil strata, *Microbiologiya*, **47**, 113–115.

NEIDLE, E.L. and ORNSTON, L.N., 1986, Cloning and expression *of Acinetobacter calcoaceticus* benABC catechol-1,2-dioxygenase structural gene catA in *Escherichia coli, J. Bacteriol.*, **168**, 815–820.

NEIDLE, E.L., HARTNETT, C., ORNSTON, L.N., BAIROCH, A., REKIK, M. and HARAYAMA, S., 1991, Nucleotide sequence of the *Acinetobacter calcoaceticus* benABC genes for benzoate-1,2-dioxygenase reveal evolutionary relationships among multicomponent oxygenases, *J. Bacteriol.*, **173**, 5385–5395.

NEL, L.H., DE HASST, J. and BRITZ, T.J., 1984, Anaerobic digestion of petrochemical effuent using and upflow anaerobic sludge blanket reactor, *Biotechnol. Lett.*, **6**, 741–746.

NIEDER, M. and SHAPRIO, J., 1975, Physiological function of the *Pseudomonas putida* PpG6 (*P. oleovorans*) alkane hydroxylase, monoterminal oxidation of alkanes and fatty acids, *J. Bacteriol.*, **122**, 93–98.

OHLENDORF, D.H., LIPSCOMB, J.D. and WEBER, P.C., 1988, Structure and assembly of protocatechuate 3,4-dioxygenase, *Nature*, **336**, 403–405.

PATEL, R.N., HOU, C.T., LASKIN, A.I., FELIX, A. and DERELANKO, P., 1980a, Microbial oxidation of gaseous hydrocarbons, production of secundary alcohols from corresponding n-alkanes by methane-utilizing bacteria, *Appl. Environ. Microbiol*, **39**, 720–726.

PATEL, R.N., HOU, C.T., LASKIN, A.I., FELIX, A. and DERELANKO, P., 1980b, Microbial oxidation of gaseous hydrocarbons, production of methylketones from corresponding n-alkanes by methane-utilizing bacteria, *Appl. Environ. Microbiol*, **39**, 727–733.

PETROLEOS MEXICANOS, 1997, *Informe anual 1996*. Mexico D.F.: PEMEX.

RAMOS, J.L., WASSERFALLEN, A., ROSE, K. and TIMMIS, K.N., 1987, Redesigning metabolic routes, manipulation of TOL plasmid pathway for catabolism of alkylbenzoates, *Science*, **235**, 593–596.

SAYE, D.J., OGUNSEITAN, O.A., SAYLER, G.S. and MILLER, R.V., 1990, Transduction of linked chromosomal genes between *Pseudomonas aeruginosa* strains during incubation *in situ* in a freshwater habitat, *Appl. Environ. Microbiol*, **56**, 140–145.

SAYLER, G.S., HOOPER, S.W., LAYTON, A.C. and KING, J.M.H., 1990, Catabolic plasmids of environmental and ecological significance, *Microb. Ecol*, **19**, 1–20.

SCHELL, M.A., 1985, Transcriptional control of the *nah* and *sal* hydrocarbon-degradation operons by the *nahR* gene product, *Gene*, **36**, 301–309.

SEUBERT, W., 1960, Degradation of isoprenoid compounds by microorganisms. V. Isolation and characterization of an isoprenoid-degrading bacterium, *Pseudomonas citronellolis, J. Bacteriol.*, **79**, 426–434.

SEXSTONE, A.J. and ATLAS, R.M., 1977, Response of microbial populations in arctic tundra soil to crude oil, *Can. J. Microbiol.*, **23**, 1327–1333.

SPOONER, R.A., LINDSAY, K. and FRANKLIN, C.H., 1986, Genetic, functional and sequence analysis of the xylR and xylS regulatory genes of the TOL plasmid pWW0. *J. Gen. Microbiol.*, **132**, 1347–1358.

STIRLING, L.A., WATKINSON, R.J. and HIGGINS, I.J., 1977, Microbial metabolism of alicyclic hydrocarbons: isolation and properties of a cyclohexane-degrading bacterium, *J. Gen. Microbiol.*, **99**, 119–125.

SUZUKI, M., HAYAKAWA, T., SHAW, J.P., REKIK, M. and HARAYAMA, S., 1991, Primary structure of xylene monooxygenase: Similarities to and differences from the alkane hydroxylation system, *J. Bacteriol.*, **173**, 1690–1695.

TAIRA, K., HAYASE, N., ARIMURA, N., YAMASHITA, S., MIYAZAKI, T. and FURUKAWA, K., 1988, Cloning and nucleotide sequence of 2,3-dihydroxybiphenyl dioxygenase gene from the PCB-degrading strain of *Pseudomonas paucimobilis* Q1, *Biochemistry*, **27**, 3990–3996.

TONGE, G.M. and HIGGINS, I.J., 1974, Microbial metabolism of acyclic hydrocarbons, growth of *Nocardia petroleophila* (NCIB 9438) on methyl cyclohexane, *J. Gen Microbiol.*, **81**, 521–524.

TSUDA, M. and IINO, T., 1988, Identification and characterization of Tn4653, a transposon covering the toluene transposon Tn4651 on TOL plasmid pWW0, *Mol. Gen. Genet.*, **213**, 72–77.

TSUDA, M. and IINO, T., 1990, Naphthalene degrading genes on plasmid NAH7 are on a defective transposon, *Mol. Gen. Genet.*, **223**, 33–39.

VAN DER LINDEN, A.C. and THIJESSE, G.J.E., 1965, The mechanisms of microbial oxidations of petroleum hydrocarbons, *Adv. Enzymol*, **27**, 469–546.

VAN DER MEER, J.R., EGGEN, R.I.L., ZEHNDER, A.J.B. and DE VOS, W.M., 1991, Sequence analysis of the *Pseudomonas* sp. strain P51 tcb gene cluster, which encodes metabolism of chlorinated catechols, evidence for specialization of catechol-1, 2-dioxygenase for chlorinated substrates, *J. Bacteriol*, **173**, 2425–2434.

VAN DER MEER, J.R., DE VOS, W.M., HARAYAMA, S. and ZEHNDER, A.J.B., 1992, Molecular mechanisms of genetic adaptation to xenobiotic compounds, *Microbiol. Rev.*, **56**, 677–694.

WALTER, J.D., AUSTIN, H.F. and COLWELL, R.R., 1975, Utilization of mixed hydrocarbon substrate by petroleum-degrading microorganisms, *J. Gen. Appl. Microbiol.*, **21**, 27–39.

WARD, D.M. and BROCK, T.D., 1976, Environmental factors influencing the rate of hydrocarbon oxidation in temperate lakes, *Appl. Environ. Microbiol.*, **31**, 764–772.

WATKINSON, R.J., 1977, *Developments in biodegradation of hydrocarbons*, London: Applied Science.

WESTLAKE, D.W.S., JOBSON, A.M. and COOK, F.D., 1978, *In situ* degradation of oil in a soil of the boreal region of the northwest territories, *Can. J. Microbiol.*, **24**, 254–260.

WORSEY, M.J. and WILLIAMS, P.A., 1975, Metabolism of toluene and xylenes by *Pseudomonas putida* (arvilla) mt-2. Evidence for a new function of the TOL plasmid, *J. Bacteriol.*, **124**, 7–13.

YEN, K.M. and SERDAR, C.M., 1988, Genetics of naphthalene catabolism in pseudomonads, *Crit. Rev. Microbiol.*, **15**, 247–268.

YEN, K.M., KARL, M.R., BLATT, L.M., SIMON, M.J., WINTER, R.B., FAUSSET, P.R. *et al.*, 1991, Cloning and characterization *of Pseudomonas mendocina* KR1 gene cluster encoding toluene-4-monooxygenase, *J. Bacteriol.*, **173**, 5315–5327.

ZOBELL, C.E., 1946, Action of microorganisms on hydrocarbons, *Bacteriol. Rev.*, **10**, 1–49.

ZYLSTRA, G.J., OLSEN, R.H. and BALLOU, D.P., 1989, Genetic organization and sequence of the *Pseudomonas cepacia* genes for the alpha and beta subunits of protocatechuate 3,4-dioxygenase, *J. Bacteriol.*, **171**, 5915–5921.

PART FOUR

Cleaner Bioprocesses

16

Clean Biological Bleaching Processes in the Pulp and Paper Industry

JUAN M. LEMA, M. TERESA MOREIRA, CAROLYN PALMA AND
GUMERSINDO FEIJOO

16.1 Introduction

Paper and cardboard production has greatly increased in recent years, being one of the sectors of the forest industry with important expectations of expansion (Table 16.1). The most important feedstock is generally wood, the main components of which are cellulose (42 % dry organic matter), hemicellulose (27 %) and lignin (25 %), and to a lesser extent wood extracts such as tannins and resin acids (6 %). Cellulose is a linear polysaccharide consisting of ß-D-glucopyranose units, which are linked by (1–4)-glucosidic bonds. Hemicelluloses are composed of different carbohydrate units and are located at the interface of cellulose and lignin. Lignin is a complex three-dimensional polymer consisting of non-repeating phenyl propanoid units linked by carbon–carbon and ether bonds. Lignins have been generally classified into three major groups – softwood lignin, hardwood lignin, and grass lignin – based on their chemical structure of monomer units (Higuchi, 1980).

Table 16.1 Predicted consumption of paper and cardboard (Source: Food and Agriculture Organization of the United Nations)

	Consumption (millions of tonnes)		
	1993	2010	Increment (%)
North America	92 388	144 101	56.0
Europe	65 373	111 916	71.2
Old URSS	4 323	8 561	98.0
Africa	3 305	8 265	150.1
Latin America	9 194	17 094	85.9
Asia	74 375	183 298	146.5
Oceania	3 244	5 943	83.2
Total	**252 202**	**479 178**	**90.0**

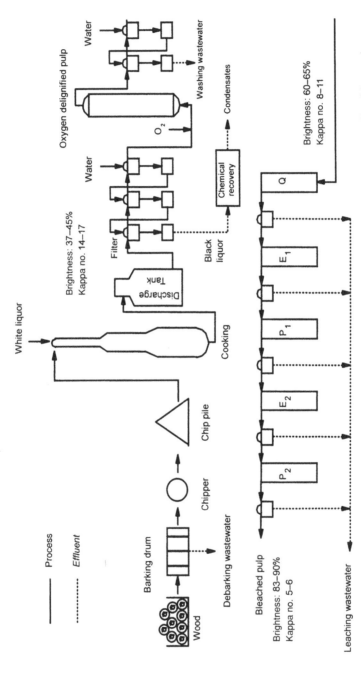

Figure 16.1 Flow diagram of Kraft pulp mill including post-digested oxygen delignification and totally chlorine free bleaching process. Symbols for each bleaching step are defined in Table 16.2

Table 16.2 Nomenclature used for the different bleaching steps in the Kraft process

Bleaching processes	Notation	Reaction
Mill Trials:		
Chlorine	C	Reaction with elemental chlorine in acid solution
Chlorine dioxide	D	Reaction with chlorine dioxide in acid solution
Mixture of chlorinated compounds	C_D or D_C	Mixture of chlorine and chlorine dioxide
Sodium hypochlorite	H	Reaction with hypochlorite in alkali solution
Oxygen	O	Reaction with pressurized oxygen in alkali solution
Ozone	Z	Reaction with ozone
Hydrogen peroxide	P	Reaction with hydrogen peroxide
Enzyme (xylanase)	X	Enzymatic hydrolysis of xylan
Alkali extraction	E	Dissolution of reaction products with sodium hydroxide
Monox-L (hypochlorous acid)	M	Reaction with hypochlorous acid
Chelating treatment	Q	Chelating treatment with EDTA
Process Pilot Plant:		
Prenox	NO_X/O_2	Acid pretreatment
Peroxyacetic acid	CH_3CO_3H	Chemical bleaching
Peroxyformic acid	HCO_3H	Chemical bleaching
Dimethyl dioxirane	DMD	Activated oxygen

The process of cellulose pulp production requires an initial step of pulping after debarking and chipping of the raw material. The purpose of pulping is to remove lignin in order to facilitate fibre separation and to improve the papermaking properties of the pulp. There are basically two methods: mechanical and chemical pulping, accounting for 33 % and a 67 % respectively of the total production. Most chemical pulping is carried out according to either the Kraft (sulphate) process (Figure 16.1) or the sulphite process. The Kraft process entails treating wood chips at 160–180°C with a liquor containing contains sodium hydroxide and sodium sulphide, which promotes cleavage of the various ether bonds in the lignin. The lignin degradation products so formed are dissolved in the alkaline pulping liquor. Depending on pulping conditions, as much as 90–95 % of the lignin is removed from wood at this stage. After separation of the pulp, the spent liquor is evaporated to a high concentration and then burned to recover energy and inorganic chemicals. Neither the Kraft nor the sulphite process removes all the lignin: about 5–10 % of the original remains in the pulp, because it cannot be removed by extended pulping without seriously damaging the polysaccharide fraction.

The pulping process plays a central role in the pollution load and composition of wastewater produced at pulp mills. Mechanical and thermomechanical pulping give high yields and accordingly low pollution loads are produced. Their wastewaters usually contain high amounts of biodegradable matter, such as carbohydrates and organic acids. Semi-chemical (e.g. neutral sulphite semi-chemical) and chemo-thermomechanical pulping wastewaters are of intermediate strength and contain higher amounts of lignin (Welander and Andersson, 1985). In the chemical pulping, a high-strength wastewater is produced with soluble carbohydrates and organic acids as the main organic components, along with hardly biodegradable wood extracts (Kringstad and Lindström, 1984). Furthermore, inorganic compounds such as sulphates, sulphites, sulphide and alkali are also present at high concentrations.

Removal of the residual lignin, which is responsible for the dark colour of Kraft pulps, calls for a further multistage bleaching process. The more common nomenclature for expressing the different bleaching stages is shown in Table 16.2. Conventionally, the

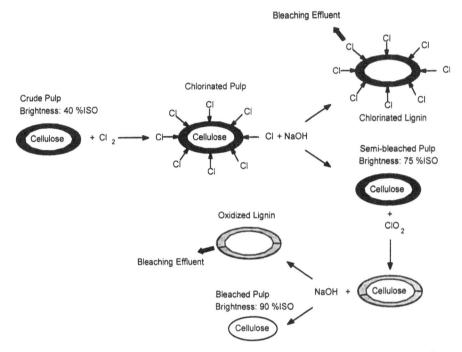

Figure 16.2   Mechanism of conventional bleaching by chlorine and chlorine dioxide

bleaching of digested pulp by Cl_2 (Figure 16.2) has implied the generation of adsorbable organic halogens (AOX), this being problematic for biological wastewater treatment due to their high toxicity and low biodegradability (Sierra *et al.*, 1994). Moreover, organohalogens have generally been regarded as undesirable xenobiotic compounds. Consequently, due to market pressures and government norms, the pulp and paper industry has made substantial changes to conventional pulping and bleaching technologies. New operating policies were developed to minimize the pernicious impact of the free disposal of liquid residues in natural courses and landfills. The proposals towards cleaner processes were focused on bleaching with chlorine-free sequences to eliminate emission of organic halogen compounds (Kinstrey, 1993).

16.2   New Bleaching Processes

The first significant change introduced in the bleaching plant has been the replacement of elemental chlorine by chlorine dioxide; the so-called 'elemental chlorine free' (ECF) bleaching process. This alternative leads to a high-brightness pulp, with acceptable strength properties and wastewaters with lower AOX concentrations and lower AOX loads, ranging from 0.7 to 0.9 kg t^{-1} air-dried pulp (ADP) for mature eucalyptus and from 0.4 to 1.0 kg t^{-1} ADP for plantation (young) eucalyptus (Nelson *et al.*, 1993).

Since 1990, 'totally chlorine free' (TCF) bleaching has been used (Figure 16.1), largely in response to market demands for non-chlorine-bleached pulp. TCF bleaching was made possible after the action of a pre-delignification step with pressurized oxygen, which leads to a pulp with considerably lower kappa number. Then the process can be based on the action of oxygen, ozone, hydrogen peroxide, or enzyme which replaces chlorine and chlorine dioxide bleaching (Byrd *et al.*, 1992; Lapierre *et al.*, 1995). Peroxide stages are

Table 16.3   Present status of bioprocessing in the pulp and paper industry

Process stage and identified problems	Biotechnological solution		
	Identified	Tested	Commercialized
Raw material treatment:			
Debarking	+	+	–
Wood preservation	+	+	–
Mechanical pulping:			
Energy saving	+	+/–	–
Pitch removal	+	+	+
Dissolved colloids	+	+	–
Chemical pulping:			
Delignification	+	–	–
Bleaching:			
Hemicellulase-aided bleaching	+	+	+
Enzymatic delignification	+	+/–	–
Paper manufacture:			
Recycled paper	+	+/–	–
Fibre modification	+	+/–	+/–
Microbial contaminants and slimes	+	+/–	+/–

+/– Results not conclusive
Source: Viikari, 1995

preceded by some treatment, such as an acid wash, and a chelating stage to remove metal ions. Peroxide-based TCF bleaching processes produce virtually no organohalogens as the AOX loads are reported to be below 0.05 kg t^{-1} ADP (Folke and Rendberg, 1994). In a recent study, Vidal *et al.* (1997) analysed the methanogenic toxicity and anaerobic biodegradability of ECF and TCF bleaching effluents from oxygen-delignified eucalyptus Kraft pulp. The effluents from chlorine and ECF bleaching sequences had similar methanogenic toxicities, with 50 % inhibiting concentrations (50 %IC) of 0.65–1.48 g of chemical oxygen demand (COD) per litre. Only the TCF bleaching effluent was distinctly less toxic, with a 50 %IC of 2.3 g COD/l. The fact that the ECF bleaching effluent was no less toxic than that of chlorine bleaching combined with the residual toxicity of TCF, indicates that other substances aside from organohalogens contribute to the high methanogenic toxicity in bleaching effluents (Stauber *et al.*, 1996).

To obtain an efficient operation both processes, ECF and TCF, have additional requirements such as an improved quality of raw materials, extended cooking, post-digested oxygen delignification, optimized washing to remove high lignin fractions and closure of the brown stock area (Johnson, 1996).

The utilization of biological or enzymatic processes in several areas of food and non-food industries has become feasible by the development of highly specialized microorganisms, which can be employed directly or through isolation of the enzymes they produce. It can be expected that progress will be made in reducing chemical inputs by the use of biotechnology, which also provides generally better environmental protection (Messner and Srebotnik, 1994). The present status of the possible biotechnical solutions at different process stages in the pulp and paper industry is summarized in Table 16.3. As it can be observed, only few of these solutions are commercial.

Enzymes (xylanases) can be used as a biotechnological alternative as a pretreatment (Viikari *et al.*, 1994; Suurnäkki *et al.*, 1996); oxidative enzymes from white-rot fungi may be used in the bleaching process itself (Hirai *et al.*, 1994; Paice *et al.*, 1995; Moreira *et al.*, 1997a).

Xylanases are being employed as enzymatic pre-treatments in Kraft pulp bleaching process at mill scale in order to improve delignification by degradation of hemicelluloses (Senior and Hamilton, 1993). Xylan forms the basic backbone polymers of wood hemicelluloses, depending on the degree of polymerization of the wood species. Elucidation of the exact mechanism of the enzyme-aided bleaching methods is key. There are different hypotheses to explain the enzymatic action on the fibre-bound substrate, but it can be concluded that two types of phenomenon are involved (Viikari *et al.*, 1994): (1) the hydrolysis of the re-precipitated xylan, formed during delignification, renders the pulp more permeable, thus facilitating the removal of residual lignin; (2) the partial hydrolysis of xylan, located in the inner layers and possibly linked to lignin, is likely to facilitate further bleaching. A pretreatment step with xylanases achieves an increase of brightness or a decrease of chemical consumption. Other positive features are the low cost of enzyme and the low investment costs if the enzyme stage is performed in the brownstock storage tower. However, the use of xylanases will always require some further chemical delignification for complete pulp bleaching and consequently will not permit large chemical savings even when the process operates at the higher enzyme dosage (Garg *et al.*, 1996).

16.3 Enzymatic Bleaching

White-rot fungi are well known for their outstanding ability to depolymerize and mineralize lignin (Kirk and Farrell, 1987) and to degrade high-molecular-weight pollutants (Field *et al.*, 1993). Several strains have been identified that cause extensive delignification or bleaching of unbleached Kraft pulps (UKPs). The first attempt to bleach Kraft pulps with white-rot fungi was made by Kirk and Yang (1979) with *Phanerochaete chrysosporium*, which lowered the lignin content of the pulp but also attacked the cellulose and significantly decreased the pulp strength. Other white-rot fungi have been found to more selectively delignify Kraft pulps; these include *Trametes versicolor* (Paice *et al.*, 1989), *Phanerochaete sordida* (Kondo *et al.*, 1994), an unidentified white-rot fungal strain called IZU-154 (Fujita *et al.*, 1993) and *Bjerkandera* sp. BOS55 (Moreira *et al.*, 1997a).

Lignin biodegradation is initiated by several extracellular oxidative enzymes excreted by white-rot fungi, including lignin peroxidase (LIP) (Tien and Kirk, 1984), manganese-dependent peroxidase (MnP) (Glenn and Gold, 1985), manganese-independent peroxidase (MIP) (De Jong *et al.*, 1992), laccase (Eggert *et al.*, 1996) and hydrogen peroxide-generating oxidases (Kersten and Kirk, 1987; De Jong *et al.*, 1994). Purified ligninolytic enzymes have been shown to cause limited delignification and bleaching of UKP, provided that the hydrogen peroxide is carefully dosed and the enzymes are coincubated with low-molecular-weight cofactors: veratryl alcohol for LIP (Arbeloa *et al.*, 1992), manganese, organic acids and surfactants for MnP (Paice *et al.*, 1993; Kondo *et al.*, 1994) and *n*-substituted aromatic compounds for laccase (Bourbonnais and Paice, 1992). The role of LIP in pulp biobleaching by whole cultures is not clear because this enzyme has generally not been detected during fungal biobleaching in many of good biobleaching strains (Archibald, 1992a; Moreira *et al.*, 1997a). Laccase and MnP, on the other hand,

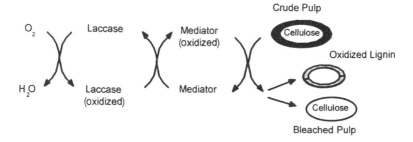

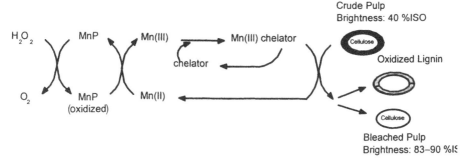

Figure 16.3 The oxidative pathway for catalytic action of laccase and manganese peroxidase on lignin (adapted from Gold and Glenn, 1988)

are excreted at varying levels by different white-rot fungal cultures when biobleaching occurs (Hirai *et al.*, 1994; Moreira *et al.*, 1997a).

Laccase is a blue copper phenoloxidase that contains four copper atoms per polypeptide chain and is capable of catalysing the four-electron transfer reaction necessary to fully reduce oxygen to water (Figure 16.3). Although by definition *p*-diphenols serve as electron donors for laccase, these enzymes have a broad substrate specificity including polyphenols, methoxy-substituted monophenols, aromatic amines and a considerable variety of other natural and synthetic substrates (Eggert *et al.*, 1995; Muñoz *et al.*, 1997). Since *P. chrysosporium*, the model white-rot fungus of choice, was believed not to produce laccase, the role of laccase in lignin degradation has been less intensely studied (Srinivasan *et al.*, 1995). Laccase can catalyse the alkyl–phenyl and C_α–C_β cleavages of phenolic dimers which are used as model lignin substructures and can catalyse the dimethoxylation of several lignin model compounds (Ishihara, 1980). The redox potentials of the laccases studied so far have not been thought sufficiently high to remove electrons from the non-phenolic aromatic substrates that must be oxidized during lignin degradation. Bourbonnais and Paice (1992) showed that an artificial laccase substrate, ABTS (2,2′-azinobis-[3-ethylbenzthiazoline-6-sulphonate]), could act as a redox mediator which enables laccase to oxidize non-phenolic lignin model compounds. The laccase mediator concept combines the action of the enzyme with a low-molecular-weight and environmentally friendly redox mediator, which generates strongly oxidizing compounds that specifically degrade lignin leaving the cellulose fibre intact (Figure 16.3). Laccase Mediator System technology was tested successfully at a pilot plant trial using pulps of different origins (Call and Mücke, 1996). The performance of this system has been further improved by optimizing the mixture of system components, protecting the enzyme against inactivation and enhancing reaction kinetics to obtain a reduction of the mediator quantity, which is the main operating cost of the process. For many pulps, a 50 % delignification could be obtained using 5 kg of mediator per tonne of pulp. The most efficient mediators found belonged to the *n*-heterocyclic compound group (Call, 1994).

Since its discovery in *P. chrysosporium* in 1984 (Kuwahara *et al.*, 1984), MnP has been found to be secreted by many white-rot fungi. Initially, MnP was thought to play only a limited role in lignin depolymerization, oxidizing only phenolic sub-units, while LIP was regarded as the main enzyme (Kirk and Farrell, 1987). MnP was found to depolymerize ^{14}C-labelled dehydrogenative polymerizate lignin (Wariishi *et al.*, 1991), to oxidize non-phenolic model compounds (Bao *et al.*, 1994), and to discolour dissolved coloured lignin in bleached Kraft mill effluent (Michel *et al.*, 1991; Feijoo *et al.*, 1995c). As MnP appears to have an important role in lignin biodegradation, factors involved in its catalytic cycle require special attention to determine their relative importance. The requirement of the Mn(II) ion in MnP expression is widely known, and has been reported for a great number of white-rot fungi (Brown *et al.*, 1990; Bonnarme and Jeffries, 1990; Mester *et al.*, 1995). Mn(II) is involved in the catalytic cycle of MnP and oxidized to Mn(III) (Gold and Glenn, 1988). Mn(III) complexed with an organic acid present in the medium is the oxidant capable of diffusing inside the matrix of the polymer and depolymerizing lignin (Figure 16.3). Accordingly, these complexes are also able to attack a great number of substrates with a phenolic structure similar to that of lignin (Glenn and Gold, 1985).

There is increasing evidence that MnP has a key role in the biological bleaching of pulp, which is supported by several facts. (1) MnP activity is generally detected in active biobleaching cultures of different strains (Paice *et al.*, 1993; Hirai *et al.*, 1994; Kaneko *et al.*, 1994; Katagiri *et al.*, 1995; Moreira *et al.*, 1997a). Moreover, the maximum level of MnP coincides with the time period in which pulp biobleaching occurs (Paice *et al.*, 1993; Hirai *et al.*, 1994). MnP activity has also been correlated with the biobleaching ability of different white rot fungal isolates (Addleman and Archibald, 1993). (2) Mutants of *Trametes versicolor* deficient in MnP do not bleach, and its bleaching activity is partially restored after addition of MnP (Addleman *et al.*, 1995). (3) Catalase, an enzyme that destroys hydrogen peroxide, inhibits bleaching (Paice *et al.*, 1993). (4) Pulp biobleaching and delignification by purified MnP in cell-free systems were recently demonstrated when Mn(II), Tween 80, malonate and hydrogen peroxide were supplied at appropiate concentrations (Kondo *et al.*, 1994). Nevertheless, experiments with the purified enzymes as well as *in vitro* incubation of UKP with cell-free extracellular fluids of biobleaching fungal cultures without cofactors have failed to reproduce the extensive biobleaching obtained by whole-culture systems (Archibald, 1992b; Kaneko *et al.*, 1995), thus indicating that effective biobleaching is achieved only by the combined action of enzyme, cofactors and stabilizing organic acids. However, the specific contributions of each component to bleaching in whole fungal cultures are still unknown (Paice *et al.*, 1995).

In a recent study, Moreira *et al.* (1997a) showed that *Bjerkandera* sp. BOS55 is able to cause extensive delignification and bleaching of UKP as well as oxygen-delignified Kraft pulp (OKP). Figure 16.4 shows the brightness and kappa number reduction achieved for both pulps over a 14-day incubation period by five selected fungal strains after oxalic acid extraction (Moreira *et al.*, 1996a). Brightness gains of up to 20 % ISO in UKP and up to 14 % ISO units in OKP were obtained, giving high final brightness values up to 68 and 80 % ISO units, respectively. The residual lignin in OKP is therefore less bleachable

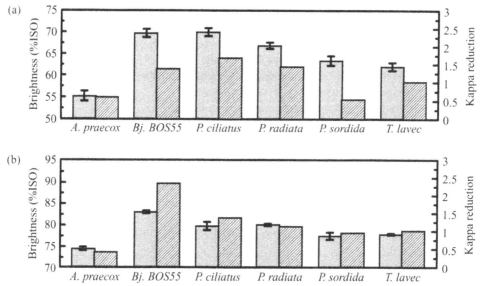

Figure 16.4 Biobleaching and delignification of UKP (a) and OKP (b) by selected white-rot fungal strains in a 14-day incubation period after oxalic acid extraction. Strains: *Agrocybe praecox, Bjerkandera* sp. BOS55, *Polyporus ciliatus, Phlebia radiata, Phanerochaete sordida* and *Trametes* sp. LAVEC94-3. The initial brightness and kappa number of UKP were 47.9(±0.5) %ISO and 12.81(±0.87); after 14 days of incubation in sterile medium, the brightness and kappa number were 49.0(±0.1) %ISO and 12.94(±1.27), respectively. The initial brightness and kappa number of OKP were 69.4(±0.4) %ISO and 9.00(±0.20); after 14 days of incubation in sterile medium, the brightness and kappa number were 69.2(±0.3) %ISO and 8.94(±0.27), respectively

by various white-rot fungal strains, which is logical because residual lignin in UKP and OKP differ in both amount and structure, as oxygen delignification removes about one-half of the residual lignin in Kraft pulp and produces lignin modifications yielding highly condensed lignin units (Reid and Paice, 1994). The specific kappa number decrease per unit of brightness gain during the biobleaching of OKP was as high as 0.20 with the best strain, *Bjerkandera* sp. BOS55. During the course of biobleaching and delignification, MnP was the major oxidative enzyme activity detected in the extracellular fluid although the biobleaching extension was not correlated to the titres of extracellular peroxidases under a wide variety of culture conditions. Therefore it seems that other components of the bleaching system, such as secondary metabolites, organic acids and hydrogen perox-ide, should be considered as possible rate-limiting factors. Surprisingly, *Bjerkandera* sp. BOS55 also presented a manganese-independent bleaching system under manganese-deficient conditions, MnP being the predominant oxidative enzyme produced (Moreira *et al.*, 1997b). Consequently, biobleaching by *Bjerkandera* sp. BOS55 in the absence of manganese must be attributed to other peroxidases or to a presently unknown cofactor for MnP other than manganese. The incomplete definition of the mechanisms which explain the bleaching process is an important obstacle for the implementation of this new process at an industrial scale. Likewise availability of enzymes; neither enzyme is currently available in sufficient quantities for mill trials, and scale-up of enzyme production from fungal cultures may be costly.

16.4   Production of Manganese-Dependent Peroxidase

Because the synthesis of ligninolytic enzymes occurs during secondary metabolism in response to nitrogen or carbon limitation (Kirk and Farrell, 1987) the productivity of the process is quite low, which constitutes one of the major obstacles to implementing bleaching sequences by using oxidative enzymes at large scale.

Several attempts to achieve production of ligninolytic enzymes by white-rot fungi in submerged or immobilized liquid cultures have been described (Feijoo *et al.*, 1995a). Most of them are completely mixed systems, with or without mechanical agitation, such as completely stirred tank reactors (Linko, 1988; Michel *et al.*, 1990) and airlift reactors or bubble columns (Bonnarme *et al.*, 1993; Laugero *et al.*, 1996). Only relatively sharp transient peaks of activity and a later decrease of its maximum level in successive batches have been observed in most cases, regardless of the strain or the carbon source used. One of the factors probably involved in the destabilization of the enzymatic complex is the action of extracellular proteases which are produced by the fungi in nutrient starvation conditions in the search of alternative nutritional sources (Dosoretz *et al.*, 1990b; Feijoo *et al.*, 1995b). The enzymes produced are quickly denatured by the action of these proteases and the final attack on the cell membrane as an ultimate nutrient source means lysis of the culture. From the analysis of results of batch experiments, other environmental factors (a high oxygen tension, gentle agitation, presence of different cofactors such as Mn(II) for MnP, veratryl alcohol for LIP or *n*-substituted aromatic compounds for laccase) are necessary to achieve maximum enzymatic titres in the cultures (Dosoretz *et al.*, 1990a; Faison and Kirk, 1985).

When operating with pellets as well as with immobilized mycelia, the branching of the hyphae favours linking of pellets or immobilized bioparticles into conglomerates, which provokes many operational problems such as fouling of the probes, blockage in the sampling and feeding lines and broth viscosity increase. Moreover, the operation with nitrogen-limited medium enhances the production of extracellular polysaccharides, which act as linking agents. All these factors cause limitations in the mass and oxygen transfer coefficients and negatively affect the overall efficiency (Bonnarme *et al.*, 1993). These drawbacks make it necessary to control mycelial growth in processes operated for long periods of time, an objective that is an important goal in the proper performance of any type of bioreactors.

Application of a perturbation in the form of gas pulsation can minimize the problems related to excessive growth of mycelia and thus avoid clogging. This pulsation is provoked by means of a pulsing device, which can be regulated in amplitude and frequency (Roca *et al.*, 1994). This pulsation system increases the overall productivity of different biotechnological processes (Sanromán *et al.*, 1994).

The effect of the pulsation of oxygen on MnP production by *Phanerochaete chrysosporium* in two types of bioreactors (fluidized-bed with pellets and packed-bed with immobilized mycelia) has been analysed (Moreira *et al.*, 1996b, 1997c). In both cases pulsation was a key factor and allowed a stable performance to be achieved. The pulsation modified the morphology of mycelial pellets because the applied shear stress on the bioparticle surface caused by the pulsation limited hyphal extension. In the case of immobilized mycelia the pulsation avoided interconnection between foam blocks and aggregation of conglomerates (Moreira *et al.*, 1997c).

The efficiency of the process largely depends on a number of environmental and operating conditions. Optimal conditions for enhanced MnP production in a fixed-bed reactor for an extended period of time correspond to a feed rate of nutrients of 250 mg

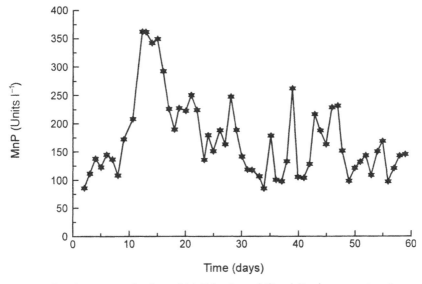

Figure 16.5 Continuous production of MnP by immobilized *P. chrysosporium* in a packed-bed bioreactor operated with pulsation at a hydraulic retention time of 24 h and a Mn^{2+} concentration of 5 mM

Table 16.4 Comparison between different systems for the production of MnP in bioreactors

Reference	Bioreactor/process	Operation time (days)	Max. activity (Units l^{-1})	Productivity (Units l^{-1} day^{-1})
Bonnarme and Jeffries (1990)	Airlift/batch	3.9	192[a]	49
Bonnarme *et al.* (1993)	Airlift/batch	3.8	365[a]	95
Laugero *et al.* (1996)	Bubble-column/batch	4	726[a]	181
Moreira *et al.* (1997b)	Packed bed/continuous	140	250	202

[a] The extinction coefficient for the dimeric dimethoxyphenol oxidation product was corrected to 49 600 M^{-1} cm^{-1}, as determined by Wariishi *et al.* (1992)

glucose l^{-1} h^{-1} and 0.6 mg ammonia l^{-1} h^{-1}, a Mn(II) concentration of 5 mM, adequate oxygen supply, a bioreactor hydraulics of a plug flow with partial mixing and an operating hydraulic retention time of 24 h (Figure 16.5). Table 16.4 shows that the results obtained following this strategy compare favourably with the best recently reported for the production of MnP, even when mutant cells of *P. chrysosporium*, strain INA-12 were used.

16.5 Future Perspectives

The application of enzymatic bleaching to the pulp and paper industry is still to be developed. The presently commercial xylanases reduce the need to bleach chemicals for high-brightness pulp, and they have relatively low cost and ease of application with great selectivity. The drawback of the hemicellulase-aided bleaching is that it is an indirect method, not directly delignifying pulp. Both laccase and MnP can achieve more substantial delignifying action than xylanase, but there are obstacles to be overcome before either enzyme can be used in a cost-effective manner.

To make the application of ligninolytic fungi and their oxidative enzymes feasible in a biobleaching process serious reduction in the time from days to hours is needed, which is only possible through knowledge of the mechanism involved. For laccase, so far no mediator is sufficiently effective and inexpensive to be commercially viable. There is currently no large-scale commercial source for enzyme production, so the costs of the process associated with the use of oxidative enzymes can not be accurately calculated. Future research should allow for a more efficient utilization of the bleaching system, adapted to each specific case, bearing in mind the importance of its being an environment-friendly clean process of low cost. Thus, oxidative enzymes, which can be regarded as catalysts for oxygen and hydrogen peroxide-driven delignification, may also find a place in the pulp bleaching process in coming years.

Acknowledgements

This work was funded by the Spanish Commission of Science and Technology (CICYT), (Projects B=O 98-0610 and IFO97-0584.

References

ADDLEMAN, K. and ARCHIBALD, F.S., 1993, Kraft pulp bleaching and delignification by dykaryons and monokaryons of *Tramestes versicolor*. *Appl. Environ. Microbiol.*, **59**, 266–273.

ADDLEMAN, K., DUMONCEAUX, T., PAICE, M.G., BOURBONNAIS, R. and ARCHIBALD, F.S., 1995, Production and characterisation of *Trametes versicolor* mutants unable to bleach hardwood Kraft pulp, *Appl. Environ. Microbiol.*, **61**, 3687–3694.

ARBELOA, M., DE LESELEUC, J., GOMA, G. and POMMIER, J.C., 1992, An evaluation of the potential of lignin peroxidases to improve pulps, *Tappi J.*, **75**, 215–221.

ARCHIBALD, F.S., 1992a, Lignin peroxidase activity is not important in biological bleaching and delignification of unbleached Kraft pulp by *Trametes versicolor*, *Appl. Environ. Microbiol.*, **58**, 3101–3109.

ARCHIBALD, F.S., 1992b, The role of fungus-fibre contact in the biobleaching of Kraft brownstock by *Trametes (Coriolus) versicolor*, *Holzforschung*, **4**, 305–310.

BAO, W.L., FUKUSHIMA, Y., JENSEN, K.A. and MOEN, M.A., 1994, Oxidative degradation of non-phenolic lignin during lipid peroxidation by fungal manganese peroxidase, *FEBS Lett.*, **354**, 297–300.

BONNARME, P. and JEFFRIES, T.W., 1990, Mn(II) regulation of lignin peroxidases and manganese-dependent peroxidases from lignin-degrading white-rot fungi, *Appl. Environ. Microbiol.*, **56**, 210–217.

BONNARME, P., DELATTRE, M., DROUET, H., CORRIEU, G. and ASTHER, M., 1993, Toward a control of lignin and manganese peroxidases hypersecretion by *Phanerochaete chrysosporium* in agitated vessels: Evidence of the superiority of pneumatic bioreactors on mechanically agitated bioreactors, *Biotechnol. Bioeng.*, **41**, 440–450.

BOURBONNAIS, R. and PAICE, M.G., 1992, Demethylation and delignification of Kraft pulp by *Trametes versicolor* laccase in the presence of 2,2′-azinobis-(3-ethylbenzthiazoline-6-sulphonate), *Appl. Microbiol. Biotechnol.*, **36**, 823–827.

BROWN, J.A., GLENN, J.K. and GOLD, M.H., 1990, Manganese regulates expression of manganese peroxidase by *Phanerochaete Chrysosporium*, *J. Bacteriol.*, **172**, 3125–3140.

BROWN, J.A., LI, D., ALIC, M. and GOLD, M.H., 1993, Heat shock induction of manganese peroxidase gene transcription in *Phanerochaete chrysosporium, Appl. Environ. Microbiol.*, **59**, 4295–4299.

BYRD, M.V., GRATZL, J.S. and SINGH, R.P., 1992, Delignification and bleaching of chemical pulps with ozone: a literature review, *Tappi J.*, **75**, 207–213.

CALL, H.P., 1994, Process of modifying, breaking down or bleaching lignin, materials containing lignin or like substances, World Patent Application, WO 94/29510.

CALL, H.P. and MÜCKE, I., 1996, The laccase-mediator system (LMS) – a new concept, in Srebotnik, E. and Messner, K. (eds) *Biotechnology in the Pulp and Paper Industry. Recent Advances in Applied and Fundamental Research*, Vienna: Facultas-Universitätsverlag, pp.27–32.

DE JONG, E., DE VRIES, E.P., FIELD, J.A., VAN DER ZWAN, R.P. and DE BONT, J.A.M., 1992, Isolation and screening of basidiomycetes with high peroxidative activity, *Mycol. Res.*, **96**, 1098–1104.

DE JONG, E., FIELD, J.A. and DE BONT, J.A.M., 1994, Aryl alcohols in the physiology of ligninolytic fungi, *FEMS Microbiol. Rev.*, **13**, 153–188.

DOSORETZ, C.G., CHEN, H.C. and GRETHLEIN, H.E., 1990a, Effect of oxygenation conditions on submerged cultures of *Phanerochaete chrysosporium, Appl. Environ. Microbiol.*, **34**, 131–137.

DOSORETZ, C.G., DASS, S.B., REDDY, A. and GRETHLEIN, H.E., 1990b, Protease-mediated degradation of lignin peroxidase in liquid cultures of *Phanerochaete chrysosporium, Appl. Environ. Microbiol.*, **56**, 3429–3434.

EGGERT, C., TEMP, U., DEAN, J.F.D. and ERIKSSON, K.E., 1995, A fungal metabolite mediates degradation of non-phenolic lignin structures and synthetic lignin by laccase, *FEBS Lett.*, **391**, 144–148.

EGGERT, C., TEMP, U. and ERIKSSON, K.E., 1996, Laccase-producing white-rot fungus lacking lignin peroxidase and manganese peroxidase. Role of laccase in lignin biodegradation, in Jeffries, T.W. and Viikari, L. (eds) *Enzymes for Pulp and Paper Processing*, ACS Symposium Series, Vol. 655, Washington, DC: American Chemical Society Press, pp.130–150.

FAISON, B.D. and KIRK, T.K., 1985, Factors involved in the regulation of ligninase activity in *Phanerochaete chrysosporium, Appl. Environ. Microbiol.*, **49**, 299–304.

FEIJOO, G., DOSORETZ, C. and LEMA, J.M., 1995a, Production of lignin peroxidase by *Phanerochaete chrysosporium* in a packed bed bioreactor operated in semi-continuous mode, *J. Biotechnol.*, **42**, 247–253.

FEIJOO, G., ROTHSCHILD, N., DOSORETZ, C.G. and LEMA, J.M., 1995b, Effect of addition of extracellular culture fluid on ligninolytic enzyme formation in *Phanerochaete chrysosporium, J. Biotechnol.*, **40**, 21–29.

FEIJOO, G., VIDAL, G., MOREIRA, M.T., MÉNDEZ, R. and LEMA, J.M., 1995c, Degradation of high molecular weight compounds of Kraft pulp mill effluents by a combined treatment with fungi and bacteria, *Biotechnol. Lett.*, **17**, 1261–1266.

FIELD, J.A., DE JONG, E., FEIJOO-COSTA, G. and DE BONT, J.A.M., 1993, Screening for xenobiotic degrading white-rot fungi, *Trends Biotechnol.*, **11**, 44–49.

FOLKE, A. and RENDBERG, F., 1994, Chlorine dioxide in pulp bleaching. An update on technical aspects and environmental effects. Literature study. Brussels, Belgium: CEFIC, Chlorate Sector Group.

FUJITA, K., KONDO, R., SAKAI, K., KASHINO, Y., NISHIDA, T. and TAHAHARA, Y., 1993, Biobleaching of softwood Kraft pulp with white-rot fungus IZU-154, *Tappi J.*, **76**, 81–84.

GARG, A.P., McCARTHY, A.J. and ROBERST, J.C., 1996, Biobleaching effect of *Streptomyces thermoviolaceus* xylanase preparations on birchwood Kraft pulp, *Enzyme Microb. Technol.*, **18**, 261–267.

GLENN, J.K. and GOLD, M.H., 1985, Purification and characterization of an extracellular Mn(II)-dependent peroxidase from the lignin-degrading basidiomycete, *Phanerochaete chrysosporium, Arch. Biochem. Biophys.*, **242**, 329–341.

GOLD, M.H. and GLENN, J.K., 1988, Manganese peroxidase from *Phanerochaete chrysosporium, Methods Enzymol.*, **161**, 258–264.

HIGUCHI, T., 1980, Lignin structure and morphological distribution in plant cell walls, in Kirk, T.K., Higuchi, T. and Chang, H. (eds) *Lignin Biodegradation: Microbiology, Chemistry, and Potential Applications*, Vol. 1, Boca Raton, FL: CRC Press, pp.1–17.

HIRAI, H., KONDO, R. and SAKAI, K., 1994, Screening of lignin-degrading fungi and their ligninolytic enzyme activities during biological bleaching of Kraft pulp, *Mokuzai Gakkaishi*, **40**, 980–986.

ISHIHARA, T., 1980, The role of laccase in lignin biodegradation, in Kirk, T.K., Higuchi, T. and Chan, H. (eds), *Lignin Biodegradation: Microbiology, Chemistry, and Potential Applications*, Vol. 2, Boca Raton, FL: CRC Press, pp.17–31.

JOHNSON, T., 1996, Pulping technologies for the 21st Century – A North American view. Part Three: Technologies for the future, *Appita*, **49**, 6–10.

KANEKO, R., IIMORI, T., YOSHIKAWA, H., MACHIDA, M., YOSHIOKA, H. and MURAKAMI, K., 1994, A possible role of manganese peroxidase during biobleaching by the pulp bleaching fungus SKB-1152, *Biosci. Biotechnol. Biochem.*, **58**, 1517–1518.

KANEKO, R., IIMORI, T., YOSHIKAWA, H., MACHIDA, M., YOSHIOKA, H. and MURAKAMI, K., 1995, Biobleaching with manganese peroxidase purified from the pulp bleaching fungus SKB-1152, *Biosci. Biotechnol. Biochem.*, **59**, 1584–1585.

KATAGIRI, N., TSUTSUMI, Y. and NISHIDA, T., 1995, Correlation of brightening with cumulative enzyme activity related to lignin biodegradation during biobleaching of Kraft pulp by white-rot fungi in the solid-state fermentation system, *Appl. Environ. Microbiol.*, **61**, 617–622.

KERSTEN, P.J. and KIRK, T.K., 1987, Involvement of a new enzyme, glyoxal oxidase, in intracellular H_2O_2 production by *Phanerochaete chrysosporium*, *J. Bacteriol.*, **169**, 2195–2201.

KINSTREY, R.B., 1993, An overview of strategies for reducing the environmental impact of bleach-plant effluents, *Tappi J.*, **76**, 105–113.

KIRK, T.K. and FARRELL, R.L., 1987, Enzymatic 'combustion': The microbial degradation of lignin, *Annu. Rev. Microbiol.*, **41**, 465–505.

KIRK, T.K. and YANG, M.H., 1979, Partial delignification of unbleached Kraft pulp with ligninlytic fungi, *Biotechnol. Lett.*, **1**, 347–352.

KONDO, R., KURASHIKI, K., SAKAI, K., 1994, In vitro bleaching of hardwood Kraft pulp by extracellular enzymes excreted from white-rot fungi in a cultivation system using a membrane filter, *Appl. Environ. Microbiol.*, **60**, 921–926.

KRINGSTAD, K.P. and LINDSTRÖM, K., 1984, Spent liquors from pulp bleaching, *Environ. Sci. Technol.*, **18**, 236–248.

KUWAHARA, M., GLENN, J.K., MORGAN, M.A. and GOLD, M.H., 1984, Separation and characterization of two extracellular H_2O_2-dependent oxidases from ligninolytic cultures of *Phanerochaete chrysosporium*, *FEBS Lett.*, **169**, 247–250.

LAPIERRE, L., BOUCHARD, J., BERRY, R.M. and van LIEROP, B., 1995, Chelation prior to hydrogen peroxide bleaching of Kraft pulps: an overview, *J. Pulp Pap. Sci.*, **21**, 268–273.

LAUGERO, C., SIGOILLOT, J.C., MOUKHA, S., FRASSE, P., BELLON-FONTAINE, M.N., BONNARME, P. *et al.*, 1996, Selective hyperproduction of manganese peroxidases by *Phanerochaete chrysosporium* I-1512 immobilised on nylon net in a bubble-column reactor, *Appl. Microb. Biotechnol.*, **44**, 717–723.

LINKO, S., 1988, Continuous production of lignin peroxidase by immobilised *Phanerochaete chrysosporium* in a pilot scale bioreactor, *J. Biotechnol.*, **9**, 163–170.

MESSNER, K. and SREBOTNIK, E., 1994, Biopulping: an overview of developments in an environmentally safe paper-making technology, *FEMS Microbiol. Rev.*, **13**, 351–364.

MESTER, T., DE JONG, E. and FIELD, J.A., 1995, Manganese regulation of veratryl alcohol in white-rot fungi and its indirect effect on lignin peroxidase, *Appl. Environ. Microbiol.*, **62**, 880–885.

MICHEL, F.C. JR., DASS, S.B., GRULKE, E.A. and REDDY, C.A., 1991, Role of manganese peroxidases and lignin peroxidases of *Phanerochaete chrysosporium* in the decolorization of Kraft bleach plant effluent, *Appl. Environ. Microbiol.*, **57**, 2368–2375.

MICHEL, F.C. JR., GRULKE, E.A. and REDDY, C.A., 1990, Development of a stirred tank reactor system for the production of lignin peroxidase (ligninases) by *Phanerochaete chrysosporium* BKM-F-1767, *J. Ind. Microbiol.*, **5**, 103–112.

MOREIRA, M.T., FEIJOO, G., LEMA, J.M., FIELD, J.A. and SIERRA-ALVAREZ, R., 1996a, Oxalic acid extraction as a post-treatment to increase the brightness of Kraft pulps bleached by white-rot fungi, *Biotechnol. Techn.*, **10**, 559–564.

MOREIRA, M.T., SANROMÁN, A., FEIJOO, G. and LEMA, J.M., 1996b, Control of pellet morphology of filamentous fungi in fluidized bed bioreactors by means of a pulsing flow. Application to *Aspergillus niger* and *Phanerochaete chrysosporium*, *Enzyme Microb. Technol.*, **19**, 261–266.

MOREIRA, M.T., SIERRA-ALVAREZ, R., FEIJOO, G., LEMA, J.M., and FIELD, J.A., 1997a, Biobleaching of oxygen delignified Kraft pulp by several white-rot fungal strains, *J. Biotechnol.*, **53**, 237–251.

MOREIRA, M.T., FEIJOO, G., SIERRA-ALVAREZ, R., LEMA, J.M., and FIELD, J.A., 1997b, Manganese is not required for biobleaching of oxygen-delignified Kraft pulp by the white-rot fungus *Bjerkandera* sp. strain BOS55, *Appl. Environ. Microbiol.*, **63**, 1749–1755.

MOREIRA, M.T., PALMA, C., FEIJOO, G. and LEMA, J.M., 1997c, Continuous production of manganese peroxidase by *Phanerochaete chrysosporium* immobilized on polyurethane foam in a pulsed packed-bed bioreactor, *Biotechnol. Bioeng.*, **56**, 130–137.

MUÑOZ, C., GUILLEN, F., MARTINEZ, A.T. and MARTINEZ, M.J., 1997, Induction and characterization of laccase in the ligninolytic fungus *Pleurotus eryngii*, *Curr. Microbiol.*, **34**, 1–5.

NELSON, P., CHIN, CH. and GROVER, S., 1993, Bleaching of Eucalyptus Kraft pulp from and environmental point of view, *Appita*, **46**, 354–360.

PAICE, M.G., JURASEK, C., BOURBONNAIS, R. and ARCHIBALD, F., 1989, Direct biological bleaching of hardwood pulp with the fungus *Coriolus versicolor*, *Tappi J.*, **75**, 217–221.

PAICE, M.G., REID, I.D., BOURBONNAIS, R., ARCHIBALD, F.S. and JURASEK, L., 1993, Manganese peroxidase, produced by *Trametes versicolor* during pulp bleaching, demethylates and delignifies Kraft pulp, *Appl. Environ. Microbiol.*, **59**, 260–265.

PAICE, M.G., BOURBONNAIS, R., REID, I.D., ARCHIBALD, F.S. and JURASEK, L., 1995, Oxidative bleaching enzymes: A review, *J. Pulp Pap. Sci.*, **21**, 280–284.

REID, I.A. and PAICE, M.G., 1994, Effect of residual lignin type and amount on bleaching of Kraft pulp by *Trametes versicolor*, *Appl. Environ. Microbiol.*, **60**, 1395–1400.

ROCA, E., SANROMÁN, A., NUÑEZ, M.J. and LEMA, J.M., 1994, A pulsing device for packed-bed bioreactors. I. Hydrodynamic behaviour, *Bioprocess Eng.*, **10**, 61–73.

SANROMÁN, A., ROCA, E., NUÑEZ, M.J. and LEMA, J.M., 1994, A pulsing device for packed-bed bioreactors. II. Application to alcoholic fermentation, *Bioprocess Eng.*, **10**, 75–81.

SENIOR, D.J. and HAMILTON, J., 1993, Xylanase treatment for the bleaching of softwood Kraft pulps: the effect of chlorine dioxide substitution, *Tappi J.*, **76**, 200–206.

SIERRA, R., FIELD, J.A., KORTEKAAS, S. and LETTINGA, G., 1994, Overview of the anaerobic toxicity caused by organic forest industry wastewater pollutants, *Water Sci. Technol.*, **24**, 113–125.

SRINIVASAN, C., D'SOUZA, T.M., BOOMINATHAN, K. and REDDY, C.A., 1995, Demonstration of laccase in the white-rot basidiomycete *Phanerochaete chrysosporium* BKM-F1767, *Appl. Environ. Microbiol.*, **61**, 4274–4277.

STAUBER, J., GUNTHORPE, L., WOODWORTH, J., MUNDAY, B., KRASSOI, R. and SIMON, J., 1996, Comparative toxicity of effluents from ECF and TCF bleaching of eucalyptus Kraft pulps, *Appita*, **49**, 184–188.

SUURNÄKKI, A., CLARK, T.A., ALLISON, R.W., VIIKARI, L. and BUCHERT, J., 1996, Xylanase- and mannanse-aided ECF and TCF bleaching, *Tappi J.*, **79**, 111–117.

TIEN, M. and KIRK, T.K., 1984, Lignin-degrading enzyme *from Phanerochaete chrysosporium*: purification, characterization and catalytic properties of a unique H_2O_2-requiring oxygenase, *Proc. Natl. Acad. Sci. USA*, **81**, 2280–2284.

VIDAL, G., SOTO, M., FIELD, J., MÉNDEZ-PAMPÍN, R. and LEMA, J.M., 1997, Anaerobic biodegradability and toxicity of wastewaters from chlorine and total chlorine free bleaching of eucalyptus Kraft pulps, *Water Res.*, **31**, 2487–2494.

VIIKARI, L., KANTELINEN, A., SUNDQUIST, J. and LINKO, M., 1994, Xylanases in bleaching: From an idea to the industry, *FEMS Microbiol. Rev.*, **13**, 335–349.

VIIKARI, L., 1995, Clean technology: biotechnology in the forest industry, in Rolz, C.E. (ed.) *Biochemical Engineering Applications in Environmental Biotechnology and Cleaner Production*, electronic course available on the Internet, http://www.icaiti.org.gt.

WARIISHI, H., VALLI, K. and GOLD, M.H., 1991, *In vitro* depolymerization of lignin by manganese peroxidase of *Phanerochaete chrysosporium, Biochem. Biophys. Res. Commun.*, **176**, 269–275.

WELANDER, T. and ANDERSSON, P.E., 1985, Anaerobic treatment of wastewater from the production of chemi-thermomechanical pulp, *Water Sci. Technol.*, **17**, 103–112.

17

The Cleaner Production Strategy Applied to Animal Production

EUGENIA J. OLGUÍN

17.1 Introduction

Animal production is one of the major economic activities throughout the world. Even the smaller countries are giving priority to the production of animal stocks of various kinds. Pigs, cattle and chickens are by far the most abundant species, reaching a world total number around one thousand million (Table 17.1). China is the world leader in pig production and the USA and Brazil appear among the first five leading countries in pig, cattle and chicken production.

With the exception of a few countries in which environmental legislation is stringently enforced, most countries are still suffering severe environmental damage due to intensive animal production units lacking proper management.

It is considered in this chapter that a cleaner production strategy should be applied to the animal production industry in the very short term in order to prevent pollution and to make it more profitable. The case of pig production has been chosen to discuss in detail.

Pig production is common in those countries in which pig meat is part of the diet of the majority of the population. However, there are interesting differences in number of pigs per country depending on the geographical region and even the country itself. Analysis of data adapted from FAO's statistics, in which only those countries registering more than one million pigs were taken into account (Table 17.1), shows that currently Asia accounts for 61.5 % of the total registered for all regions. Furthermore, the number of pigs increased steadily in Asia during the period 1995 to 1998, while in other regions (e.g. Europe) the number of pigs has remained more or less stable or has even slightly decreased over the same period.

A closer analysis of the major producers at the global level during 1998 (Table 17.2) shows that China has by far the largest number of pigs (approximately 486 million), followed by the USA (with approximately 60 million), Brazil (36 million) and Germany (around 25 million). Other countries – such as Spain, Poland, Vietnam, India, Russia, Mexico, France, the Netherlands, Denmark, Canada, Philippines and Indonesia – have a range of 10–20 million animals.

It is important to mention that the total world production of pigs reported by the FAO for 1998 (FAOSTAT, 1998) was 956 523 230 (Table 17.2), only slightly higher than the

Table 17.1 Potential market for recycling pig waste: pigs per country and per region from 1995 to 1998

Region	Country	1995	1996	1997	1998
Asia	China	424 680 600	452 200 500	468 055 400	485 698 400
	Vietnam	16 306 827	16 921 400	17 635 800	18 060 000
	India	14 311 000	14 855 000	15 419 000	16 005 000
	Philippines	8 941 190	9 025 950	9 752 180	10 210 000
	Indonesia	7 824 827	7 897 210	8 638 269	10 201 000
	Japan	10 250 000	9 900 000	9 809 000	9 800 000
	Republic of Korea	6 461 000	6 516 770	7 096 000	6 700 000
	Thailand	4 507 000	4 023 000	4 209 000	4 209 000
	Malaysia	3 282 314	3 400 000	3 400 000	3 400 000
	Myanmar	2 944 000	3 400 000	3 400 000	3 400 000
	Democratic People's Republic of Korea	3 150 000	3 100 000	3 100 000	3 100 000
	Laos	1 723 590	1 772 000	1 900 000	2 000 000
	Sub total	**504 382 348**	**533 011 830**	**552 414 649**	**572 783 400**
Europe	Germany	24 698 120	23 736 560	24 282 980	24 782 200
	Spain	18 161 000	18 625 000	19 269 000	19 597 000
	Poland	20 417 820	17 963 910	18 134 780	19 240 000
	Russian Federation	24 589 000	22 631 000	19 500 000	16 579 000
	France	14 593 000	14 800 000	14 976 000	15 430 000
	Netherlands	14 397 000	13 958 000	14 253 000	11 438 000
	Denmark	11 083 910	10 842 000	11 100 000	11 400 000
	Ukraine	13 946 000	13 144 000	11 236 000	9 479 000
	Italy	8 023 400	8 060 700	8 090 000	8 155 000
	United Kingdom	7 534 000	7 496 000	7 992 000	7 959 000
	Belgium–Luxembourg	7 053 000	7 225 000	7 313 000	7 300 000

Belarus	7 300 000	7 313 000	7 225 000	4 004 500
Romania	7 272 900	8 234 500	7 959 500	7 758 000
Hungary	4 931 000	5 289 000	5 032 000	4 356 000
Yugoslavia	4 216 000	4 216 000	4 446 000	4 192 000
Czech Republic	3 995 273	4 079 590	4 016 250	3 866 570
Austria	3 737 000	3 679 876	3 663 747	3 706 185
Sweden	2 386 000	2 338 000	2 348 800	2 313 137
Portugal	2 151 000	2 365 000	2 344 000	2 402 000
Slovakia	1 900 000	1 985 223	2 076 439	2 037 370
Ireland	1 717 000	1 664 500	1 542 000	1 498 300
Bulgaria	1 700 000	1 500 000	2 140 010	1 986 180
Switzerland	1 521 000	1 521 100	1 580 100	1 610 700
Finland	1 394 000	1 394 000	1 394 000	1 295 100
Lithuania	1 205 200	1 127 600	1 270 000	1 259 800
Croatia	1 175 460	1 175 461	1 196 285	1 174 602
Sub total	**197 961 033**	**204 029 610**	**204 639 862**	**207 956 694**
Latin America and Caribbean				
Brazil	35 900 000	35 800 000	36 600 000	36 062 100
México	15 500 000	15 020 000	15 405 000	15 923 000
Venezuela	3 200 000	3 200 000	3 100 000	3 200 000
Argentina	3 200 000	3 200 000	3 100 000	3 100 000
Ecuador	2 795 000	2 708 000	2 621 000	2 618 000
Bolivia	2 655 530	2 568 767	2 481 930	2 404 833
Paraguay	2 525 000	2 525 000	2 525 000	2 525 000
Peru	2 481 000	2 481 000	2 533 200	2 401 400
Colombia	2 480 000	2 480 000	2 431 000	2 500 000
Chile	1 962 000	1 722 403	1 485 615	1 489 990
Cuba	1 500 000	1 500 000	1 500 000	1 750 000
Sub total	**74 198 530**	**73 205 170**	**73 782 745**	**73 974 323**

Environmental Biotechnology and Cleaner Bioprocesses

Table 17.1 *(cont'd)*

Region	Country	1995	1996	1997	1998
North America	United States of America	59 990 000	58 264 000	56 171 000	60 250 000
	Canada	11 290 500	11 588 000	11 482 700	11 482 500
	Sub total	**71 280 500**	**69 852 000**	**67 653 700**	**71 732 500**
Africa	Nigeria	7 150 000	7 400 000	7 600 000	7 600 000
	Madagascar	1 592 000	1 628 620	1 662 000	1 670 000
	South Africa	1 627 985	1 602 578	1 617 404	1 630 000
	Cameroon	1 410 000	1 410 000	1 410 000	1 410 000
	Democratic Republic of Congo	1 170 000	1 410 000	1 410 000	1 410 000
	Sub total	**12 949 985**	**13 451 198**	**13 699 404**	**13 720 000**
Oceania	Australia	2 652 810	2 526 412	2 684 000	2 490 000
	Papua New Guinea	1 400 000	1 450 412	1 500 000	1 500 000
	Sub total	**4 052 810**	**3 976 824**	**4 184 000**	**12 213 634**
Asia		504 381 921	532 841 030	552 372 149	573 024 400
Europa		208 226 694	203 386 001	200 431 510	194 343 233
Latin America and Caribbean		73 974 323	73 782 745	73 205 170	74 198 530
North America		71 280 500	69 852 000	67 653 700	72 092 500
Africa		12 949 985	13 198 198	13 472 404	13 480 000
Oceania		4 052 810	3 976 412	4 184 000	3 990 000
TOTAL		**874 866 233**	**897 036 386**	**911 318 933**	**931 128 663**

Adapted from FAOSTAT DATA 1990–1998; only countries with more than 1 million pigs were taken into account

Table 17.2 World major producers of confined animal species (1998)

Pigs		Cattle		Chickens	
Country	Number of animals	Country	Number of animals	Country	Stock (×1000)
China	485 698 400	India	209 084 000	USA	1 553 000
USA	60 250 000	Brazil	162 000 000	Indonesia	1 195 000
Brazil	35 900 000	USA	99 501 000	Brazil	908 000
Germany	24 782 000	China	96 192 530	Mexico	408 800
Spain	19 597 000	Russian Federation	31 700 000	Russian Federation	405 000
World total	**956 523 230**		**1 030 927 390**		**13 623 865**

Source: FAOSTAT DATA 1990–1998

total production of pigs for countries registering more than 1 million animals, which was 931 128 663. Thus, data from Table 17.2 allows valid conclusions as to the world status and potential market for recycling pig waste.

Several environmental problems are caused by poorly or non-managed pig farms in which wastewater and solid wastes are discharged to water bodies or fields without proper treatment. Release of noxious odours and nutrients – such as nitrogen and phosphorus, which may accumulate in water bodies and cause eutrophication – are the most immediate negative impacts. Nitrogen content of pig waste is 5.4–6.3 kg t^{-1} of waste and phosphorus content in the range 2.2–3.1 kg t^{-1} (Newell, 1980; Sweeten, 1992). Furthermore, the problem becomes acute because pig waste is produced in a proportion of around 5 % of animal weight per day and contains a very high biochemical oxygen demand (BOD); in the range of 10–90 g l^{-1} (Jewell and Loehr, 1977). Longer term and more severe impacts are caused by release of methane into the atmosphere during natural biodegradation or degradation of organic carbon inside anaerobic ponds. Methane is a greenhouse gas which has a persistence time of approximately 10 years (Masera, 1991).

17.2 Cleaner Pig Production Units

Because pig production is one of the most polluting agroindustries, extensive efforts should be oriented towards the design and implementation of cleaner pig production units (CPPUs). Application of the assessment methodology described by van Berkel (1995) allows a systematic and detailed analysis of the current way of operation of the PPU aimed at designing cleaner operations and processes. The following discussion is a brief account of the application of such methodology. Once modifications have been introduced to reduce wastes and wastewater at source, other complementary options may be applied for recyling and recovering nutrients from pig waste (Laliberte *et al.*, 1997). A full discussion of an integrated biosystem for such purposes, which has been investigated by the author for several years, is presented below.

The first step involves *source identification*. A flow diagram of the various stages involved in a full pig production cycle is required. Inputs such as energy, feed, water to clean facilities and outputs such as dissipated heat, pig waste, wastewater and noxious odours need to be specified.

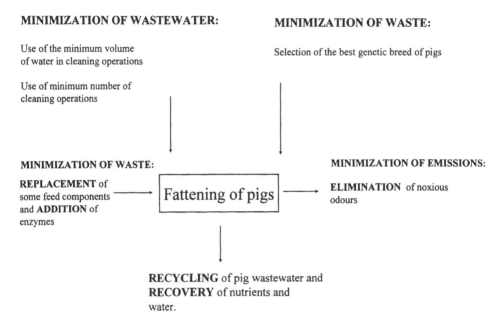

Figure 17.1   Minimization of waste, emissions and wastewater volume together with recycling are part of the *option generation* stage

Once sources of wastes and emissions have been identified, an assessment of their volume and composition is made, as part of the second stage (called *cause evaluation*). Feed composition in terms of protein, fat and energy content is evaluated. Chemical additives such as copper and zinc salts have to be quantified. Feeding procedures are evaluated as cause of feed losses.

One of the most relevant factors controlling the wastewater volume and composition is the volume of water utilized to clean facilities. It has been reported that water consumption may vary from 18 to 40 litres of water per animal unit (100 kg) per day in tropical climates (Taiganides, 1992) or from 20 to 45 litres of water per pig per day in Asia (Ng *et al.*, 1989). The number of cleaning operations is important too. These two factors, together with the design for channelling wastewater out of facilities, may also be the cause of the noxious odours.

Market demands and capital availability influence the chosen breed. Genetic characteristics influence metabolism and size, and therefore the quantity and composition of waste generated per animal. Finally, another factor influencing the volume and composition of waste and emissions is the management procedure used to handle it.

The third stage of the assessment methodology is critical, because it involves the *option generation*. At this stage, minimization of waste and wastewater volume at the source (i.e. within the process) may be achieved by applying at least three modifications (Figure 17.1):

1   *Water saving*, especially during the cleaning of the facilities. This modification may be a key factor in terms of the cost-benefit ratio of CPPUs. A clear illustration of this statement is presented later this chapter. The conclusion drawn from a comparison of running the same system under two options, one in which only 17 litres of water/pig/ day and another in which 45 litres of water/pig/day were utilized for cleaning the

facilities, was that the cost of the infrastructure in the second option was approximately 2.5 times larger and the investment recovery time nearly double that of the first.

2 *Selection of the best genetic combination.* Better feed conversions may be attained when choosing the best breed for a particular production purpose. Thus, less waste per pig unit should be produced.

3 *Replacement* of some feed components, especially those chemical additives which cause major negative impact to the environment. This exercise is not easy, because a careful balance between cost of feed components and benefit in terms of weight gain and quality of the produced meat has to be worked out.

A very attractive approach for the minimization of the waste volume and change of composition is the addition of enzymes to the feed with the purpose of promoting digestibility of the starch and protein in the cereals used as raw materials. A decrease of 30 % in waste volume has been described, with the added advantage of producing waste with less nitrogen than the one which results after using conventional feed (Novo Nordisk, 1994). This Danish company commercializes a xylanase for wheat-based feeds, a beta glucanase for barley-based feeds, an alpha amylase for improved starch digestion, a protease for improved digestibility of protein and a carbohydrate for improving digestibility of other cereals and vegetable protein carbohydrates.

Once modifications have been introduced to reduce wastes and wastewater at the source, other complementary options may be applied for recycling pig waste and for the recovery of nutrients contained in the waste. A full discussion of an integrated biosystem for such purposes, which has been investigated by the author for several years, is presented below.

17.3 Integrated System for Recycling Pig Wastewater, and Recovering Biogas *Spirulina* and *Lemna* Biomass (Bio-Spirulinema System)

17.3.1 *General Aspects*

An integrated system for the recycling of animal waste consisting of two stages, starting with anaerobic digestion followed by high-rate oxidation ponds (HROPs) with recuperation of biogas and the valuable cyanobacterium *Spirulina*, has been reported as an option for small-scale units (Olguín, 1982) and for arid environments (Olguín and Vigueras, 1981; Olguin, 1986). The recuperation of *Lemna* biomass as source of protein in poultry diets was reported for the recovery of nutrients from ponds fertilized with pig waste (Olguín *et al.*, 1986).

More recent developments, in which diluted seawater has been utilized to cultivate *Spirulina* in HROPs with the double purpose of recovering nutrients and treating the anaerobic effluents, have illustrated the usefulness of such systems (Olguín *et al.*, 1993, 1994b). Evaluations of the system under temperate climatic conditions (Olguín *et al.*, 1997) or under tropical conditions (Olguín *et al.*, unpublished results) and evaluation of the recuperation of *Lemna* biomass under temperate climatic conditions (Hernández *et al.*, 1997), have provided enough information to indicate that operation of the integrated system as illustrated in Figure 17.2 is feasible.

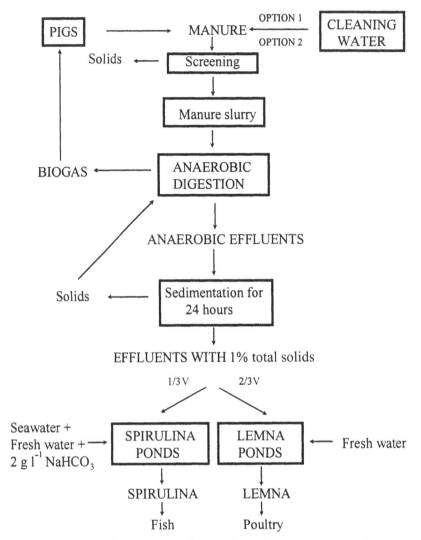

Figure 17.2 The bio-spirulinema system for recycling of pig wastewater and recovery of biogas, *Spirulina* and *Lemna*

17.3.2 *Main Processes and Operation Conditions*

The system includes several unit operations, starting with the cleaning of the pig facilities. A key factor in the economic viability of CPPUs, and therefore of this integrated system, is the volume of water used during such operation, as demonstrated below. After dilution of the pig manure, a second operation unit (Figure 17.2) consisting of the screening of solids which could block the anaerobic filter, is required.

The use of anaerobic filters as the first process in degrading the high content of organic matter contained in the pig waste has several advantages. The first one is that 55–60 % of the original chemical oxygen demand (COD) of the waste can be removed (Olguín and Castillo, unpublished results) and, simultaneously, some of the carbon can be recycled and transformed from complex polymeric molecules such as cellulose into methane and carbon dioxide (biogas). Secondly, organic nitrogen and phosphorus are converted into inorganic molecules, which can be recovered in the form of valuable *Spirulina* and *Lemna* biomass in the second stage of the process. Thirdly, some carbon which remains as dissolved bicarbonate in the anaerobic effluents can be recovered when utilized as the main carbon source by *Spirulina*. Last, but not least, the anaerobic conditions and high temperatures which predominate during the process result in the inhibition of aerobic pathogens and 90 % reduction in the numbers of helminth eggs present (Dixo *et al.*, 1995).

Due to intrinsic process characteristics, anaerobic digestion alone is not able to produce effluents that can meet the discharge standards (Tilche *et al.*, 1996). The new standards recently introduced in the countries of the European Union (Directive no. 271/91) are very difficult to meet with anaerobic digestion alone, particularly for nutrients such as ammonia-nitrogen and phosphorus which may cause eutrophication of lakes and seas. Thus, treatment of anaerobic effluents is required. In the system described in Figure 17.2, the HROPs inoculated with *Spirulina* and the ponds inoculated with *Lemna* are used to recover most of the nutrients contained in such effluents.

The HROPs have been investigated over some decades for the treatment of various effluents (Oswald, 1988); they are important for nutrient recycling and will grow in importance in the twenty-first century (Oswald, 1995), because they offer the great advantage of high dissolved oxygen levels with a very low energy input and therefore operate at a very low cost. The oxygen is produced through photosynthesis by microalgae, which can grow actively at the expense of the nitrogen, phosphorus and other micronutrients contained in the effluents.

HROPs are quite efficient reactors for the removal of contaminants. According to Oswald (1988), removal of more than 90 % of the BOD can be attained with a flow rate of 15 cm s^{-1} and with an oxygenation factor of 1.55. In fact, Yang and Nagano (1983) reported 97 % BOD removal in HROPs in which a native mixture of microalgae established naturally was used to treat anaerobic effluents from pig waste. These researchers also reported that this kind of pond consumes a lot less energy than other oxidation ponds: 0.49 kWh kg^{-1} of BOD removed in contrast to consumptions of 5.6 and 1.6 kWh kg^{-1} of BOD removed in oxidation ditches and oxidation ponds, respectively. HROPs inoculated with *Spirulina* for the treatment of anaerobic effluents of pig waste were found to remove 99 % of phosphates and 100 % of ammonia-nitrogen when the effluents were added at the beginning of the growth cycle either in batch (Olguín *et al.*, 1994a) or semicontinuous cultures (Olguín *et al.*, 1997).

In the system described here (Figure 17.2), the HROPs inoculated with *Spirulina* operate with a mixture of seawater (in a ratio of 1:3) plus fresh water and should receive the supernatant fraction of the anaerobic effluents (after being settled for a period of 24 h), in a proportion of 2 % of the total volume. Semi-continuous or continuous cultures of *Spirulina* can be established operating at flow velocities in the range of 15–25 cm s^{-1} and pond depths of 6–30 cm. According to Belay (1997), who has described the experience of clean *Spirulina* cultures at the Earthrise Farms, the depth of the pond depends on the season, desired algal density and, to a certain extent, the desired biochemical composition of the biomass, particularly pigment concentration.

Although information is available concerning design and operation of tubular photobioreactors for *Spirulina* cultivation (Richmond *et al.*, 1993; Torzillo, 1997), there is no information on the use of this kind of reactor for the cultivation of *Spirulina* with the purpose of recycling nutrients contained in anaerobic effluents. Work is in progress on this research line (Olguín and Olmos, 1998). Such reactors are a lot more expensive than HROPs but they are more productive; their use is thus justified only for the production of high added value products from *Spirulina*, such as pigments or poly-unsaturated fatty acids (PUFAs).

The recovery of nutrients from anaerobic effluents in the Bio-Spirulinema system is also achieved through ponds inoculated with *Lemna* sp. These aquatic floating plants are quite efficient for removing nitrogen and phosphorus and have been promoted for waste management for two decades (Culley *et al.*, 1981) due to their high protein content and potential use in aquaculture as an animal feed. Although their growth characteristics have been described in nutrient-enriched water (Reddy and DeBusk, 1984; Guy *et al.*, 1990a,b) and other important parameters such as the effect of harvest rate, waste loading and stocking density on the productivity of this useful plant have been evaluated (Said *et al.*, 1979), very little work has been done regarding its capacity for recycling the nutrients contained in anaerobic effluents (Balasubramanian *et al.*, 1991; Hernández *et al.*, 1997).

According to the flow diagram of the integrated system (Figure 17.2), two-thirds of the total anaerobic effluents can be treated through *Lemna* ponds. The reason for such design is that *Lemna* may tolerate total nitrogen concentrations as high as 450 mg l^{-1} when it is previously adapted to anaerobic effluents (Hernández *et al.*, 1997) and these effluents may be added in a proportion of 4–5 % of the total pond volume. The ponds for *Lemna* cultivation do not require mixing or any specific length-width ratio as in the case of *Spirulina* ponds. Also, because *Lemna* fronds float on the surface, harvesting is less complex than *Spirulina*.

According to the evaluation done by our research group under temperate climatic conditions (Hernández *et al.*, 1997), *Lemna* has the capacity of high removal efficiencies for total nitrogen, ammonia nitrogen, phosphates and COD (77 %, 80 %, 92 % and 90 % respectively) and thus competes effectively with other aquatic plants for this purpose.

17.3.3 *Productivity and Chemical Composition of* Spirulina

Production of *Spirulina* biomass is the result of a combination of factors. In cultures in which nutrients are not limiting and the pH is in the optimum range (8.3–9.8), productivity is a function mainly of solar irradiance and temperature. Thus, it is also a function of the reactor design. It is well accepted now that these two factors are better managed in tubular photobioreactors and that productivity is higher in such reactors than in HROPs (Richmond *et al.*, 1993).

Kinetic studies carried out in HROPs mixed with paddle wheels (Richmond *et al.*, 1990) showed that *Spirulina* productivity can be as high as 60–70 t ha^{-1} $year^{-1}$, but only if certain conditions are met: (a) temperature near the optimum (35°C) to mitigate the photoinhibition effect, which is more acute at a suboptimal temperature (around 20°C); (b) use of shades to mitigate the photoinhibition effect; (c) harvest at the end of the day to mitigate the loss of biomass through nocturnal respiration; (d) reduction of contaminants by addition of ammonia ions.

Photoinhibition is one of the factors which affect productivity in a significant way (Vonshak, 1997) and the temperature of the culture is a determinant for the expression of this negative effect (Jensen and Knutsen, 1993). A *Spirulina* culture growing in a turbidostat at 25°C showed a 50 % reduction in its photosynthetic rate when exposed to an irradiance of 1720 mmol m^{-2} s^{-1}, but was very little affected when it was cultivated at 35°C at the same irradiance.

The effect of temperature on productivity is related to the fact that the loss of biomass through nocturnal respiration is higher at a suboptimal temperature (Torzillo *et al.*, 1991). Temperature also affects the protein content; higher contents are related to temperatures near the optimum (35°C) and lower contents are observed in cultures at suboptimal temperatures (Torzillo *et al.*, 1991). It has also been shown (Tomaselli *et al.*, 1988) that temperatures higher than 42°C cause a significant reduction in the protein content.

Protein content of *Spirulina* biomass is also affected strongly by light intensity. Air-lift *Spirulina* cultures at the lab level (Tadros *et al.*, 1993) were found to show a higher protein content (in the range of 51–61 %), when the culture was exposed to the lower irradiance tested (30 μEm^{-2} s^{-1}). A similar effect has been shown in the case of the phycobiliproteins, which may account for as much as 60 % of the total protein content (Garnier and Thomas, 1993). The effect of the light intensity on the protein content of *Spirulina* biomass is probably also related to the effect of the mixing rate of the culture (attained by paddle wheels in HROPs) on the protein content, reported by Olguín *et al.* (1994b). A maximum protein content (in the range of 71 %) was attained when the mixing rate was 20 rpm, equivalent to a flow rate of 14.5 cm s^{-1} in small HROPs.

Protein content of *Spirulina* cultures can also be affected by other operating factors such as dissolved oxygen and ammonia nitrogen concentration in solution. An excess of dissolved oxygen caused a decrease in the protein content (Márquez *et al.*, 1995) and concentrations of ammonia nitrogen in the range 30–50 mg l^{-1} favoured higher protein contents than in semi-continuous cultures in which ammonia nitrogen was already depleted (Olguín *et al.*, 1997).

17.3.4   *Productivity and Chemical Composition of* Lemna

The expected productivity and protein content of *Lemna* sp. growing on anaerobic effluents (providing 200–350 mg l^{-1} of total nitrogen) may be in the range of 6–7 gm^{-2} day^{-1} and 40–45 %, respectively, according to the evaluation done by our research group under temperate climatic conditions (Hernández *et al.*, 1997). Harvesting should be performed every 3 days and nitrogen concentration should be kept around 350 mg l^{-1} in order to maintain the highest protein content.

17.3.5   *Qualities of the Major Products of the Bio-Spirulinema System*

The major products of the integrated system for the recycling of animal waste, and especially for the recycling of pig waste, are biogas, *Spirulina* and *Lemna* biomass, as mentioned earlier. The recovery of the nutrients contained in the animal waste and their transformation in these three valuable products is the basis of the high potential of application of such systems world-wide.

Biogas

Biogas (a mixture of methane and carbon dioxide) has been recognized as an extremely useful renewable energy source for many years, since anaerobic digestion started to be

Table 17.3 Calorific value of gases from batch anaerobic digestion of various organic materials at 30°C

Organic matter digested	Specific production (l kg^{-1} organic matter)	Percentage of methane in digester gas	Calorific value (MJ m^{-3})
Pig waste	415	80.8	28.9

Source: Wheatley, 1980

investigated and proposed as a feasible process to treat wastes and wastewaters of different origin (Jewell and Loehr, 1977). It contains a high calorific value (Table 17.3) and it can be utilized to replace part of the fossil fuel utilized in the PPU, a fact that improves considerably the economic viability of the treatment of animal wastes through anaerobic digestion (Polprasert *et al.*, 1994).

Spirulina *biomass*

The cyanobacteria of the genus *Spirulina (Arthrospira)* have great potential as a source of several chemical compounds (Richmond, 1987; Vonshak, 1990). Its cultivation has been promoted on the basis that, under appropriate conditions, its protein content can attain 60–70 % of its dry mass (Ciferri, 1983). More recently, its market has grown as a 'food for health' and emphasis is given to compounds other than protein, such as pigments (phycocyanin and β-carotene) and PUFAs, which have been shown to have therapeutic effects in humans (Belay *et al.*, 1993, 1994).

The *Spirulina* biomass produced in the integrated system described in Figure 17.2 can be utilized only as a feed because it is cultivated on nutrients recovered from animal waste. However, this market is also growing due to the great attributes this cyanobacterium has as an animal feed supplement. Belay *et al.* (1996) have done an extensive review of the discovered effects of *Spirulina* on growth, survival and tissue quality in a whole range of animals and of more recent studies on its immunomodulatory, anti-viral and anti-cancer effects. They conclude that, from the point of view of cost-effectiveness, the most promising application may be its immune enhancement effects and, through this, its antiviral and anti-bacterial properties because these effects are exhibited at very low supplemental concentrations in feed. It is expected that the *Spirulina* biomass produced within integrated systems such as Bio-Spirulinema may be produced at a cost competitive with those of other conventional protein sources and, therefore, *Spirulina* could also be effectively used as a protein source. Thus, the market of such recovered biomass could be oriented towards various uses, including the production of high-added-value products such as pigments and PUFAs.

Lemna *biomass*

Lemna biomass is a very attractive source of protein for fish and poultry (Culley *et al.*, 1981) and may be used as a feed for cattle (Russof *et al.*, 1977). In the author's experience, fresh *Lemna* counteracted the problems of cannibalism in poultry which appeared when there was a deficiency in protein intake (Olguín *et al.*, 1986).

17.3.6 *Water Saving as Key Factor of the Economic Viability of the Bio-Spirulinema System*

To illustrate that the volume of water utilized to clean pig facilities is the key factor determining the economic viability of the system, two different exercises are presented. *Option 1* assumes a water-saving practice in which a consumption of 17 litres per pig per day is performed. *Option 2* assumes the use of an excess of water (45 litres per pig per day). In each case, calculation of the cost of infrastructure for different sizes of pig farm, from 10 to 20 000 pigs, are provided. In this way the analysis covers a wide range, from a typical micro-enterprise (10 pigs), to small (100 pigs), medium (1000 pigs) and large (20 000 pigs) enterprises.

The wastewater volume was calculated as the equivalent to the 92 % of the volume used to clean the pig facilities (17 and 45 litres per pig per day, respectively), assuming some loss for evaporation and other minor causes (Table 17.4). Assuming a 10-day hydraulic retention time (HRT) and a 30 % of extra volume for the packing material, the total volume of the anaerobic filters was calculated. About 20 % of the anaerobic effluents volume was assumed to be lost during the sedimentation operation for 24 h. The volume of anaerobic effluents (containing around 1 % of total solids) to be treated in the HROP inoculated with *Spirulina* was assumed to be only one-third of the total volume after sedimentation; the other two thirds were assumed to be treated in *Lemna* ponds.

The total infrastructure cost was calculated assuming a cost of US$100.00 dollars per m^3 for the anaerobic filters, US$400 dollars per m^2 for the *Spirulina* ponds, and US$100 per m^2 for the *Lemna* ponds.

The main conclusion of the whole exercise (Table 17.4) is that the total infrastructure cost of option 2 (in which an excess of water was assumed to be used) is nearly 2.5 times larger than that of option 1, in which a considerable saving of water was assumed to be performed. A complete cost-benefit analysis, in which the investment recovery time is calculated for the two options, is currently in progress. Final calculations are expected to draw a similar conclusion in the sense that a strategy of saving water at the source (i.e. during the cleaning operation) results in a shorter investment recovery time (approximately half).

It should be mentioned that these costs may be higher than those reported for other countries, because they were calculated on the basis of the current cost of building materials in Mexico, and the exchange rate peso to dollar is constantly increasing. However, option 1 (in which a minimum of water is utilized) is quite feasible because it the water volume is similar to that allowed or already utilized in various countries. At Berrybank in Victoria, Australia, water use per pig is limited to 9 litres per day for stock watering and washing and 12 litres per day reclaimed water for manure washdown. As 4 litres are utilized per pig per day in the growth of the animal, the total effluent volume from the piggery is equivalent to 17 litres per pig per day (Heath, 1998).

17.4 Final Remarks

The animal production industry is a global major economic activity which provides food, raw materials for other industries (such as the leather industry), pharmaceutical products and jobs for several millions of people around the world. Hence, the industry should be constantly modernized and updated to lower costs of production, to keep already existing markets and open new ones.

Table 17.4 Infrastructure cost of the 'Bio-Spirulinema System' in relation to the water volume utilized to clean facilities

	Option 1 (17 l pig/day)				Option 2 (45 l pig/day)			
	10 pigs	100 pigs	1 000 pigs	20 000 pigs	10 pigs	100 pigs	1 000 pigs	20 000 pigs
Wastewater volume[a] ($m^3 day^{-1}$)	0.16	1.6	16.6	332	0.41	4.14	41.4	828
Anaerobic filter volume[b] (m^3)	2.15	21.5	215.8	4 316	5.38	53.8	538.2	10 764
Anaerobic effluent volume after sedimentation ($m^3 day^{-1}$)[c]	0.133	1.33	13.3	266	0.33	3.30	33	660
Spirulina ponds surface (m^2)[d]	21.7	217.2	2 172.3	43 446.6	53.9	539	5 390	107 800
Lemna ponds surface (m^2)[e]	11.2	112.3	1 123.1	22 462.2	27.8	278.6	2 786.6	55 733.3
Total infrastructure cost (US$)[f]	1 195	11 950	119 500	2 390 000	2 972	29 720	297 200	5 944 000

[a] Assuming that 8.0 % of cleaning water is lost due to evaporation
[b] Assuming an HRT of 10 days and that a 30 % extra volume is allowed for packing material
[c] Assuming 20 % of anaerobic effluent volume is lost during sedimentation of solids for 24 h
[d] Assuming only one-third of anaerobic effluent volume is used to fertilize *Spirulina* ponds in a proportion of 2 % (v/v). A column water of 10 cm is also assumed
[e] Assuming two-thirds of anaerobic effluent volume is utilized to fertilize *Lemna* ponds in a proportion of 2 % (v/v). A column water of 10 cm is also assumed
[f] Assuming the following costs: anaerobic filters US$100.00 m^{-3}; *Spirulina* ponds US$40.00 m^{-2} (20 cm high); *Lemna* ponds US$10.00 m^{-2} (30 cm high)

On the other hand, one of the major challenges for the twenty-first century is to introduce new strategies and policies in order to enhance true industrial sustainable development in which the social, economic and environmental dimensions may coexist in harmony. The application of the cleaner production strategy to animal production may help such an important industry to become an important part of the construction of the so urgently needed sustainable development. However, more research and development is required in this field, in conjunction with as a comprehensive promotion programme, including training of industry personnel and human resource formation at various levels.

References

BALASUBRAMANIAN, P.R. and BAI, R.K., 1991, Recycling of biogas-plant effluent through aquatic plant (*Lemna*) culture, *Bioresource Technol.*, 213–215.

BELAY, A., 1997, Mass culture of *Spirulina* outdoors – The Earthrise Farms experience, in Vonshak, A. (ed.) *Spirulina platensis (Arthrospira): Physiology, Cell-biology and Biotechnology*, London: Taylor & Francis, p.131.

BELAY, A., OTA, Y., MIYAKAWA, K. and SHIMAMATSU, H., 1993, Current knowledge on potential health benefits of *Spirulina*, *J. Appl. Phycol.*, **5**, 235.

BELAY, A., OTA, Y., MIYAKAWA, K. and SHIMAMATSU, H., 1994, Production of high quality *Spirulina* at Earthrise farms, in Siew Moi, P., Yuan Kun, L., Borowitzka, M.A. and Whitton, B.A. (eds) *Algal Biotechnology in the Asia-Pacific Region*, Malaysia: Institute of Advanced Studies, University of Malaya, p.92.

BELAY, A., KATO, T. and OTA, Y., 1996, *Spirulina (Arthrospira)*: potential application as an animal feed supplement, *J. Appl. Phycol.*, **8**, 303–311.

BERKEL, R. VAN, 1995, Introduction to cleaner production assessments with applications in the food processing industry, *UNEP Industry and Environment*, **January–March**, 8–15.

CIFERRI, O., 1983, *Spirulina*, the edible microorganisms, *Microbiol. Rev.*, **47**, 551.

CULLEY, D., REJMANKOVA, E., KVET, J. and FRYE, J.B., 1981, Production chemical quality and use od duckweeds (*Lemnaceae*) in aquaculture, waste management and animal feeds, *World Maricult. Soc.*, **12**(2), 27–49.

DIXO, N.G.H., GAMBRILL, M.P., CATUNDA, P.F.C. and VAN HAANDEL, A.C., 1995, Removal of pathogenic organisms from the effluent of an upflow anaerobic digester using waste stabilization ponds, *Water Sci. Tech.*, **31**, 275–284.

FAOSTAT, 1998, http://apps.fao.org.

GARNIER, F. and THOMAS, J.C., 1993, Light regulation of phycobiliproteins in *Spirulina maxima*, in Doumenge, F., Duranch-Chastel, H. and Toulemont, A (eds) *Spiruline, algue de vie*, Monaco Musée Océanographique, pp.7–10.

GUY, M., GRANOTH, G. and GALE, J., 1990a, Cultivation of *Lemna gibba* under desert condition. I: Twelve months of continuous cultivation in open pond, *Biomass*, **21**, 145–156.

GUY, M., GRANOTH, G. and GALE, J., 1990b, Cultivation of *Lemna gibba* under desert condition. II: The effect of raise winter temperature, CO_2 enrichment and shading on productivity, *Biomass*, **23**, 1–11.

HERNÁNDEZ, E., OLGUÍN, E.J., TRUJILLO, S. and VIVANCO, J., 1997, Recycling and treatment of anaerobic effluents from pig waste using *Lemna* sp. under temperate climatic conditions, in Wise, L.D. (ed.) *Global Environmental Biotechnology*, Amsterdam: Elsevier, pp.293–304.

HEATH, L., 1998, Total waste management system for the pig industry, Water Research Foundation of Australia, Berrybank, Australia (http://cyberone.com.au/~enviro/berrybank.html).

JENSEN, S. and KNUTSEN, G. 1993, Influence of light and temperature on photoinhibition of photosynthesis in *Spirulina platensis, J. Appl. Phycol.*, **5**, 495–504.

JEWELL, W.J. and LOEHR, R.C., 1977, Energy recovery from animal wastes, in Taiganides, E.P. (ed.) *Animal Wastes*, London: Applied Science, pp.273–294.

LALIBERTE, G., OLGUIN, E.J. and DE LA NOUE, J., 1997, Mass cultivation and wastewater treatment using *Spirulina*, in Vonshak, A. (ed.) *Spirulina (Arthrospira): Physiology, Cell-Biology and Biotechnology*, London: Taylor & Francis, p.159.

MÁRQUEZ, F.J., SASAKI, K., NISHIO, N., and NAGAI, S., 1995, Inhibitory effect of oxygen accumulation on the growth of *Spirulina Platensis*, *Biotechnol. Lett.*, **17**, 225–228.

MASERA, 1991, México y el cambio climático global, *Ciencia y Desarrollo*, **XII**, 52–57.

NEWELL, P.J., 1980, The use of high rate contact reactor for energy production and waste treatment from intensive livestock units, in Vogt, F. (ed.) *Energy Conservation and Use of Renewable Energies in the Bio-Industries*, New York: Pergamon Press, p.395.

NG, W.J., CHIN, K.K. and WONG, K.K., 1989, Biological treatment alternatives for piggery wastes, in Wise, D.L. (ed.) *International Biosystems* Vol. 2. Boca Raton, FL: CRC Press.

Novo Nordisk, 1994, *Novo Nordisk A/S and Products for 'Cleaner Production'*.

OLGUÍN, E.J., 1982, Some concepts and projects of appropriate biotechnology as applied to the mexican situation, *Resource Conserv. Recycl.* **5**, 79–83.

OLGUÍN, E.J., 1986, Appropriate biotechnological systems in the arid environment, in Doelle, H.W. and Heden, C.G. (eds) *Applied Microbiology*, Dordrecht: Reidel Publishing Co. UNESCO, pp.111–134.

OLGUÍN, E.J. and OLMOS, P., 1998, Recycling of pig wastewater with recovery of *Spirulina* under tropical conditions. 2. Utilizing tubular photoreactors, in Olguín, E., Sánchez, G., Galicia, S. and Hernández, E. (eds) *Memorias del Tercer Simposium Internacional Bioprocesos más limpios y Desarrollo Sustentable*, Veracruz, Ver. WAITRO, CITELDES, Instituto de Ecología, pp.67–72.

OLGUÍN, E.J. and VIGUERAS, J., 1981, Unconventional food production at the village level in a desert area of Mexico, in *Proceedings of the 2nd World Congress of Chemical Engineering*, Montreal, Canada: pp.332–335.

OLGUÍN, E.J., BENÍTEZ, J.R. and ARIAS, E., 1986, Desarrollo agropecuario integral con la participación de una cooperativa de mujeres, *Desarrollo y Medio Ambiente*, **1**, 2–8.

OLGUÍN, E.J., HERNÁNDEZ, B., ARAUS, A., GALICIA, S., RAMÍREZ, M.E., CAMACHO, R. and MERCADO, G., 1993, Aspectos del cultivo de *Spirulina* en agua de mar suplementada con efluentes anaeróbicos en lagunas de oxidación de alta tasa, *Biotecnología*, **3**, 16–23.

OLGUÍN, E.J., HERNÁNDEZ, B., ARAUS, A., CAMACHO, R., GONZÁLEZ, R., RAMÍREZ, M.E. *et al.*, 1994a, Production of *Spirulina* on sea water supplemented with anaerobic effluents from pig waste, in De la Noue, J. and Laliberté, G. (eds) *Microalgae: from the Laboratory to the Field*, Quebec, Canada: Université Laval, pp.145–154.

OLGUÍN, E.J., HERNÁNDEZ, G., ARAUS, A., CAMACHO, R., GONZÁLEZ, R., RAMÍREZ, M.E. *et al.*, 1994b, Simultaneous high biomass protein production and nutrient removal using *Spirulina maxima* in sea water supplemented with anaerobic effluents, *World J. Microbiol. Biotechnolnol.*, **10**, 576–578.

OLGUÍN, E.J., GALICIA, S., CAMACHO, R., MERCADO, G. and PÉREZ, T.J., 1997, Production of *Spirulina* sp in sea-water supplemented with anaerobic effluents in outdoor raceways under temperate climatic conditions, *Appl. Microbiol. Biotechnol.*, **48**, 242–247.

OSWALD, W.J., 1988, Microalgae and waste-water treatment, in Borowitzka, A. and Borowitzka, J. (eds) *Microalgae Biotechnology*, New York: Cambridge University Press, pp.306–328.

OSWALD, W.J., 1995, Ponds in the twenty-first century, *Water Sci. Tech.*, **31**, 1–18.

POLPRASERT, C., YANG, P.Y., KONGSRICHAROERN, N. and KANJANAPRAPIN, W., 1994, Productive utilization of pig farm wastes: a case study for developing countries, *Resource Conserv. Recycl.*, **11**, 245–259.

REDDY, K. and DEBUSK, F.W., 1984, Growth characteristics of aquatic macrophytes cultured in nutrient-enriched water:II.Azolla, Duckweed and Salvinia, *Econ. Bot.*, **39**, 200–208.

RICHMOND, A., 1987, *Spirulina*, in Borowitzka, M.A. and Borowitzka, L.J. (eds) *Microalgal biotechnology*, Cambridge: Cambridge University Press, pp.86–121.

RICHMOND, A., LICHTEMBERG, E., STAHL, B. and VONSHAK, A., 1990, Quantitative assessment of major limitations on productivity of *Spirulina platensis* in open raceways, *J. Appl. Phycol.*, **2**, 195–206.

RICHMOND, A., BOUSSIBA, S., VONSHAK, A. and KOPEL, R., 1993, A new tubular reactor for mass production of microalgae outdoors, *J. Appl. Phycol.*, **5**, 327–332.

RUSSOF, L.L., GANTT, D.T., WILLIAMS, D.M. and CHOLSON, J.H., 1977, Duckweed as a potential feedstuf for cattle, *J. Dairy Sci.*, **60**, 160–161.

SAID, Z., CULLEY, D., STANDIFER., EPPS, E., MYERS, R. and BONEY, S., 1979, Effect of harvest rate, waste loading, and stocking density on the yield of duckweeds. *Proc. World Maricult. Soc.*, **10**, 769–780.

SWEETEN, J.M., 1992, Livestock and poultry waste management: a national overview, in Blake, J., Donald, J. and Magette, W. (eds) *National Livestock, Poultry and Aquaculture Waste Management, Proceedings of the National Workshop, 29–31 July 1991*, ASAE Publication 03-92, 1992, 4.

TADROS, M.G., SMITH, W., JOSEPH, B. and PHILLIPS, J., 1993, Yield and quality of cyanobacteria *Spirulina maxima* in continuous culture in response to light intensity, *Appl. Biochem. Biotechnol.*, **39/40**, 337–347.

TAIGANIDES, E.P., 1992, Pig waste management and pollution control: ideas for Mexico, Presented at the Conference on 'Pig waste management', Acapulco, Mexico, July 8–12.

TILCHE, A., BORTONE, G., GARUTI, G. and MALASPINA, F., 1996, Post-treatments of anaerobic effluents, *Antonie van Leeuwenhoek*, **69**, 47–59.

TOMASELLI, L., GIOVANNETTI, L., SACCHI, A. and BOCCI, F., 1988, Effects of temperature on growth and biochemical composition in *Spirulina platensis* strain M2, in Stadler, T., Mollion, J., Verdus, M.C., Karamanos, Y., Morvan, H. and Christiaen, D. (eds) *Algal Biotechnology*, London: Elsevier Applied Science, pp.305–314.

TORZILLO, G., 1997, Tubular Bioreactors, in Vonshak, A. (ed.) *Spirulina(Arthrospira): Physiology, Cell-Biology and Biotechnology*, London: Taylor & Francis, pp.101–115.

TORZILLO, G., SACCHI, A., MATERASSI, R. and RICHMOND, A., 1991, Effect of temperature on yield and night biomass loss in *Spirulina platensis* grown in tubular photobioreactors, *J. Appl. Phycol.*, **3**, 103–109.

VONSHAK, A., 1990, Recent advances in microalgal biotechnology, *Biotechnol. Adv.*, **8**, 709–727.

VONSHAK, A., 1997, Outdoor mass production of *Spirulina*: The basic concept, in Vonshak, A. (ed.) *Spirulina (Arthrospira): Physiology, Cell-Biology and Biotechnology*, London: Taylor & Francis, pp.79–99.

WHEATLEY, B.I., 1980, The gaseous products of anaerobic digestion – biogas, in Stafford, D.A., Wheatley, B.I. and Hughes, D.E. (eds) *Anaerobic Digestion*, London: Applied Science, pp.415–428.

YANG, P.Y. and NAGANO, S.Y., 1983, Integrating anaerobic digestion and algal-biomass treatment processes for swine wastewater, *in Proceedings of the 38th Purdue Industrial Waste Conference*, Michigan: Ann Arbor Science.

Clean Technologies through Microbial Processes for Economic Benefits and Sustainability

HORST W. DOELLE, ARAN HANPONGKITTIKUN AND
POONSUK PRASERTSAN

18.1   Introduction

All technological developments are aimed at improving the quality of life of a community of people (DaSilva *et al.*, 1992). Developing countries are looking for programmes reducing the risk to health and achieving sustainable, economical growth conducive to a higher per capita income (Doelle, 1996a,b). The introduction of clean technologies using microbial processes requires therefore a complete change from the presently existing industrialized economic to a more socio-economic approach. Waste management must become an integrated part of our new clean technology systems, which are often referred to as *integrated biosystems* (Chan, 1997a) or *socio-economic biotechnology systems* (Doelle, 1989, 1993, 1994).

This chapter will describe some aspects of such an approach in the sugarcane, palm oil mill and fish processing industries.

18.2   The Sugarcane and Sugar Processing Industry

The processing of sugar from sugarcane is practised in many sub-tropical and tropical countries such as Australia, Brazil, Cuba, Fiji, India, Malawi, Mexico and Zimbabwe. In most cases, sugarcane farmers and sugarcane processors are different entities; the farmer works under contract with the processor to produce a certain amount of sugarcane. In some countries, however, sugarcane farmer cooperatives may have their own sugarcane-processing facilities. The sugarcane industry, from the farmer to the processor, has suffered – and still is suffering – from the instability of income due to fluctuations on the overseas market demand and prices as well as weather fluctuations affecting the yield of sugarcane. The instability of the income causes in many cases a reduction in waste management. Pollution is therefore a major concern because of the enormous variation in harvesting and sugar-processing procedures and their efficiency. There is no doubt that the introduction of microbial process systems into the sugarcane industry would not only offer a more stable income through diversifying product formation but also would help to eliminate pollution and transform the sugarcane industry into a clean-technology industry.

18.2.1 *Present Practices*

Sugarcane harvest

In surveying first the sugarcane harvesting methods, one has to realize that these methods vary from hand-cutting (Africa) to highly mechanized techniques (Australia). Furthermore, burning of the cane before harvest is a common practice in many countries. As the sugarcane is a vulnerable crop, it has to be processed ideally within 8–10 h after cutting. A longer delay causes formation of the polymer dextran, complicating the sugar extraction process and reducing the yield. Furthermore, sugarcane should have a sucrose content of 10–14 % for efficient economic sugar production, but this is often not achievable owing to abnormal weather conditions. A low-sugar-content sugarcane becomes a waste product, costing the farmer valuable revenue. Even an acceptable sugarcane produces agricultural waste products because tops and trash has to be removed before the sugarcane can be delivered to the mill. In many countries, tops and trash are burned. The introduction of stringent air pollution laws increasingly prohibits the burning of sugarcane residues before or after the harvest.

Sugarmill operation

Sugarmill operations, like harvesting methods, vary widely in different countries. The extraction of the sugarcane juice produces large amounts of fibre (bagasse) and the clarification a mud cake (only clarified juice can be used for further mill processing). The fibre is usually used to generate the energy required for the sugarmill operation. Excess bagasse and the mud cake are problems in many regions of the world. If inhaled, bagasse is known to cause severe asthma and lung diseases.

The clarified juice is now concentrated to a sugarcane syrup (approximately 82 % sugar) and crystallized in evaporators. Depending on the efficiency of the mill, there will be only one or up to three crystallization steps – and the larger the number of crystallization steps, the lower the value of the residual molasses. One therefore categorizes the molasses into A (approx. 72 % sugar), B (approx. 60 % sugar) and C or blackstrap molasses (approx. 45 % sugar). The additional evaporation – crystallization steps also concentrate the salt (up to 15 % final content), form furfurals and other melanin-like brown-coloured substances.

18.2.2 *Possibilities for Cleaner Technologies Through Introduction of Microbial Processes*

In order to overcome the income fluctuations and pollution problems, the sugarcane industry must become more flexible, allowing other products to be produced for the local market to reduce the overseas market dependence, to compensate for the losses and return to the farmer a more stable and predictable income. A socio-economic cooperative system, as outlined in Figure 18.1, combining farming, milling, processing and waste re-use could solve these problems. Microbial process industries combine very well with chemical and agricultural engineering systems.

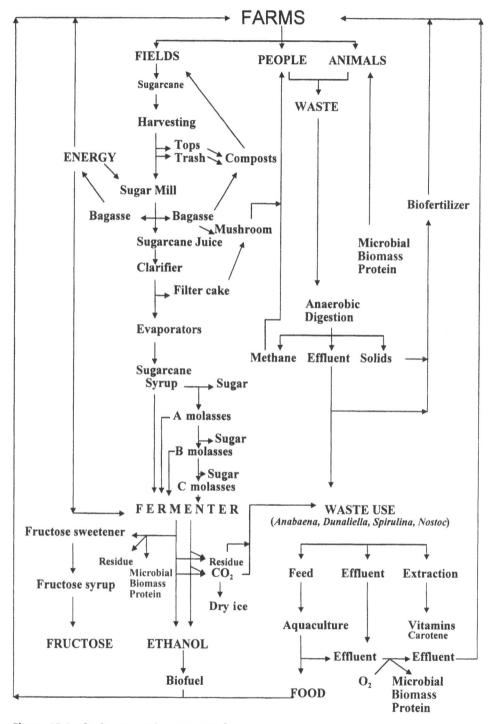

Figure 18.1 Socio-economic sugarcane farm cooperative system

Microbial processes on the farm

On the farm itself, tops and trash from the sugarcane harvest should be collected for composting (Stentiford and Dodds, 1992). In many areas of Australia, for example, it is prohibited to burn garden trash. Such a law should also be extended to the farms. Composting is an excellent way to produce biofertilizer with a relatively high mineral salt content. As will be outlined below, this compost can be converted into rich humus using supplementation with animal waste or anaerobic digester sludge and earthworm cultivation (Rodriguez, 1997a,b). The rich humus compost can be sold to the local community or spread back onto the fields. This microbial process not only would eliminate air pollution but would also reduce the costs of expensive chemical fertilizer used by the farmer, which only harm the soil microbial mineralization process.

Microbial processes attached to the sugarmill operation

The energy requirement for the mill should be supplemented with biogas, which has a much higher heat value compared to the present use of bagasse. This would free the bagasse, which is an excellent resource for mushroom production. As will be outlined in Section 18.3, lignocellulosic fibres supplemented with solids from the anaerobic digester would give an ideal carbon:nitrogen ratio. Mushrooms are the only microorganisms capable of excreting the enzymes needed to separate the lignin economically from cellulose and hemicellulose. The residue from mushroom production will enrich the composting of the tops and trash or can be used as biofertilizer.

Some countries have already started with the diversification of their sugarcane industry, allowing better health standards in the community. For example, the National Alcohol Programme in Brazil helped the national economy of Brazil significantly through the production of ethanol from sugarcane juice and molasses. The ethanol produced by the microbial yeast process was used to substitute petrol (gasoline) in the car fleet, reducing the need for importation of oil. It is possibly true to say that Brazil was one of the first, or *the* first country to demonstrate that the economy of a country can substantially benefit by a reduction of imports. Most of our industrialized economic thinking is orientated solely towards the export of commodities. It was very unfortunate that the programme had to be halted due to lack of microbial waste management. Other countries, such as Malawi and Zimbabwe, followed the way of the Brazilians, but with better results regarding the distillery waste. In Malawi (and in India) biomethanation experiments with distillery waste have shown excellent results. In India (Karhadkar *et al.*, 1990), mesophilic biomethanation of distillery spentwash was very successful in a diphasic anaerobic process. The first acid phase with an organic loading of 30 kg COD m^{-3} day^{-1} was followed by a methane phase operation enabling a 65 % reduction in COD and 0.3 m^3 biogas produced per kg COD using an organic loading rate of 3.25 kg COD m^{-3} day^{-1}. With a methane phase at 5 kg COD m^{-3} day^{-1} and a mean retention time of 27 days, a 70 % reduction in COD was achieved with the same amount of biogas formed.

As anaerobic digestion does not reduce the chemical oxygen demand (COD) to the required minimum level for discharge into the environment, the effluent should be guided into a high rate oxidation pond (Olguín *et al.*, 1995) as a polishing step. In this case the production of the protein-rich microalgae *Spirulina* or other algae may be a key factor for the economic viability of the whole system. This would be even more the case if fish breeding or any aquaculture ponds could be added to the system (see Section 18.3). Aquaculture could, of course, be of various types.

If the microbial yeast technology for ethanol production is replaced by a bacterial *Zymomonas* technology (Doelle *et al.*, 1991), cleaner ethanol production can significantly reduce the biochemical oxygen demand (BOD) and COD in the effluent (Doelle *et al.*, 1993). A lower cell biomass, negligible by-product formation and faster time of fermentation would further contribute to economic efficiency. In addition, special strains of *Zymomonas* are capable of producing not only fructose and ethanol (Doelle and Doelle, 1991) but also other sugar products (Johns *et al.*, 1991). The important factor in this consideration for flexibility is that low-, medium- and high-yielding sugarcane can be used for fermentation, composting and mushroom production.

Microbial processes can help in introducing a clean technology if appropriately introduced and maintained. They could convert the sugarcane industry into a clean industry with much greater viability and sustainability. It is the flexibility and diversification of product formation, together with the mixing of animal and human waste with stillage or other residues during anaerobic digestion (Hobson and Wheatley, 1992; Kositratana, 1990), or composting (Stentiford and Dodds, 1992) and mushroom production (Chang and Miles, 1989; Zadrazil, 1992; Zadrazil *et al.*, 1992), which secure not only the ecological environment but also the sustainability and lifting of living standards. This introduction of clean technologies may also attract joint venture capital investment (Doelle, 1996b); however, it requires additional infrastructure training, which could be provided with the help of the global network of Microbiological Resources Centres (Doelle, 1996a).

18.3 The Palm Oil Industry

Oil palm plantations in south-east Asia are predominantly found in Malaysia, Thailand and Indonesia. As Figure 18.2 indicates, the important oil yield represents only 20 % of the original fresh fruit bunches, including the kernel oil. The rest of the fresh fruit bunches, together with water effluents, are 'waste products'. Whereas the solid contents consist mainly of lignin and cellulosic substances, the palm oil mill effluent (or POME) can easily reach a BOD of 40 000 ppm. In Malaysia, the leading world producer of palm oil, it has been estimated that the effluents from all the palm oil mills throughout the country are responsible for the same pollution as that generated by a population of 32 million people, about twice the population of Malaysia. In southern Thailand, the waste accumulation of the liquid residue alone is equivalent to the BOD generated by 3 million people.

In Thailand, the oil palm plantations had a production yield of 1 530 000 t of fresh fruit bunches (FFB) for a production of 255 910 t of crude palm oil (Chainuwat, 1991). Each tonne of FFB produces 1 t of liquid residue, 230 kg empty fruit bunches (EFB), 145 kg pericarp fibre, 60 kg shells and approximately 30 kg of palm cake. The 1 t liquid residue has a BOD load of 27 kg, a COD of 52 kg, 13 kg suspended solids and oil and grease of 8 kg (Hanpongkittikun *et al.*, 1996). Both Governments, Malaysia and Thailand, have now in place regulations prohibiting the release of untreated effluent.

18.3.1 *Present Waste Treatment Practices*

In order to overcome the massive solid waste problem, the technology of utilizing EFB, fibre and shells in boilers and incinerators was adopted (Sulaiman and Shafii, 1987). To produce 1 million t of palm oil, 5 million t FFB are required, producing 500 000 t EFB,

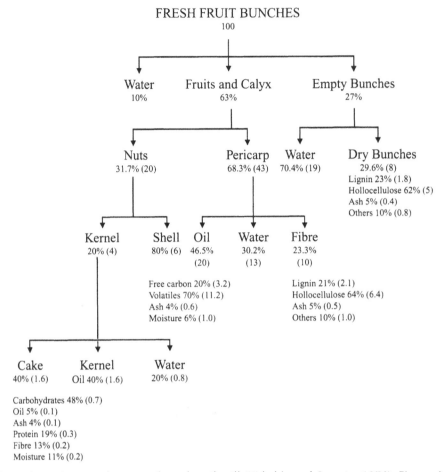

Figure 18.2 Input and output of a palm oil mill (Kirkaldy and Sutanto, 1976). Figures in parentheses indicate weight portions of original fresh fruit bunches

500 000 t pericarp fibre and 800 000 t palm shells, which produce altogether 211 MW electricity (Kirkaldy and Sutanto, 1976). The ash (5 %) contains a substantial amount of potash fertilizer and is usually recycled to the palm oil plantations, that is, about 20 000 t of it. However, boilers emit black smoke and the incinerators release large particles causing severe air pollution, which contravenes the clean air acts of these countries (Kume *et al.*, 1990).

Palm oil mill effluent (POME) has been used in various ways. Some oil palm planta-tions use the anaerobically treated effluent for irrigation, which may spill into the water supply and may not be used during the rainy season. Investigations in Malaysia and Thailand indicated that using POME in an oil palm plantation without adding any other fertilizer increases the yield by up to 13 %. Furrow irrigation is not recommended due to possible overloading and spillage into waterways as most of the plantations are in regions of heavy tropical rainfalls.

In Thailand at present pond treatment is very popular, whereby an anaerobic pond is followed by an aerobic and a polishing pond. As the temperature of POME is 75–90°C, a cooling process is first required; this is carried out by letting the effluent flow through a long channel or into cooling ponds before entering the anaerobic pond. Anaerobic digestion with open pond systems have the disadvantages that often a heavy scum develops, closing the system, and that biogas can not be recovered, instead adding to air pollution and greenhouse effect. After the anaerobic digestion system, certain additional aerobic pond systems are needed to improve the quality of the effluent. The constraint which inhibits the discharge of the final effluent into water courses is its brownish colour. This requires additional use of a number of stabilization ponds resulting in an expansion of the treatment area.

18.3.2   *Prospects of Microbial Processes in the Palm Oil Industry*

The situation described above is undoubtedly intolerable for ecological sustainability. The biogas is not collected as an energy source because of surplus energy from burning the solid material; both cause severe air pollution and aggravate the greenhouse effect. Furthermore, valuable carbon is wasted, which could bring further significant income to the industry and the farmers on the oil palm plantations. Any change in the practises presently used, however, requires a significant if not total change of attitude, mainly by the consultants to the industry. Alternatives have been suggested, such as closed anaerobic biodigesters for biogas energy production (Petitpierre, 1982), mulching (Singh *et al.*, 1989), use of solids as subsurface mulch (Vikineswary and Ravoof, 1990) and composting (Jeris and Regan, 1968, 1973). The EFB has a carbon:nitrogen ratio of 38:1, which has to be lowered to at least 25:1 in order to be useful for plants; 3 % ammonium sulphate addition has been suggested to achieve this (Atan *et al.*, 1995). In a socio-economically integrated system, this nitrogen could come from an anaerobic digester.

A socio-economic integrated system has been outlined in Figure 18.3, indicating how the introduction of various microbial processes could help the move towards a clean palm oil mill technology.

Solid wastes

The first priority must be cessation of the use of incinerators, which burn the surplus solid material not required for energy production in the mill, and thus elimination of the worst air polluter. The solid material could be divided into two groups on their chemical composition: whereas EFB and the fibre are mainly lignocellulosic in nature, the main components of the shell and cake are volatile materials and carbons. EFB and the pericarp fibre, possibly with the addition of some cake or shells, are very good resources for mushroom production. Mushroom production is a well established industry in many countries around Thailand and is rising because of increasing demand (Chang, 1980; Chang and Miles, 1991; Zadrazil, 1992; Zadrazil *et al.*, 1992). Mushrooms can convert waste materials into human food. They are rich in protein and contain several vitamins and minerals, and should be regarded as high-protein vegetables to enrich human diets. Mushrooms can be grown on materials resistant to natural biological degradation (cellulose, lignocellulose, hemicellulose) because the mycelium secretes extensive enzyme complexes which can directly attack/degrade these compounds (Chang, 1997a). Mushroom production can be carried out in furrows between trees (oil palm or rubber) and trays stacked on top of each other using primitive farming techniques; they are thus very cost effective. After improving culture techniques, they should be cultivated as widely and cheaply as other common vegetables (Chang and Hayes, 1978; Chang and Quimio, 1982; Chang and Miles, 1989; Quimio *et al.*, 1990; Chang *et al.*, 1993). The second

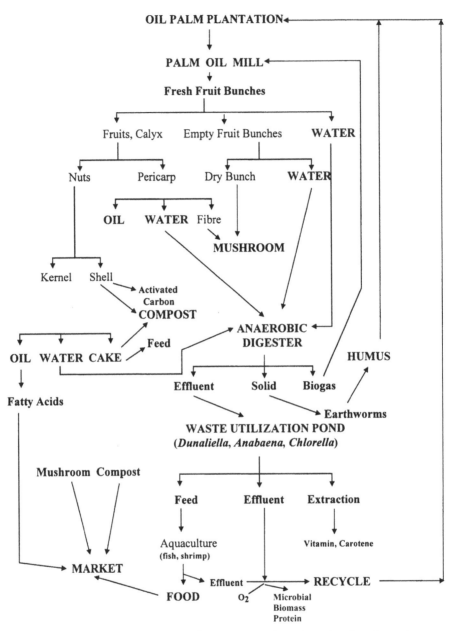

Figure 18.3   The use of clean microbial processes in the palm oil mill industry for a sustainable and pollution-free industry with significant economic benefits

major attribute of mushrooms is their medicinal properties, which have long been recognized in China, Korea and Japan (Chang, 1997b). A recent upsurge of interest in traditional remedies for various physiological orders has led to the coining of the term 'mushroom nutraceuticals' (Chang and Buswell, 1996). A mushroom nutraceutical is a refined, partially defined mushroom extract which is consumed in the form of capsules or tablets as a dietary supplement (not food) and which has potential therapeutic applications. A regular intake may enhance the immune response of the human body, thereby increasing resistance to disease and, in some cases, causing regression of the disease state (Liu *et al.*, 1995; Chang, 1997b).

The residue from the mushroom industry can now be used for composting, together with the shells or some of the palm cake. In contrast to mulching of the cellulosic/lignocellulosic material, which is very labour intensive, composting can easily be mechanized on a large scale (Manderson, 1997), as has been shown in New Zealand. Compost is a much better soil improver and stabilizer than the EFB and fibre itself. Experiments in Malaysia have shown (Atan *et al.*, 1995) that composting EFB ensures a perfect nutrient balance for efficient microbial activity at 30°C, in particular when the EFB is treated with liquid sludge (POME). With weekly turning of the pile the compost is ready within 9 weeks, when the carbon:nitrogen ratio has dropped from 30:1 to about 20:1 (Huan, 1989). The optimal ratio for composting is in the range of 25–30:1 (Manderson, 1997). The necessary attributes of successful composting must be kept in mind when assessing the quality of the final material:

- it must be safe and free of pathogens
- the process must conserve nutrients of value to green plants
- production economics must be favourable.

Composting technologies can be categorized on temperature of operation (mesophilic, thermophilic), oxygen availability (aerobic, anaerobic) and mixing (tumbled system) or non-mixing (static piles or windows).

The humus obtained from composting can be further improved through earthworm addition, as will be described later.

The palm cake can first be cleaned up by hydrolysing the residual traces of oil with lipases (Pichaiyut *et al.*, 1996), followed by saccharification of the major complex carbohydrate components using cellulolytic microorganisms, steam explosion (Oi *et al.*, 1994) or solid substrate fermentation for the production of cellulases (Prasertsan *et al.*, 1992, 1996a and b; Prasertsan and Oi, 1992). Palm cake may, of course, also be added for composting as mentioned above.

These microbial process methods not only eliminate very severe air pollution caused by burning and incineration but also produce additional products of value, mainly mushrooms and compost or humus for soil improvement.

Liquid wastes

In order to keep the palm oil industry viable, after removing burning and incineration from their practices, one has to look for alternative energy from the liquid waste.

It was mentioned earlier that for every tonne of FFB processed, approximately 1 tonne of liquid effluent is produced with an approximately 52 kg COD or 52 000 ppm COD. At present a three-stage ponding system is being used in Malaysia (Chooi, 1981) with no further product formation. The liquid waste could, however, produce enough biogas, and thus energy, to run the plant mill operation, which would balance the elimination of incinerators.

The basic calculations are as follows (Petitpierre, 1982). If one realizes that 1 m^3 biogas is equivalent to 0.7 l fuel oil, 0.8 l petrol, 2.7 kg dry firewood or 0.3 m^3 propane gas, and that the processing of 1 t FFB produces 1 t or 1 m^3 liquid waste (POME) with a COD of 52 kg, the production of biogas is $52 \times 0.3 = 15.6$ m^3 t^{-1} FFB, which is equivalent to 325 MJ. The average palm oil mill has a capacity of processing 20 t FFB h^{-1} or

50 000 t FFB per year. If all the effluent is treated by closed anaerobic digestion, 780 000 m^3 biogas equivalent to 546 000 l fuel can be generated. Converted into electricity, this biogas would operate a 200 kW (250 kva) generator for 9828 h or in excess of 24 h day^{-1} (fuel consumption being 0.275 l kW^{-1}). In basic terms it means that the liquid waste from 1 t FFB can supply about 39 kWh, or more electricity than required to treat the same weight of FFB.

In comparing various anaerobic digestion treatment methods, Edewor (1986) found a tank digester most economical for POME with a break-even cost after 2 years of operation. An organic loading of 2.8 kg COD m^{-3} day^{-1} gave a 90 % reduction in COD after an HRT of 18.5 h and a 0.57 m^3 biogas production kg^{-1} COD day^{-1} (Chua and Giau, 1986). Using a temperature of 44–52°C, an organic loading of 4.8 kg volatile solids [VS] m^{-3} day^{-1} was very successful with an HRT of only 10 days (Quah and Gillies, 1981) using 3700 m^3 digesters. The biogas production was 28.3 m^3 m^{-3} POME digested or 0.59 m^3 kg^{-1} VS added. With a retention time of 10 days, biogas production would be 2.8 m^3 m^{-3} effective digester volume day^{-1}. Very similar results were obtained by other research groups (Chin, 1981; Chin and Wong, 1983; Cheah *et al.*, 1988; Borja-Padilla and Banks, 1993).

The sediment sludge from the anaerobic digester is still rich in nitrogen and minerals and this can easily be used as fertilizer or added to the composting of the solid waste to improve the carbon:nitrogen ratio. The sludge sediment from anaerobic digesters has also been used for earthworm production in Colombia, which is now being successfully implemented in Vietnam (Rodriguez, 1997a,b). The transformation of manure by earthworms is an interesting process that brings benefits from three points of view: economical, biological and ecological. *Eisenia foetida* transforms the material swallowed within a few hours to material composed of organic substances, mainly humic acid. Humus is the main reserve of organic components in the soil. Manure that has been transformed by earthworms supplies readily assimilable nutrients to the soil because of the capacity of the earthworm to promote the production of humus.

The liquid from the digester can now be transferred into shallow ponds for aquaculture (Moi, 1987; Chan, 1993; Li, 1997). Algae are excellent nitrogen and phosphorus scavengers (Olguín *et al.*, 1988, 1994, 1995; Aziz and Ng, 1992; Israel, 1997a,b). Algae are an excellent protein source and can further be used in a second deeper pond as food for fish production (Rajagopalan and Webb, 1975; Moi, 1987; Chan, 1997a,b). The design and cost of such systems can be relatively low (Khan, 1996; Chan, 1997b) and of significant benefit (Chan, 1997c).

The introduction of clean technologies using microbial processes can lead immediately to a sustainable industry with significant economic benefits to the industry and no doubt also to the oil palm farmer. It completely eliminates the pollution of air, soil and water by gaining additional products for the market place as well as palm oil.

18.4 The Seafood Processing Industry

The seafood processing industry is one of the largest export industries in Thailand and has earned more than US$2 billion annually from exports of seafood and seafood products (Prasertsan *et al.*, 1994) over the period 1988–1993. There are more than 40 seafood processing factories in Southern Thailand, with the major factories employing up to 2000 people each. The quantitites of solid and liquid waste produced by these fish-processing plants are enormous, and a great variety of raw materials (especially tuna, shrimp, crab, squid, clam) is used. A survey in the Songhkla region of Southern Thailand revealed that the amount of wastewater discharged from each plant was 300–500 m³ day⁻¹ (Prasertsan *et al.*, 1988) with a BOD content of 7000–12 000 ppm and a COD up to 20 000 ppm. The BOD/COD ratio of wastewater from all sources ranges between 0.47 and 0.61 (Jung *et al.*, 1990). More recent data indicate that the quantity of wastewater is up to 3000 m³ day⁻¹. Wastewater treatment methods vary from activated sludge, oxidation ponds, aerated lagoons and combinations of these. Different raw materials and processing methods, together with the biological variability, account for the variation in BOD/COD ratio. The pH of the wastewater ranges between 6 and 7, total solids 5000–15 000 ppm, volatile solids 2000–6500 ppm, total nitrogen 100–700 ppm, phosphorus 30–44 ppm and the chloride content 500–4000 ppm NaCl. Every plant, however, also produces solid waste, which can be as high as 70 % in the case of the shrimp-processing plants.

18.4.1 Present Operating System

Solid waste

The solid waste in a seafood processing plant ranges from whole fish, off-cuts of fish, bone, skin and intestinal organs to heads, shells and pieces of shrimp waste. The non-shrimp solid wastes are mainly sold to fishmeal factories, while shrimp wastes are used for the production of chitin and chitosan, as well as feed for chickens.

In fishmeal factories the solids are screened, wet milled and dried (Middlebrooks, 1979; Valle and Aguilera, 1990). Fish meal is a highly concentrated animal feed supplement consisting of high-quality protein, minerals and B-complex vitamins.

Shellfish (shrimp head and shell) are rich in the polysaccharide chitin, which can be recovered and deacylated to chitosan. This process has been introduced most recently in Southern Thailand. Chitosan has a wide spectrum of applications in medical as well as biotechnological fields. It is of particular interest for the production of biodegradable packaging material (Knorr, 1991).

Liquid waste

The large amount of liquid waste is preferably treated in a pond or aerated lagoon systems (Prasertsan *et al.*, 1988). An example of such a system is presented in Figure 18.4 (Prasertsan and Choorit, 1988). Although this combination of pond and aerated lagoon is a popular wastewater treatment method because of low operating and maintenance costs, a high investment on very large areas of land is required, which is larger than the processing plant itself. Problems have been observed in some of these wastewater treatment plants with the appearance of a red colour (photosynthetic bacteria) and bad smell. The discharge of red effluent into streams would most certainly have an impact on the quality of natural waterways or reservoirs. The bad smell may be no problem at the moment, but an increasing population will certainly perceive it differently. Furthermore, the land required for the pond system is very expensive in some areas. Use of activated sludge (AS) is becoming more popular although it has high investment cost (approximately 20 million Baht or US$0.8 million) and high cost of operation because it is an energy-intensive process. Dissolved air flotation (DAF) has also been implemented in some factories before the activated sludge to remove oil, grease and protein. Both AS and DAF processes create high quantities of sludge, which need to be treated or disposed of.

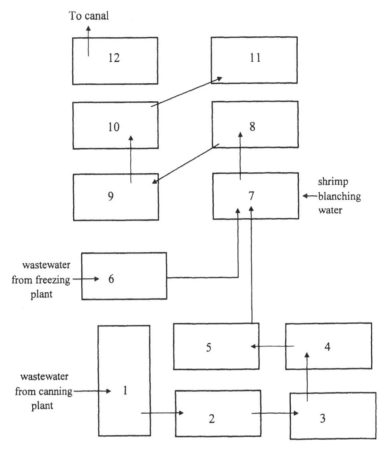

Figure 18.4 Flow chart of the wastewater treatment system of a seafood processing plant (Prasertsan and Choorit 1988). Ponds 1 and 2: anaerobic; pond 3: facultative; ponds 4, 7 and 8: aerated; ponds 5, 6, 9–12: facultative

18.4.2 *Clean Technologies Based on the use of Microorganisms*

Solid waste

The solid wastes from the fish-processing industries are very rich in protein, one of the reasons why fish meal and fish sauce production has been known for a long time. Because of the relatively high hydrolytic activities in these solid wastes, fish silage, fish sauce and protein hydrolysate production, chitin and chitosan isolation and production of many enzymes, microbial biomass protein (MBP), and oil flavour compounds could form a new waste re-use industry.

Fish silage The solids are acidified to pH 3–4, which is optimal for protein digestion by the fish enzymes (Gildberg, 1993). Because no salt is added, autolysis is much faster than in the case of fish sauce. The bitter aqueous phase makes the liquid unsuitable for human consumption. However, it is an excellent alternative to fish meal production (Raa and Gildberg, 1982) and very suitable as feed to immature young animals (Wignall and Tatterson, 1976; Stone and Hardy, 1986).

Fish sauce Fish sauce production is a very simple and ancient process (Gildberg, 1993). The small fish is mixed with 20–40 % sea salt and stored for 6–12 months. High-quality fish sauce contains about 10 % peptides and amino acids and approximately 25 % salt. The whole process makes use of trypsin-like enzymes (Orejana and Liston, 1982) in the fish solids. In commercial processes the addition of enzymes such as papain and bromelain reduces the storage time to a few weeks (Beddows *et al.*, 1976; Beddow and Ardeshir, 1979; Kumalaningshi, 1990).

Protein hydrolysate Protein hydrolysates are excellent nitrogen sources for microbial growth media (Beuchat, 1974; Gildberg *et al.*, 1989; Rebeca *et al.*, 1991), but fish peptones are commercially produced so far only in Japan and Norway (Ahmas *et al.*, 1990; Uchida *et al.*, 1990) with a nutritive value similar to that of casein (Rebeca *et al.*, 1991). A further application of protein hydrolysates are as protein and flavour supplements in fish feed and pet foods (Pigott, 1982). They can also be used as fertilizer on vegetables (Wyatt and McGourty, 1990), retarding the ageing of lettuce and peas and delaying flowering and fruiting of tomatoes. Much of the fish flavouring, fish soup and fish paste products available on the market are in fact fish protein hydrolysates (In, 1990; Shoji, 1990).

Protein hydrolysate manufacture requires the mincing of the solid waste, small fish or cheap fish raw material and the addition of proteases from *Bacillus subtilis* (Gildberg, 1993), *Bacillus thermoproteolyticus* (Nakajima *et al.*, 1992), papain and bromelain or the acidic proteases from fish itself (Reece, 1988).

Enzyme and MBP production The enzyme chitin deacetylase required for the production of chitosan is present in a number of fungal species (Kafetzopoulos *et al.*, 1991). It is also possible to produce the enzyme chitinase from *Serratia marcescens* grown on chitin (Cosio *et al.*, 1982), which hydrolyses chitin to *N*-acetylglucosamine, an excellent raw material for the production of MBP (*Pichia kudriavzenii*). Shrimp waste is also rich in enzymes such as alkaline phosphatase, hyaluronidase and acetylglucosaminidase (Olsen *et al.*, 1990). A large-scale process for the recovery of these enzymes has been very successful, producing up to 2300 U g^{-1} alkaline phosphatase, 260 U g^{-1} *N*-acetylglucosminidase and 5 U g^{-1} hyaluronidase. These enzymes can be used in crude form as a supplement in starter diets in aquaculture.

Liquid waste

The most useful microbial process for the treatment of the liquid waste, including all the wastes from the solid waste bioconversion processes, is the anaerobic digester system. Although such systems exist in Thailand for the treatment of wastewaters from tapioca (= cassava = manihot) starch and distillery factories (Tantichareon *et al.*, 1986) using the biogas produced, the anaerobic digester has not been employed as yet to treat the wastewater from seafood processing plants. A clean technology for disposal of both solid and liquid waste in the fish processing industry is presented in Figure 18.5.

Experiments have already shown (Prasertsan *et al.*, 1994) that a higher than 75 % COD reduction could be obtained up to an organic loading rate of 1 kg m^{-3} day^{-1} with an HRT of 11 days. Increasing the organic loading to 1.3 kg m^{-3} day^{-1} corresponds to an HRT of 6.6 days with a maximal biogas production of 1.5 m^3 m^{-3} day^{-1} or a 1.3 m^3 biogas kg^{-1} COD with a 65 % COD reduction. For the treatment of tuna condensate, a constant 11 day HRT and an organic loading of 1.3 kg COD m^{-3} day^{-1} achieved a maximal biogas

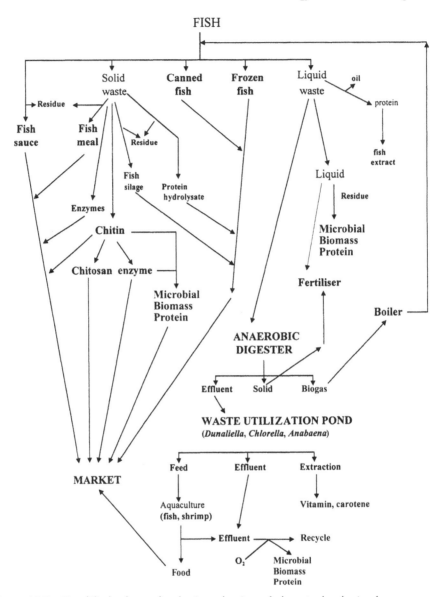

Figure 18.5 Simplified scheme for the introduction of clean technologies for a sustainable fish-processing industry

productivity of 1.1 m³ m⁻³ day⁻¹ or 0.75 m³ kg⁻¹ COD with a reduction of 60 % COD. If the organic loading is increased to 2.5 kg COD m⁻³ day⁻¹, biogas production stops. It is suspected that a relatively high carbon:nitrogen ratio may be responsible for this cessation.

Tuna condensate could also be used first for the production of MBP such as the photosynthetic bacterium *Rhodocyclus gelatinosus* (Prasertsan *et al.*, 1993a,b), *Bacillus subtilis*, *Candida tropicalis* F-129 (Sangsri *et al.*, 1996) or other yeasts (Sujarit *et al.*, 1996).

The effluent from the biodigester may then be transferred into a shallow pond for the production of the alga *Chlorella* (Ratanapradit *et al.*, 1996) and photosynthetic bacteria.

Both microorganisms have a high chlorophyll, carotenoid, protein, vitamin and fatty acid content, and therefore make an excellent feed for fish production in a second deeper pond, or could be sold to the large shrimp-farm industry along the South Thailand coast. Liquid waste also contains a large variety of enzymes, in particular proteases, which could be recovered by simple biochemical purification steps.

Fish processing waste is one of the richest resource materials for further enzymatic and microbial process development. The introduction of all or part of the mentioned processes would not only eliminate the problems in the presently used wastewater treatment methods but would also undoubtedly raise the economic benefits, viability and sustainability of the industry in a clean ecological environment.

18.5 Concluding Remarks

The sugarcane, palm oil and seafood processing are important industries in the developing countries, particularly in South-east Asia. The present technologies employed in these and other agroindustries generate large quantities of solid and liquid wastes, most of which not only remain unused but also cause environmental problems. Ever-increasing competition in the world market, together with the introduction of the ISO 14000 Series of Environmental Management Standards, will force industries into the adoption of pollution-preventing technologies in treating wastes as valuable resource material for value-added product production. These clean technologies can only be achieved through the implementation of integrated biosystems. It is our belief that such systems would help significantly in achieving sustainability.

References

AHMAS, K.A., 1990, Utilisation of marine biomass for production of microbial growth media and biochemicals, in Voigt, M.N. and Botta, R.J. (eds) *Advances in Fisheries Technology and Biotechnology for Increased Profitability*, Lancaster, USA: Technomic Publ. Co, Inc, pp.361–372.

ATAN, Y., AWANG, M.R., OMAR, M., HASHIM, A., KUME, T. and HASHIMOTO, S., 1995, Biodegradation of oil palm empty fruit bunch into compost by composite microorganisms, in *Proceedings of the EU-ASEAN Conference on Combustion of Solids & Treatment Products*, February.

AZIZ, M.A. and NG, W.J., 1992, Feasibility of wastewater treatment using the activated-algae process, *Biosource Technol.*, **40**, 205–208.

BEDDOWS, C.G. and ARDESHIR, A.S., 1979, The production of soluble fish protein solution for use in fish sauce manufacture. II. The use of acids at ambient temperature, *J. Food Technol.*, **14**, 613–623.

BEDDOWS, C.G., ISMAIL, M. and STEINKRAUS, K.H., 1976, The use of bromelain in the hydrolysis of mackerel and the investigation of fermented fish aroma, *J. Food Technol.*, **11**, 379–388.

BEUCHAT, L.R., 1974, Preparation and evaluation of a microbial growth medium formulated from catfish waste peptone, *J. Milk Food Technol.*, **37**, 277–281.

BORJA-PADILLA, R. and BANKS, C.J., 1993, Thermophilic semi-continuous anaerobic treatment of palm oil mill effluent, *Biotechnol. Letts*, **15**, 761–766.

CHAINUWAT, C., 1991, Status of palm oil production in the year 1990/91. In *First Workshop on Research on Oil Palm and Palm Oil*, Prince of Songkla University, Hat Yai, Thailand.

CHAN, G., 1993, Aquaculture, Ecological Engineering: Lessons from China. *AMBIO*, **22**, 491–494.

CHAN, G., 1997a, Integrated biomass system for sustainability. Presented at the UNDP/UNU ZERI Indo-Pacific Workshop, Montford Boys Town, Suva, Fiji, May 1997. Institute of Advanced Studies, UNU, Tokyo at www.ias.unu.edu/proceedings/icibs/ibs/info/fiji/.

CHAN, G., 1997b, Biotechnology and design of waste treatment plants. Presented at the UNDP/ UNU ZERI Indo-Pacific Workshop, Montford Boys Town, Suva, Fiji, May 1997. Institute of Advanced Studies, UNU, Tokyo at www.ias.unu.edu/proceedings/icibs/ibs/info/fiji/.

CHAN, G., 1997c, Benefits of integrated development. Presented at the UNDP/UNU ZERI Indo-Pacific Workshop, Montford Boys Town, Suva, Fiji, May 1997. Institute of Advanced Studies, UNU, Tokyo at www.ias.unu.edu/proceedings/icibs/ibs/info/fiji/.

CHANG, S.T., 1980, Mushroom production in SE Asia, *Mushroom Newsletter for the Tropics*, **1**, 18–22.

CHANG, S.T., 1997a, Practical mushroom culture in the Tropics, Presented at the UNDP/UNU ZERI Indo-Pacific Workshop, Montford Boys Town, Suva, Fiji, May 1997. Institute of Advanced Studies, UNU, Tokyo at www.ias.unu.edu/proceedings/icibs/ibs/info/fiji/.

CHANG, S.T., 1997b, Nutritional and medicinal properties of mushrooms, Presented at the UNDP/ UNU ZERI Indo-Pacific Workshop, Montford Boys Town, Suva, Fiji, May 1997. Institute of Advanced Studies, UNU, Tokyo at www.ias.unu.edu/proceedings/icibs/ibs/info/fiji/.

CHANG, S.T. and BUSWELL, J.A., 1996, Mushroom Nutriceuticals, *World J. Microbiol. Biotechnol.*, **12**, 473–476.

CHANG, S.T. and HAYES, W.A., 1978, *The Biology and Cultivation of Edible Mushrooms*, London: Academic Press.

CHANG, S.T. and MILES, P.G., 1989, *Edible Mushrooms and their Cultivation*, Florida, USA: CRC Press.

CHANG, S.T. and MILES, P.G., 1991, Recent trends in wld production of cultivated edible mushrooms, *Mushroom J.*, **504**, 15–18.

CHANG, S.T. and QUIMIO, T.H., 1982, *Tropical Mushrooms – Biological Nature and Cultivation Methods*, Hong Kong: Chinese University Press.

CHANG, S.T., BUSWELL, J.A. and MILES, P.G., 1993, *Genetics and Breeding of Edible Mushrooms*, Philadelphia, USA: Gordon and Breach.

CHEAH, S.C., MA, A.N., OOI, L.C.L. and ONG, A.C.H., 1988, Biotechnological applications for utilisation of wastes from Palm Oil Mills, *Fat Sci. Technol.*, **50**, 536–540.

CHIN, K.K., 1981, Anaerobic treatment kinetics of palm oil sludge, *Water Res*, **15**, 199–202.

CHIN, K.K. and WONG, K.K., 1983, Thermophilic anaerobic digestion of POME effluent, *Water Res*, **17**, 993–995.

CHOOI, C.F., 1981, Ponding system for palm oil mill effluent treatment, in *Proceedings of a Workshop on Review of Palm Oil Mill Effluent Technology*, Kuala Lumpur, pp.53–62.

CHUA, N.S. and GIAU, H.L., 1986, Biogas production and utilisation, in *Proceedings of a National Workshop on Recent Development in Pam Oil Milling Technology and Pollution Control*, Kuala Lumpur.

COSIO, T.G., FISHER, R.A. and CARROAD, P.A., 1982, Bioconversion of shellfish chitin waste: waste pretreatment, enzyme production, process design, and economic analysis, *J. Food Sci.*, **47**, 901–905.

DASILVA, E.J., RATLEDGE, C. and SASSON, A., 1992, *Biotechnology: Economic and social aspects*, Cambridge: Cambridge University Press.

DOELLE, H.W., 1989, Socio-economic biotechnology development for developing countries, *MIRCEN – J. Appl. Microbiol. Biotechnol.*, **5**, 391–410.

DOELLE, H.W., 1993, Use of socio-economical principles for the advancement of fermentation technologies, in *International Cooperation and Education in Applied Microbiology*, Osaka: Osaka University Press, pp.69–75.

DOELLE, H.W., 1994, *Microbial Process Development*, Singapore: World Scientific Publishers.

DOELLE, H.W., 1996a, The role of MIRCENs in Technology Transfer, *ASM News*, **62**, 334–335.

DOELLE, H.W., 1996b, Joint venture capital investment for clean technologies and their problems in developing countries, *World J. Microbiol. Biotechnol.*, **12**, 445–450.

DOELLE, M.B. and DOELLE, H.W., 1991, High fructose formation from sugarcane syrup and molasses using *Zymomonas mobilis* mutants, *Biotechnol. Letts*, **13**, 875–879.

DOELLE, H.W., KENNEDY, L.D. and DOELLE, M.B., 1991, Scale-up of ethanol production from sugarcane using *Zymomonas mobilis, Biotechnol. Letts*, **13**, 131–134.

DOELLE, H.W., KIRK, L., CRITTENDEN, R., TOH, H. and DOELLE, M.B., 1993, *Zymomonas mobilis* – Science and Applications, *Crit. Rev. Biotechnol.*, **13**, 57–98.

EDEWOR, J.O., 1986, A comparison of treatment methods for palm oil effluent (POME), *J. Chem. Technol. Biotechnol.*, **36**, 212–218.

GILDBERG, A., 1993, Enzymic processing of marine raw materials, *Process Biochem.*, **28**, 1–15.

GILDBERG, A., BATISTA, I. and STROM, E., 1989, Preparation and characterisation of peptones obtained by a two-step enzymatic hydrolysis of whole fish, *Biotechnol. Appl. Biochem.*, **11**, 413–423.

HANPONGKITTIKUN, A., PRASERTSAN, P. and SRISUWAN, G., 1996, *Minimum Environmental Requirements and Environmental Management Guidelines for the Palm Oil Mill Industry*, Report to Department of Industrial Works, Ministry of Industry, Thailand.

HOBSON, P.N. and WHEATLEY, A.D., 1992, *Anaerobic Digestion – Modern Theory and Practice*, London: Elsevier Applied Science.

HUAN, K.K., 1989, Trials on composting EFB of oil palm with and without prior shredding and liquid extraction, in *Proceedings of the 1989 PORIM International Palm Oil Development Conference Module II: Agriculture*, pp.217–244.

IN, T., 1990, Seafood flavourants produced by enzymatic hydrolysis, in Keller, S. (ed.) *Making Profits out of Seafood Wastes*, Alaska Sea Grant Programme, Report No. 90-07, Fairbanks, AK, USA, pp.197–201.

ISRAEL, A., 1997a, Uses and processing of algae and seaweeds, *UNDP/UNU ZERI Indo-Pacific Workshop*, Montford Boys Town, Suva, Fiji, May 1997. Institute of Advanced Studies, UNU, Tokyo at www.ias.unu.edu/proceedings/icibs/ibs/info/fiji/.

ISRAEL, A., 1997b, Algoculture in the Tropics, *UNDP/UNU ZERI Indo-Pacific Workshop*, Montford Boys Town, Suva, Fiji, May 1997. Institute of Advanced Studies, UNU, Tokyo at www.ias.unu.edu/proceedings/icibs/ibs/info/fiji/.

JERIS, J.S. and REGAN, R.W., 1968, Progress report on cellulose degradation in composting, *Compost Sci.*, **9**, 20–22.

JERIS, J.S. and REGAN, R.W., 1973, Controlling environmental parameters for optimum composting. I. Experimental procedures and temperature, *Compost Sci.*, **14**, 10–15.

JOHNS, M.R., GREENFIELD, P.F. and DOELLE, H.W., 1991, By-products from *Zymomonas mobilis, Adv. Biochem. Bioeng./Biotechnol.*, **44**, 97–120.

JUNG, S., PRASERTSAN, P. and BUCKLE, K., 1990, Fish cannery wastewater characteristics, *ASEAN Food J.*, **5**, 82–83.

KAFETZOPOULOS, D., MARTINOU, A. and BOURITIS, V., 1991, Isolation and characterisation of chitin deacetylase from *Mucor rouxii*.

KARHADKAR, P.P., HANDA, B.K. and KHANNA, P., 1990, Pilot-scale distillery spentwash biomethanation, *J. Environ. Eng.*, **116**, 1029–1045.

KHAN, S.R., 1996, Low cost biodigesters. Programme for Research on Poverty Alleviation. *Grameen Trust Report*, February.

KIRKALDY, J.L.R. and SUTANTO, J.B., 1976, Possible utilisation of by-products from palm oil industry, *Planter*, **52**, 118–126.

KNORR, D., 1991, Recovery and utilisation of chitin and chitosan in food processing waste management, *Food Technol.*, **45**, 114–120.

KOSITRATANA, N., 1990, Thermophilic anaerobic contact process for distillery wastewater treatment, *ASEAN Biotech. for Agro-Industry Waste Management*.

KUMALANINGSHI, S., 1990, Accelerated method of fish sauce production from lemuru fish (*Scardinella* sp.), in Reilly, P.J.A., Parry, R.W.H. and Barile, I.I. (eds) *Postharvest Technology, Preservation and Quality of Fish in South-East Asia*, Echanis Press, Manila: p.21.

KUME, T., HO, H., ISHIGAKI, I., MUHAMMAD, L.J., ZAINON, O., FOZIAL, A. *et al.*, 1990, Effect of gamma irradiation on microorganisms and components in empty fruit bunch and palm press fibres of oil palm wastes, *J. Sci. Food Agriculture*, **52**, 147–157.

LI KANGMIN, 1997, Fish polyculture in China. *UNDP/UNU ZERI Indo-Pacific Workshop*, Mont-ford Boys Town, Suva, Fiji, May 1997. Institute of Advanced Studies, UNU, Tokyo at www.ias.unu.edu/proceedings/icibs/ibs/info/fiji/.

LIU, F., OOI, V.E.C. and CHANG, S.T., 1995, Antitumour components of the culture filtrates from *Tricholoma* sp., *World J. Microbiol. Biotechnol.*, **11**, 486–490.

MANDERSON, G., 1997, Biosolids processing with particular reference to composting, *UNDP/ UNU ZERI Indo-Pacific Workshop*, Montford Boys Town, Suva, Fiji, May 1997. Institute of Advanced Studies, UNU, Tokyo at www.ias.unu.edu/proceedings/icibs/ibs/info/fiji/.

MIDDLEBROOKS, E.J., 1979, *Industrial Pollution Control, Vol. 1: Agro-Industries*, New York: John Wiley & Sons, pp.202–235.

MOI, P.S., 1987, The potential of integrating fish culture into an algal pond system treating palm oil mill effluent, in *Proceedings of an International Oil Palm/Palm Oil Conference*, June, Kuala Lumpur.

NAKAJIMA, M., SHOJI, T. and NABETAMI, H., 1992, Proteas hydrolysis of water soluble fish proteins using a free enzyme membrane reactor, *Process Biochem.*, **27**, 155–160.

OI, S., MATSUZAKI, K., TANAKA, T., IIZUKA, M., TANIGUCHI, M. and PRASERTSAN, P., 1994, Effect of steam explosion treatment on enzymatic hydrolysis of palm cake and fiber as solid wastes and natural resources, *J. Fermentation Bioeng.*, **77**, 326–328.

OLGUÍN, E.J., DOELLE, H.W. and MERCADO, G., 1995, Resource recovery through recycling of sugar processing by-products and residuals, *Resources Conserv. Recycl.*, **15**, 85–94.

OLGUÍN, E.J., TELLEZ, P. and GONZALEZ, J., 1988, Evaluacion de algunas alternativas biotecnologicas para la diversificacion de la industria azucarera, *Rev. Asoc. Tecn. Azucar. Mexico*, **2**, 18–22.

OLGUÍN, E.J., HERNANDEZ, B., ARAUS, A., CAMACHO, R., GONZALEZ, R., RAMIREZ, M.E., *et al.*, 1994, Simultaneous high biomass protein production and nutrient removal using *Spirulina maxima* on sea water supplemented with anaerobic effluents, *World J. Microbiol. Biotechnol.*, **10**, 576–578.

OLSEN, R.L., JOHANSEN, A. and MYRNES, B., 1990, Recovery of enzymes from shrimp waste, *Process Biochem.*, **25**, 67–68.

OREJANA, F.M. and LISTON, J., 1982, Agents of proteolysis and its inhibition in patis (fish sauce) fermentation, *J. Food Sci.*, **47**, 198–203.

PETITPIERRE, G., 1982, Palm oil effluent treatment and production of biogas, *Oleagineux*, **37**, 367–373.

PICHAIYUT, W., HANPONGKITTIKUN, A. and PRASERTSAN, P., 1996, Immobilisation of lipase and application to the hydrolysis of Palm Oil, *Biotechnology: Prospect for the Future*, Nov. 1996, Thailand, II-12.

PIGOTT, G.M., 1982, Enzyme modification of fishery by-products, in Martin, R.E., Flick, G.F., Hebard, C.E and Ward, D.R. (eds), *Chemistry and Biochemistry of Marine Food Products*, CT, USA: AVI Publ., pp.447–452.

PRASERTSAN, P. and CHOORIT, W., 1988, Problem and solution of the occurrence of red color in wastewater of seafood processing plant, *Songklanakarin J. Sci. Technol.*, **10**, 439–466.

PRASERTSAN, P. and OI, S., 1992, Production of cellulolytic enzymes from fungi and use in the saccharification of palm cake and palm fibre, *World J. Microbiol. Biotechnol.*, **8**, 536–538.

PRASERTSAN, P., CHOORIT, W. and SUWANO, S., 1993a, Cultivation of isolated photosynthetic bacteria in wastewater from seafood processing plant, in *Proceedings of the 9th International Biotechnology Symposium*, Crystal City, Virginia, USA.

PRASERTSAN, P., CHOORIT, W. and SUWANO, S., 1993b, SCP production and treatment of seafood processing effluent by isolated and identified photosynthetic bacteria, in *Proceedings of the 4th Annual Meeting of th Thai Society of Biotechnology*, October, Bangkok.

PRASERTSAN, P., JATURAPORNPIPAT, M. and SIRIPATANA, C., 1996a, Utilisation and treatment of tuna condensate by photosynthetic bacteria, in *Proceedings of an International Conference on Environmental Biotechnology*, September, Palmerston North, New Zealand.

PRASERTSAN, P., JITBUNJERDKUL, S. and HANPONGKITTIKUN, A., 1990, Process, waste utilisation, and wastewater characteristics of palm oil mill, *Songklanakarin J. Sci. Technol.*, **12**, 169–176.

PRASERTSAN, P., JITBUNDERKUL, S. and OI, S., 1992, Thermostable CMCase of thermophilic fungus, *Myceliophthora thermophila*, *Annu. Rep. IC Biotechnol.*, **15**, 401–403.

PRASERTSAN, P., JUNG, S. and BUCKLE, K., 1994, Anaerobic filter treatment of fishery wastewater, *World J. Microbiol. Biotechnol.*, **10**, 11–14.

PRASERTSAN, P., KUNGHAE, A., MANEESRI, J. and OI, S., 1996b, Solid Substrate Cultivation of *Aspergillus niger* ATCC 6275 on palm cake, *Annu. Rep. ICBiotechnol.*

PRASERTSAN, P., WUTTIJUMMONG, P., SOPHANODORA, P. and CHOORIT, W., 1988, Seafood processing industries within Songhkla-Hat Yai region: the survey of basic data emphasis on wastes, *Songklanakarin J. Sci. Technol.*, **10**, 447–451.

QUAH, S.K. and GILLIES, D., 1981, Practical experience in the production and uses of biogas, in *Proceedings of a National Workshop on Oil Palm By-Product Utilisation*, Kuala Lumpur.

QUIMIO, T.H., CHANG, S.T. and ROYSE, D.J., 1990, *Technical Guidelines for Mushroom Growing in the Tropics*. Rome: FAO.

RAA, J. and GILDBERG, A., 1982, Fish silage: A review. *CRC Crit. Rev. Food Sci. Nutr.*, **16**, 383–419.

RAJAGOPALAN, K. and WEBB, B.H., 1975, Palm oil mill waste recovery as a by-product industry. II. Biological Utilisation, *Planter*, **51**, 126–132.

RATANAPRADIT, K., PRASERTSAN, P., TUNSKUL, P. and AUGSUPANIT, S., 1996, Cultivation of *Chlorella* sp. T9 in effluent from seafood processing plant. *Biotechnology and its Global Impact*, Gesellschaft fur Biotechnologische Forschung, Braunschweig, Germany.

REBECA, B.D., PENA-VERA, M.T. and DIAZ-CASTANEDA, M., 1991, Production of fish protein hydrolysates with bacterial proteases; yield and nutritional value, *J. Food Sci.*, **56**, 309–314.

REECE, C., 1988, Recovery of proteases from fish waste, *Process Biochem.*, **23**, 62–66.

RODRIGUEZ, J.L., 1997a, Use of earthworms: A case study. *UNDP/UNU ZERI Indo-Pacific Workshop*, Montford Boys Town, Suva, Fiji, May 1997. Institute of Advanced Studies, UNU, Tokyo at www.ias.unu.edu/proceedings/icibs/ibs/info/fiji/.

RODRIGUEZ, J.L., 1997b, Recycling in integrated farming systems. *UNDP/UNU ZERI Indo-Pacific Workshop*, Montford Boys Town, Suva, Fiji, May 1997. Institute of Advanced Studies, UNU, Tokyo at www.ias.unu.edu/proceedings/icibs/ibs/info/fiji/.

SANGSRI, R., PRASERTSAN, P. and CHOORIT, W., 1996, Effect of nutrient concentration on growth of *Rhodocyclus gelatinosus* R7 in tuna condensate, in *Proc. Biotechnology: Prospect for the Future*, November, Prachuap Khiri Khan, Thailand.

SHOJI, Y., 1990, Creamy fish protein, in Keller, S. (ed.) *Making Profits out of Seafood Wastes*, Alaska Sea Grant Programme, Report No. 90-07, Fairbanks, AK, USA, pp.87–93.

SINGH, G., MANOHARAN, S. and SAN, T.T., 1989, United plantations' approach to palm oil mill by-product management and utilisation. *PORIM Intern. Palm Oil Develop. Conf.*, Kuala Lumpur.

STENTIFORD, E.I. and DODDS, C.M., 1992, Composting, in Doelle, H.W., Mitchell, D.A. and Rolz, C.E. (eds) *Solid Substrate Cultivation*, London: Elsevier Applied Science, pp.211–247.

STONE, F.E. and HARDY, R.W., 1986, Nutritional value of acid stabilised silage, *J. Sci. Food Agric.*, **37**, 797–803.

SUJARIT, C., THANAGOSES, P. and HANPONGKITTIKUN, A., 1996, Single cell protein production from tuna condensate, in *Proceedings of the 8th Annual Meeting of the Thai Society for Biotechnology*, IV-3, Prachnap Khiri Kan, Thailand.

SULAIMAN, N.M. and SHAFII, A.F., 1987, *Maximising waste utilisation in Palm Oil Mills*, Institute of Chemical Engineers Symposium Series No. 100.

TANTICHAROEN, M., LERTTRILUCK, S., BHUMIRATANA, S. and SUPAJANYA, N., 1986, Biogas production from tapioca starch wastewater, in *Proceedings of a Regional Training Workshop on Energy from Biomass*, Yhonburi, Thailand, pp.523–544.

UCHIDA, Y., HUKUHARA, H., SHIRAKAWA, Y. and SHOJI, Y., 1990, Bio-fish flour, in Keller, S. (ed.) *Making Profits out of Seafood Wastes*, Alaska Sea Grant Programme, Report No. 90-07, Fairbanks, AK, USA, pp.187–195.

VALLE, J.M. DEL and AGUILERA, J.M., 1990, Recovery of liquid by-products from fish meal factories: a review, *Process Biochem.*, **25**, 122–130.

VIKINESWARY, S. and RAVOOF, A.A., 1990, Abstract of decomposition of empty fruit bunches in a sub-surface mulch, *Regional Seminar on Management and Utilisation of Agricultural and Industrial Wastes*, Kuala Lumpur.

WIGNALL, J. and TATTERSON, I., 1976, Fish silage, *Process Biochem.*, **11**, 17–19.

WYATT, B. and McGOURTY, G., 1990, Use of marine by-products on agricultural crops, in Keller, S. (ed.) *Making Profits out of Seafood Wastes*, Alaska Sea Grant Program, Report 90-07, Fairbanks, AK, USA, pp.187–195.

ZADRAZIL, F., 1992, Conversion of lignocellulose into feed with white-rot fungi, in Doelle, H.W., Mitchell, D.A. and Rolz, C.E. (eds) *Solid Substrate Cultivation*, London: Elsevier Applied Science, pp.321–340.

ZADRAZIL, F., OSTERMANN, D. and DEL COMPARE, G., 1992, Production of edible mushrooms, in Doelle, H.W., Mitchell, D.A. and Rolz, C.E. (eds) *Solid Substrate Cultivation*, London: Elsevier Applied Science, pp.283–320.

19

Cleaner Biotechnologies and the Oil Agroindustry

AGUSTÍN LÓPEZ-MUNGUÍA

19.1 Technology and Raw Materials

During the last few decades, biotechnology has been a priority both for developed countries to increase and consolidate their dominant position in the international markets and for underdeveloped countries in their strategies to reach self-sufficiency and sustainability. However, it is now clear that a particular regional strategy in biotechnology must be defined, as there is a biotechnology that allows higher yields of traditional and new agricultural crops, new products or clean technologies; at the same time, biotechnology favours the substitution of raw materials, without necessarily contributing to a more sustainable society. Several cases have illustrated this situation, such as the substitution of cane sugar by corn fructose syrups (Sasson and Costarini, 1991; Santoyo and Muñoz, 1993). Mexico, which has been self-sufficient in sucrose from cane sugar since 1993, with an average production of more than 4 million t/year, now faces not only financial problems but also increasing competition in the market by fructose, both as part of the free trade agreement and as a consequence of dumping strategies of fructose producers. More than 250 000 t of fructose were consumed in 1996. Another example concerns potato production in Mexico and the changes necessary in economic policy, in order to compensate for the negative effects of new technologies (Marks *et al.*, 1991).

It has also during the last few decades become clear that industrial development requires a radical transformation and new paradigms to avoid the risk of irreversibly damaging our ecosystems. From a sustainable balance, living standards of cities in developed countries require up to 200 times their surface to produce the energy they require to degrade their residues, recycle the carbon dioxide produced by human and industrial activities, satisfy their need for food and raw materials, etc. If all countries in the world reach those standards, we would require an Earth three times its present size. Technology in general, and biotechnology in particular, is a valuable tool in the changes required to move into a better world.

In the specific case of the fat and oil industry, we can clearly identify the traditional paradigm, based exclusively in productivity, substitution, business opportunities and export criteria, and the new paradigm, to which the world moves with enormous difficulties in spite of the conclusions of the 1992 meeting in Rio de Janeiro. It is the paradigm of sustainability that has been extensively documented (Verstraete and Top, 1992; Moser, 1992; Lopez-Munguía *et al.*, 1994). However, as in the case of other industrial sectors where biotechnology has had an impact, the fat and oil industry is facing a complex combination of circumstances derived not only from the lack of research and technology in poor countries but also from the acceptance of new biotechnologies by society, as well as some trends in consumption. Actually, substitution has been present since the first technological development. Cotton, for instance, despite the multiplicity of substitutes derived from synthetic fibres, is still appreciated in the textile industry; the pharmaceutical industry has considerably depended on raw materials from plants, and although many active principles are now synthesized in industrial laboratories, plants are still the main source of drugs for people all over the world. In some cases, substitution is avoided or regulated by national trade policies, designed to protect the loss of natural resources at national or regional levels. However, this practice is now unacceptable to those countries following neoliberal policies. In the following sections the impact of biotechnology in production and the situation of the fat and oil industry is described in this context.

19.2 The Market

The fat and oil industry is very dynamic and has suffered important transformations in the last few years. World production increased from 32 million tonnes in 1983 to 80 million tonnes in 1993, mainly at the expense of a growing vegetal fat (shortening) sector which was 60 % of the market in 1963 (Gurtler, 1994). There is an important reduction in animal fat consumption, mainly butter, which was the main source of animal fat third in importance after soybean and palm oil (Moses and Cape, 1991) at the end of the 1980s. In data published by the USDA for the 1997/1998 production cycle, of the total world oil production of 101 million tonnes, fish oil accounted for a little more than 1.3 million tonnes, while 1.8 and 0.5 million tonnes of butter were produced in the EU and USA, respectively.

Soybean has for years been the most important source of oil, in spite of the growth in palm oil and rape-seed production. In 1997/1998, 22.6 million tonnes of soya oil were produced world-wide, 22.3 % of the total oil produced. Palm oil was only 4 % of the market in 1963; it increased to 10 % in 1985 12.4 % by the end of the 1980s. In 1997/1998, with a total production of 17.4 million tonnes, palm oil became the second largest oil produced, and the most-traded vegetable oil in the world, as sales made up roughly 40 % of world oil trade. Nevertheless, the supply has been characterized by instability and competition. The USA, Brazil and Argentina dominate the market of soybean and soya oil, while Malaysia and Indonesia dominate that of palm oil, as a result of aggressive technological policies of companies such as Unilever. At the national level, the effort of countries like Canada, in the case of rapeseed production, are worth mentioning. Now considered a national crop, rapeseed had to be transformed by traditional plant breeding techniques in order to reduce the amount of erucic acid, which is toxic to humans, from 25.5 to only 0.2 %. This transformed rapeseed is known as canola (Abraham and Hron, 1992).

Although the main market is for human consumption, 5.7 million tons of oil are used for cattle feed and 10.6 million tons directed to a growing lipochemical industry (Ceccaldi, 1995). In this industry more than 20 companies participate, although four of them control 75 % of the market: Henkel, Unichema, Oleofina and Akao. It is estimated that 4 million tonnes of oil derivatives are annually produced by the lipochemical industry, including fatty acids, fatty acid methyl esters, fatty alcohols and glycerol (Morin *et al.*, 1994).

19.3 Structure and Application of Fats and Oils

The application of fats and oils is a direct function of their structure; the type and distribution of fatty acids, which are classified according to their size as short chain ($< C_6$), medium chain (C_6–C_{12}) and long chain ($> C_{12}$). The fatty acids are also classified according to the level and position of insaturation: saturated, unsaturated and polyunsaturated. Soya oil is highly unsaturated (24 % of oleic acid $C_{18:1}$ and 54.5 % of linoleic acid $C_{18:2}\omega_6$). Cotton, corn and sunflower oils have similar structures. They are all industrially hydrogenated, transforming linolenic acid ($C_{18:3}\omega_3$) into linoleic ($C_{18:2}\omega_6$), linoleic into oleic acid ($C_{18:1}$) and finally oleic acid into the saturated stearic acid ($C_{18:0}$). This chemical process increases the melting point, a condition required to maintain commercial oil liquid, and increases stability, reducing rancidity and off-flavours produced by oxidation. However, hydrogenated fat is now considered inconvenient due to the formation of *trans* isomers during hydrogenation, and to the requirement for essential fatty acids (Abraham and Hron, 1992).

Coconut oil also has a characteristic composition, as 54 % of the fatty acid present is lauric acid ($C_{12:0}$), highly appreciated for its surfactant properties in the production of soaps and lotions, in addition to applications in the food industry. Only the oil from palm kernel has a similar composition. Other examples of fats and oils highly appreciated by their unique structure are olive oil, with 72.5 % oleic acid, and cocoa butter, with a high stearic acid content and a structure that results in a solid fat that melts at mouth temperature, a desirable characteristic for chocolates.

19.4 Fat and Oil Biotechnology

19.4.1 *Genetic Engineering*

The impact of genetic engineering on agricultural production has been the subject of intense research and intense debate. Nevertheless, since 1986, when the first field trials was conducted, thousands of experiments have been carried out in more than 30 countries with almost any crop; in 1998, sales of genetically engineered seeds were estimated at up to $1.5 billion. About 69.5 million acres were planted with transgenic plants; 74 % in the USA and 10 % in Canada and, although most of the trials were conducted in industrialized countries, 15 % of the area cultivated with transgenics was planted in Argentina (James and Krattiger, 1996; Thayer, 1999). Soybean is the major transgenic crop, accounting for 52 % of the land, followed by corn (30 %) and cotton (9 %). In the particular case of oil crops, genetic engineering techniques allow the possibility of redesigning the metabolic pathway in order to produce specific oils with a given type and distribution of fatty acids. This will be of tremendous impact in industry, particularly if we consider that 95 % of the oil produced world-wide is made of only six fatty acids, while hundreds of other plants have numerous fatty acids with potential interest in industry – such as vernonia, castor bean, flax and jojoba (Battey *et al.*, 1989).

The synthesis of fatty acids and their later incorporation to triglycerides (TG) is carried out in several compartments within plant cells. Carbon chains are produced in the

Table 19.1 Target enzymes for various strategies in the design of transgenic plants with new fatty acid or lipid contents

Strategy	Enzyme overexpressed	Reduce Expression (Antisense)
Increase oil content	Acyl carrier protein Acetyl-CoA carboxylase Acetyl-CoA transacylase ß-ketoacyl-ACP synthase III Malonyl-CoA ACP transacylase	
Increase oil content and palmitate	ß-ketoacyl-ACP synthase I	
Increase stearate and decrease palmitate	ß-ketoacyl-ACP synthase II	
Decrease saturated fatty acids Increase petroselinic acid	Stearoyl-ACP desaturase	
Lower oil content		Acetyl-CoA carboxylase
Increase stearic acid (cocoa butter)		Stearoyl-ACP desaturase
Increase palmitic acid (margarine)		ß-ketoacyl-ACP synthase II
Increase oleic acid		Oleoyl-ACP desaturase
Increase lauric acid (coconut)		Lauroyl-ACP thioesterase (manipulation)

From Parveez *et al.* (1994) and Liu and Brown (1996)

plastids where an enzymatic complex of at least six enzymes (the fatty acid acyl synthase complex), assembles two carbon units up to 16- (palmitate) and 18- (stearate) carbon chains, their primary products. Fatty acids may be transferred to glycerol phosphate by the same plastids, or exported to the cytoplasm, transformed, and finally stored in lipid bodies.

The knowledge of lipid biochemistry, one of the major limiting factors in this field, is rapidly growing as new genes and enzymes are being discovered and isolated, and therefore the strategies and targets for genetic engineering programmes have also increased. In Table 19.1 a review of key enzymes in the production of transgenic plants is presented. For instance, the unstable polyunsaturated linolenic acid has been reduced in transgenic plants using antisense technology to downregulate the activity of the enzyme stearate desaturase. Companies such as DuPont have cloned most of the genes which encode the enzymes of oil biosynthesis, from fatty acid elongation and desaturation through TG assembly. By regulating the expression both up and down they are able to change the fatty acid composition of soybean at will and have products of the type 'low linolenic' (80 % oleic, 3 % linolenic); usually wild soybean has 15–20 % oleic and 8–10 % linolenic acid. The company Calgene also has a portfolio of edible and industrial oils, produced mainly from rapeseed. For instance the high-stearate (40 %) canola oil its useful in the manufacture of non-hydrogenated margarines and spreads. The attraction here is a huge market of vegetable oils used in production of premium margarine, such as corn oil, that are sold at up to US$0.1 dollars per pound over the commodity vegetable oil. The high-stearate rapeseed does not require hydrogenation to have the physical properties of margarine.

It is now clear that the first target of genetic engineering products in agriculture is resistance to herbicides, pests, and infections in general. Now, after the approval by the FDA of the transgenic tomato, AgrEvo, Calgene, and DuPont (InterMountain in Canada) have started to introduce in the fields canola crops with high content of specific fatty acids, such as lauric, stearic, erucic and oleic for the already-mentioned purposes, as well as high content of medium-chain fatty acids for a market requiring fat with reduced calories.

19.4.2   *Fermentation Processes*

Although it is difficult to expect an oil production process derived from fermentation, at least in theory it is possible to obtain a microbial oil derived from bacteria, yeast, mould, or algae (Vázquez-Duhalt and Arredondo, 1991). Yeasts such as *Lipomyces starkei* or *Rhodotorula gracilis* and fungi such as *Aspergillus terreus* or *Mortiella vinacea* may accumulate up to 60–70 % of their dry weight as oil.

19.4.3   *Enzyme Technology*

Several enzymes have been used to ameliorate or modify extraction processes and to design lipids with improved physical and nutritional properties, using raw materials of high availability.

The use of 1,3-specific lipases in acidolysis reactions has resulted in the production of lipids with improved melting behaviour (for application in spreads, non-dairy creams and confectioneries), with improved organoleptic properties (reduced waxiness) and reduced total saturated or non-saturated fat. In particular, structured TGs result from the inter-esterification of medium and long-chain fatty acids, and offer a variety of benefits in nutrition and the treatment of disease. These TGs provide rapid delivery of energy via oxidation and are very useful in clinical nutrition. Medium-chain TGs, containing C_6 to C_{12} fatty acids are not likely to be stored in the adipose tissue of the body but oxidized for energy in the liver, providing a dense source of calories which the body can readily use. However, essential fatty acids are required and therefore a minimum amount of ω_3, ω_6 and ω_9 fatty acids are included. The essential linolenic acid for instance is desaturated and elongated to arachidonic acid, a precursor of important metabolites, and 2–5 % in the diet enhances the immune function, reduces blood clotting and reduces serum TGs (Kennedy, 1991). Fish oils, rich in γ-linolenic and eicosapentanoic acid, are also used in combination with structured TGs in patients administered enteral feeding, and have recently received a lot of attention for their role in nutrition (Daly *et al.*, 1991). Now that butter use has decreased considerably in developed countries following the introduction of 'light milk', we learn about the importance of butyric acid, as the main source of energy for the colonic epithelium, accounting for about 70 % of total energy consumption. Butyric acid is obtained from milk fat but is also a primary metabolite of fibre, and may help to explain the advantage of high-fibre diets in the prevention of intestinal cancers (Smith and German, 1995).

A product developed by Nabisco Foods and commercialized by Pfizer as 'Salatrim' is becoming popular as a fat substitute. It combines short-chain fatty acids (acetic, propionic and butyric) with fewer calories per unit weight and the long-chain fatty acid (stearic), only partially absorbed. The result is a TG with reduced energy content (Kosmark, 1996).

An important market for another lipase product is the substitution of cocoa butter. Unilever has developed a process in which a TG profile very similar to that of cocoa butter is obtained from palm oil and stearic acid in n-hexane, after acidolysis. Similar results have been obtained by Calgene using a mixture of cotton and sunflower oil, after transesterification with lipase. The actual market of the substitute is 225 000 t per year, representing US$500 million. This means that less and less chocolate is now made from cocoa, and more and more from rapeseed and other oil seeds. From its aztec origins, only the name remains.

Another area of application has been the use of enzymes to improve or modify the agroindustries involved in oil extraction, recently reviewed by Rosenthal *et al.* (1994) and Domínguez *et al.* (1994). Different hydrolytic enzymes are used in pretreatments to enhance oil extraction, for instance of soybean and sunflower (Domínguez *et al.*, 1993), or in aqueous extraction processes, avoiding the use of organic solvents, in the case of sunflower kernels (Domínguez *et al.*, 1995), avocado (Buenrostro and López-Munguía, 1986) and coconut (Cintra *et al.*, 1986), or oleoresins from flowers (Rubio *et al.*, 1990).

19.5 The Coconut Industry: A Case Study

Our group, among others, has been involved development of in a cleaner enzyme technology for coconut oil extraction (Cintra *et al.*, 1986). The advantages of using enzymes have long been recognized as higher yields and fewer by-products are obtained. The product obtained after the enzymatic treatment of an aqueous solution of coconut copra does not require refining because it is of high quality.

The type and concentration of enzymes to use was the subject of a particular study, as the enzyme cost is the main deterrent for the application of enzymes in this field. Therefore the process was further optimized in order to reduce water consumption and recycle the enzymes to reduce their impact on production costs (Barrios *et al.*, 1990). It was also found that, as in other vegetable extraction processes, the type of enzyme depends not only on the fruit structure but also on the state of maturity at which the fruit is processed. For the particular case of coconut, the enzymatic process has several advantages: the product can be treated fresh or dried, avoiding the nutritional damage that takes place when sun dried; the investment in equipment is low – only a mill, a mixing tank and a centrifuge are required. Actually, this process was scaled up in an expeller extraction coconut plant.

The coconut industry faces an economical crisis derived from multiple factors that clearly illustrate some of the threats and opportunities already mentioned. In Table 19.2 an overall picture of this agroindustry is presented. The situation in the Philippines has been recently reviewed (Manicad, 1995). Mexico produces around 100 000 t of coconut oil per year, but the amount imported from the Philippines increases each year, including non-refined oil. The situation in both countries as far as agronomic and agroindustrial conditions is concerned is bad, but is worst in the Philippines where 30 % of the agronomic surface is used in coconut production, satisfying 70 % of the international market. Coconut also faces increasing competition from palm oil, which renders the highest yield per unit area of all vegetable crops, and its growth in the Asian agroindustries has been supported by Unilever, particularly in Malaysia and Indonesia. In 1998, a massive programme for production of palm oil was announced as part of the 'Alianza para el Campo' governmental project. Will it be the beginning of the end of this agroindustry?

Table 19.2 Problems faced by the agroindustries involved in coconut production and oil extraction

- Use of traditional varieties with enormous losses due to infections such as lethal yellowing and the viroid disease *Cadamg-Cadamg*. Most of the coconut palms are old, and the introduction of resistant varieties such as the dwarf coconut palm has been slow
- Progressive decrease in crop yields
- Deficient production structure, controlled by the processor with most of the labour force falling below the poverty line
- Lack of a national policy on agriculture and trade. The market is influenced by preferential trade agreements and lower-priced substitutes (e.g. palm kernel)
- The industry in general has not been modernized. Coconut copra is either smoke or sun dried, with high rates of oxidation and risks of aflatoxin contamination. Oil obtained by extraction in expellers is of low quality, requiring intense refining processes

19.6   Fat Substitutes

Finally, in any discussion of factors affecting this industry, it is not possible to ignore the dramatic changes in the beliefs and controversies surrounding the role of oil and fat in human health and nutrition. It is a fact that fat has become the first enemy in the diet of people in developed countries, as well as in the most-favoured sectors of the under-developed countries. There is enormous concern about saturated and unsaturated fat, about individual fatty acids and their configuration and about serum cholesterol. But little attention is paid to the beneficial aspects of fat. It has been forgotten that fat is a necessary part of our diet, and that it is the *amount* and *type* of fat we are eating that are important.

As we have mentioned, some oil plants are being designed by genetic engineering with these objectives in mind. However, the main stream of fat substitutes in the market are derived from the industrial sector using chemical and biotechnological techniques, to compete for a growing market of around US$800 million in 1996. In fact, to more than a dozen fat substitutes already approved, the FDA has just added the first substitute of oil that can be used in frying: Olestra, produced by Procter & Gamble. It is a sucrose molecule esterified with fatty acids in all its hydroxyl groups. This product behaves like a TG, but is not recognized by lipases during digestion. It was found that the product dissolves liposoluble vitamins and carotenes, and that it causes stomach cramps if consumed in excess. However, the product is safe, according to the FDA, and the public is 'protected' by saturating Olestra with the vitamins the organism may lose (Klis, 1996). In the same category is the already-mentioned 'Salatrim', and 'Caprenin' (a structured TG combining capric C_{10}, caprylic C_8 and behenic C_{22} fatty acids), both with a low caloric content (5 instead of 9 kcal mol^{-1}). Other substitutes include glucans obtained from yeast cell wall and texturized proteins. The increased consumption of these products will affect without doubt the evolution of the oil markets.

19.7   Conclusions

The situation of the fat and oil sector is strongly affected by the actual biotechnological advances on one hand, combined with the trends in consumption, globalization and free trade policies on the other. Although most of the information of future plans in agriculture is now in the hands of private companies and in many cases kept confidential, several changes are possible and probable.

- The introduction into the market of margarine obtained by enzymatic processes or directly from oil seeds, substituting the chemical process. Saturated fats will be produced directly from transgenic plants, avoiding the catalytic process of hydrogenation.

- The substitution of cocoa butter will continue, using products obtained by enzymatic processes, but in the long term fat directly obtained from oil seeds will reach the market.

- Some private companies with patent protection will start to gain control of this sector. This is already the case with DuPont and Calgene, which have transgenic plants already on the market and patents covering important properties and parts of their genome.

- Some products and technologies may face a dark future. Such is the case of coconut, threatened by competition from palm kernel oil. However, soon the competition will be strengthened by the production of oil rich in lauric acid from transgenic plants. In terms of economics, this may not be important as coconut accounts for only 4–5 % of the market, but this is an evaluation made by our actual economic system, losing the social perspective.

- Unconventional fatty acids are being introduced into oil crops to meet specific non-food demands. This is the case of lauric acid, but also of high-erucic acid oil, which serves as liquid wax required by the chemical industry. Petroselinic acid, an isomer of oleic acid with the double bond in ω_6, can be converted by oxidation into lauric and adipic acids, with their own specific markets (Liu and Brown, 1996). Ricinoleic and vernolic acids are also industrially appreciated.

- Although we have not mentioned the application of oil in the production of biofuels, in the future TGs may also find an enormous market in the production of methyl esters of fatty acids that may be used as biocarburants and for lubrication.

- In underdeveloped countries there is little or no control at all in the introduction of transgenic products from developed countries. To the various threats one has to add the involuntary introduction of modified genomes to the local biodiversity, as may happen to corn in Mexico.

We are therefore forced to reflect on the technological model and market trends in which we are immersed and their consequences in the short and long term. Modern biotechnology may be used to conduct our industries in several directions. The case of enzyme technology in the development of clean technologies is an example: as new and more efficient enzymes become available, non-sustainable and contaminating processes may be substituted by biological alternatives, reducing the risk of environmental damage. This is the case in some oil extraction and oil modification processes described here. As stated by Hinkelamert, man cuts the tree branch on which he sits: he does it quickly and efficiently. He gives the example of cars, constructed to go faster and to be more comfortable. However, we live in traffic jams and contamination, wondering each time if a bicycle wouldn't be better. It is important therefore to insist on the enormous opportunities biotechnology offers to increase yields, to develop new products, to solve problems and to promote regional products and key strategies to face the twenty-first century. But the same tools may result in more dependence, loss of regional products and traditional technologies and more poverty. We should not forget that many of our hopes are placed in science and technology, to which we have to give back the sense, social and human aspects they have lost before we find ourselves in a traffic jam.

References

ABRAHAM, G. and HRON, R.J., 1992, Oilseeds and their oils, in Hui Y.H. (ed.) *Encyclopedia of Food Science and Technology*, Vol. 3, New York: John Wiley & Sons, pp.1901–1946.

BARRIOS, V.A., OLMOS, D.A., NOYOLA, A. and LÓPEZ-MUNGUÍA, A., 1990, Optimization of an enzymatic process for coconut oil extraction, *Oeagineux*, **45**, 35–42.

BATTEY, J.F., SCHMID, K.M. and OHLROGGE, J.B., 1989, Genetic engineering for plant oils: potential and limitations, *Trend Biotechnol.*, **7**, 122–126.

BUENROSTRO, M. and LÓPEZ-MUNGUÍA, A., 1986, Enzymatic extraction of avocado oil, *Biotechnol. Letts*, **8**, 505–506.

CECCALDI, P., 1995, Huiles Vegetables: Pour une valorisation industrielle, *Biofutur*, **September**, 16–18.

CINTRA, M.O., LÓPEZ-MUNGUÍA, A. and VERNON, C.J., 1986, Coconut oil extraction by a new enzymatic process, *J. Food Sci.*, **51**, 695–697.

DALY, J.M., LIEBERMAN, M., GOLDINE, J., SHOU, J., WEINTRAUB, F.N., ROSATO, E.F. and LAVIN, P., 1991, Enteral nutrition with suplemental with arginine, RNA and ω_3 fatty acids, *J. Parent. Ent. Nutr.* **15** (Suppl.).

DOMÍNGUEZ, H., NÚÑEZ, M.J. and LEMA, J.M., 1993, Oil extractability from enzymatically treated soybean and sunflower: range of operational variables, *Food Chem.*, **46**, 277–284.

DOMÍNGUEZ, H., NÚÑEZ, M.J. and LEMA, J.M., 1994, Enzymatic pretreatment to enhance oil extraction from fruits and oilseeds: a review, *Food Chem.*, **49**, 271–286.

DOMÍNGUEZ, H., NÚÑEZ, M.J. and LEMA, J.M., 1995, Aqueous processing of sunflower kernels with enzymatic technology, *Food Chem.*, **53**, 427–434.

FRIEND, G., 1996, The New Bottom Line, *Los Angeles Times*, February 26.

GURTLER, J.L., 1994, Marché mondial des oléagineux. Production et consommation dans le monde. *Oleagineux Corps gras lipides*, **3–1**, 185–188.

JAMES, C. and KRATTIGER, A.F., 1996, *Global Review of the Field Testing and Commercialization of Transgenic Plants: 1986–1995*. Ithaca, USA: ISAAA AmeriCenter.

KENNEDY, J.P., 1991, Structured lipids: Fats of the future. *Food Technol.*, **45**, 76–83.

KLIS, J.B., 1996, FDA approves fat substitute, Olestra. *Food Technol.*, **50**, 124.

KOSMARK, R., 1996, Salatrim: properties and applications, *Food Technol.*, **50**, 98–101.

LIU, K. and BROWN, E.A., 1996, Enhancing vegetable oil quality through plant breeding and genetic engineering, *Food Technol.*, **50**, 67–71.

LOPEZ-MUNGUÍA, A., WACHER, R.C., ROLZ, C. and MOSER, A., 1994, Integración de Tecnologias indígenas y biotecnologías modernas: Una utopía?, *Interciencia*, **19**, 177–182.

MANICAD, G., 1995, Biotechnology and the Philippine Coconut Farmers, *Biotechnol. Dev. Monitor.*, **23**, 6–10.

MARKS, L.A., KLEIN, K.K. and KERR, W.A., 1991, Efectos Económicos de la Biotecnología: la Industria Mexicana de la Papa, in Alter R. Jaffe (ed.) *Análisis de impacto de las biotecnologías en la Agricultura*, San José, Costa Rica: Instituto Interamericano de Cooperación para la Agricultura, IICA, pp.131–187.

MORIN, L., DROUNE, Y. and REQUILLART, V., 1994, La demande non alimentaire des huiles et graisses, *Oleagineux Corps gras lipides*, **3–1**, 188–191.

MOSER, A., 1992. Ecological biotechnology, the new dimension in technology, *Acta Biotechnol.*, **12**, 69–78.

MOSES, V. and CAPE, R.E., 1991, *Biotechnology. The Science and the Business*, New York: Harwood Academic Publishers.

PARVEEZ, G.K., CHOWDHURY, M.K. and SALEH, N.M., 1994, Current status of Genetic Engineering in oil bearing crops. Asia-Pacific, *J. Mol. Biol. Biotechnol.*, **2–3**, 174–192.

ROSENTHAL, A., PYLE, D.L. and NIRANJAN, K., 1994, Aqueous and enzymatic processes for edible oil production. Review, *Enzyme Microb. Technol.*, **19**, 402–420.

RUBIO, H.D., BARZANA, G.E. and LÓPEZ MUNGUÍA, A., 1990, Mexican Patent 176018.

SANTOYO CORTÉS, H. and MUÑOZ RODRÍGUEZ, M., 1993, *Alternativas para el Desarrollo Agroindustrial*, México: Universidad Autónoma de Chapingo.

SASSON, A. and COSTARINI, V., 1991, *Biotechnologies in Perspective*, France: UNESCO.

SMITH, J. and GERMAN, B., 1995, Molecular and genetic effects of dietary derived butyric acid, *Food Technol.*, **45**, 87–90.

THAYER, A.M., 1999, Transforming agriculture, *Chem. Eng. News*, **77**, 21–35.

VÁZQUEZ-DUHALT, R. and ARREDONDO VEGA, P.O., 1991, Microalgas: fuente de aceites comestibles y terapéuticos, *Biotecnología*, **6**, 19–33.

VERSTRAETE, W. and TOP, E., 1992, Holistic environmental biotechnology, in Fry, J.C., Gadd, G.M., Herbert, R.A., Jones, C.W. and Watson, I.A. (eds) *Microbial Control of Pollution*, Vol. 40, Society for General Microbiology, Cambridge University Press, pp.1–17.

In Search of Novel and Better Bioinsecticides

ARGELIA LORENCE AND RODOLFO QUINTERO

20.1 Introduction

The human population is facing a great challenge to produce more food at a faster rate but in a sustainable way. In our discussion here, 'sustainable' means to have methods of production that protect the environment and the biodiversity and also keep natural resources for future generations. In this context, it is important to analyse the international situation: the world population keeps growing and it is estimated that the mark of 8000 million people will be reached by the year 2005. Most of the increase, 2500 million, will occur in developing countries, and nearly 85 % of the total will be living there. In order to cover the food needs, agricultural production will have to be doubled around year 2025; however hunger will still leave about one billion malnourished people.

During 1996 in Mexico the situation of food production was as follows: we imported 10 million tonnes of basic grains, 7 million of which were corn and 250 000 beans, which represents an expenditure of $3 billion. This is a worrying prospect, not only due to the lack of rain in many agricultural sectors of the country but also because it indicates structural problems in the Mexican agriculture. In 1994 the FAO constructed a simulation model to estimate domestic food demand to the year 2010 (FAO-México, 1994); the model takes into account all uses of agricultural products (direct human food, industrial uses, animal feed, etc.) and considers that the NGP will grow at 4.8 % while population will increase at 1.9 % between 1989 and 2000 and 1.6 % in the period 2000–2010 (Table 20.1). In these conditions the total internal demand for agricultural products will grow annually 2.6 % in the period 1989–2010, which represents an annual increase of 0.4 % of calories available for the population. We can anticipate today that these estimations will never occur, because the growth of the NGP will be smaller than estimated. Even though the FAO's model indicates that future growth of demand will be smaller than in the last three decades, the demand will grow faster than the domestic production, as shown in Table 20.2.

For these reasons the only way to cover the food demand will be through food imports. The yearly rates of imports will be high: 6.7 % for wheat, 5.2 % for rice and 3.9 % for corn. The agricultural commercial deficit will grow 57 and 49 % for the periods 1989–2000 and 2000–2010, respectively.

Table 20.1 Future demand for food products in Mexico (thousand t/year)

Product	1988–1990	2010	Increase
Rice	786	1 349	563
Corn	16 395	23 675	7 280
Wheat	4 731	7 613	2 882
Beans	1 384	1 523	139
Barley	597	1 451	854
Sorghum	7 851	12 582	4 731
Vegetable oils	1 248	2 414	1 166
Milk	8 988	13 759	4 771
Bovine meat	1 931	3 129	1 198
Eggs	1 058	1 727	669
Pork meat	835	1 750	915
Bird meat	768	1 793	1 025

Table 20.2 Trends, future demand and importation of foods products: annual growth rate, Mexico

Product	Domestic demand			
	Historical trends		Projection 1988–2010	Importation 1988–201
	1961–1990	1980–1990		
Wheat	5.1	2.1	2.3	6.7
Corn	3.4	2.7	1.8	3.9
Rice	3.6	3.0	2.6	5.2
Sorghum	11.2	0.0	2.3	2.3
Cereals	4.8	1.8	2.1	4.0
Meats	5.3	3.5	3.1	0.9
Total foods	4.8	2.6	2.7	3.0
Total (foods/other products)	4.6	2.6	2.6	3.0

For several years domestic production has been increased by using new agricultural methods developed by adoption of technology of the type known as 'Green Revolution' and also by the opening of large areas for irrigation, which at the same time had important negative environmental impacts. The contamination related to agricultural inputs has been growing mainly due to the extended use of pesticides. These products have grown 5 % annually, from 14 000 t in 1960 to 60 000 t in 1986, as a result of the diffusion of this type of technology and also due to the fact that many pathogenic agents have developed resistance to chemical products and new pests have appeared. From 1988 to 1992 the import of plaguicides to Mexico increased from 30 000 t to 60 000 t, but how much of this is applied domestically, and the amount of pesticides produced and applied in Mexico, is not known (Ortega *et al.*, 1994).

It is thus necessary to increase food production but how could it be done keeping the environment and the biodiversity safe and conserving the natural resources for the future? A satisfactory answer faces problems of large technological complexity: agricultural productivity must increase but at the same time consumption of agrochemicals should decrease. This goal can be reached at least partially through

- replacing traditional agrochemicals (insecticides, fungicides, herbicides, etc.) by novel and less toxic products, susceptible to short-term biodegradation in the environment;
- national policies to reduce the input of agrochemicals;
- support of research and development directed towards integrated pest management.

In this chapter, we will describe some of the most interesting aspects of the most widely commercial used bioinsecticide, *Bacillus thuringiensis* (*Bt*), because it is a good model to understand the problems of replacing toxic agrochemicals with biodegradable products. The areas that will be covered are:

- the search for novel *Bt* activities against different kind of insects;
- study of the mechanism of action of the *Bt* δ-endotoxins;
- generation of transgenic plants containing the δ-endotoxins *Bt* genes (*cry* genes)
- development of new systems for use of *Bt* toxins.

20.2 Bioinsecticides based on *Bt*

Bacillus thuringiensis is a Gram-positive soil bacterium that produces insecticidal crystal proteins (ICP), also known as δ-endotoxins, during the sporulation process; these are toxic to the larvae of a number of destructive insect pests in forestry and agriculture. The spores and the crystals produced by this bacterium were discovered at the beginning of the twentieth century, and commercial products based on it were used until 1956 in France. However, the δ-endotoxins (also called cry proteins), the main constituent of the crystals, have recently attracted the attention of several research groups and also of some international companies for two reasons: δ-endotoxins with novel activities have been discovered, and the arrival of molecular biology to agriculture has generated transgenic plants containing the *cry* genes.

In 1990 it was reported that *Bt* represents around 95 % of the total biopesticides market, its sales volume being $260 million. It has been estimated that the demand for these products will grow at least 10 % annually until the end of the twentieth century. The principal producers of *Bt* biopesticides are Abbott, Sandoz, Du Pont and Novo Nordisk. The market distribution by regions is: USA and Canada 50 %, Orient 18 %, China 10 %, 8 % Central and South America, rest of the world 14 % (Bravo and Quintero, 1993).

If we follow the historical development of the discovery of δ-endotoxins with novel activities we can see that its growth rate has been practically exponential since the 1980s and today we can foresee that with new biological methodologies available more toxins will be found with novel activities. Particularly it must be pointed out that *Bt* toxins are not toxic to humans and other mammals after 40 years of use and this characteristic has made them very attractive because it is not necessary to carry out long and costly clinical trials to probe its lack of toxicity and, at the same time, the past experience of commercial use makes easier and faster the introduction of similar products.

At the present time several *Bt* strains have been identified with different insecticidal activities; most of them are toxic to lepidopteran larvae (caterpillars), but there are also strains with toxic activity agains dipterans (flies and mosquitoes), coleopterans (beetles), mites and ants (hymenopterans), platyhelminthes (flat worms), protozoans (amoebae) and nematodes (roundworms) (Feitelson, 1993). It is estimated that the list of susceptible organisms will keep growing and this area, the search of toxins with novel activities, is the principal goal of many research groups around the world, in academia as well as in private companies (Bravo *et al.*, 1995b).

20.3   Mode of Action of *Bt* δ-Endotoxins

The elucidation of the mechanism of action of the *Bt* toxins is, besides the search for novel activities, the big area of interest in the field of the *Bt* bioinsecticides. This knowledge will allow the design of more potent molecules or manufacture of molecules with a novel and/or higher spectrum of action. The current knowledge about the mode of action of the *Bt* δ-endotoxins is illustrated in Figure 20.1.

It is known that cry proteins accumulate within the cell, forming crystals during the sporulation process. In the crystals, the proteins are immature products (protoxins); in order to be active the crystal must be solubilized in the extreme pH conditions of the insect midgut and in the same place also must be activated by proteolytic processing. The mature products or toxins bind to specific receptors in the brush border membrane of the midgut columnar cells where, after a substantial conformational change caused by interaction with the receptor, it is inserted into the membrane and, presumably by oligomerization, forms pores. These pores disturb the endogenous permeability properties of the target membrane. After that, there is a massive entry of water into the cells, causing the destruction of the epithelial tissue and finally the death of the larvae (Yamamoto and Powell, 1993).

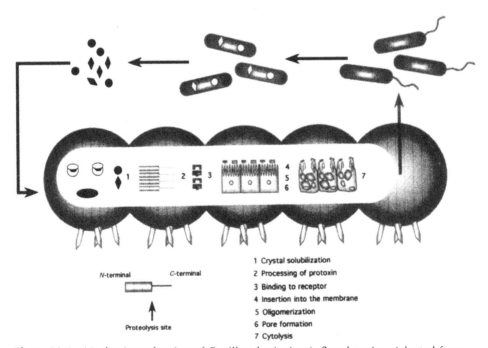

1 Crystal solubilization
2 Processing of protoxin
3 Binding to receptor
4 Insertion into the membrane
5 Oligomerization
6 Pore formation
7 Cytolysis

N-terminal   *C*-terminal

Proteolysis site

Figure 20.1   Mechanism of action of *Bacillus thuringiensis* δ-endotoxins. Adapted from Bravo *et al.*, 1995b, by permission

The putative receptor has recently been isolated and characterized. It has been reported that the Cry1Ac receptor in *Manduca sexta*, *Heliothis virescens* and *Lymantria disparis* is an aminopeptidase N that is linked to the cell membrane by a glycosyl–phosphatidylinositol anchor.

Using several methods *in vitro* to know with more detail the effect of the δ-endotoxins in the membrane that contains the receptor, several authors have suggested that the primary action of the cry proteins in the membrane is to induce some endogenous proteins to allow the flux of ions through the membrane – in other words, to open an ionic channel selective to cations, or forms (alone or with the receptor) a cationic channel (Lorence *et al.*, 1995). Some δ-endotoxins have been studied incorporating them into planar bilayers formed by synthetic lipids, where it has been shown that these toxins could form ionic channels *per se*. The ionic channels formed by these proteins allow the preferential flux of monovalent cations; other important characteristic of these channels is that apparently they are formed by one or more molecules, because its conductance (the property that describes how much resistance must an ion overcome to pass through the channel) is very variable (500–4000 pS). It is important to point out that in order to observe these channels experimentally micromolar concentration of toxins have to be used, while the effect of them in the apical membrane of the epithelial cells occur at nanomolar or picomolar concentration. This fact suggests that the receptor reduces the toxin concentration required for the protein or the protein–receptor complex to insert and form pores into the membrane (Bravo *et al.*, 1995a). There is still an open question: do the channels observed in the black lipid bilayers in the absence of the receptor correspond to the channels formed by the cry proteins *in vivo*?

At least 60 different *cry* genes have been described and classified considering the similarity of the amino acid sequence in 19 different groups (1, 2, . . . , 19) and subgroups (A, B, . . . , etc.) (Crickmore *et al.*, 1997). Doing alignments of the primary structure of the cry protein toxic regions it has been possible to identify five very conserved regions (blocks 1, 2, 3, 4 and 5) that are separated from others of low similarity and variable length.

20.4 Structure and Function of δ-Endotoxins

The tridimensional structure determined by X-ray diffraction studies of the toxic portion of one of these proteins, Cry3A, has shown that it is organized in three domains, to each of which a specific function has been assigned. Domain I (residues 1 to 290) is constituted of six amphipathic helices surrounding another hydrophobic one, α-helix 5; these are possible constituents of a structure that forms pores. Domain II (residues 290–500) is formed by three β-antiparallel layers that end in loops in the vertex of the molecule; this conformation is called β-prisms. The region that forms the loops is the less conserved (hypervariable) region; the interchange of fragments of the hypervariable region between three closely related proteins allows interchange of specificities, and for this reason the domain is associated with the interaction with the receptor. Domain III is composed of β-sheet arrays in the form of a β-sandwich (this topology is called double helix β); toxins with mutations in this domain are very unstable to protease treatment, and the role of this domain is to be responsible for the structural stability of the whole molecule (Li *et al.*, 1991). It is important to note that the five conserved regions are in the central part of the molecule, and it is proposed that the cry proteins have a common conformation and similar mechanism of action. This proposal was corroborated recently when the structure

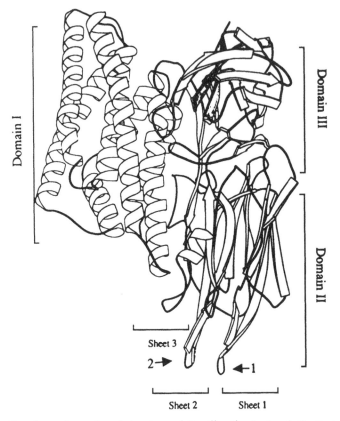

Figure 20.2 The three-dimensional structure of *Bacillus thuringiensis* Cry3a toxin. Domain I contains seven α-helixes, six of them arranged the seventh around, α-helix 5. Domain II is formed by three β-sheets (loops 1 and 2 are denoted by arrows). Domain III has a β-sandwich structure with a topology typical of double β-helix

of another member of the cry family Cry1Aa, was published (Grochulski *et al.*, 1995). There is only one significant difference between Cry3A and Cry1Aa: the loops of domain II of the latter toxin are longer. This structural information supports the idea that the hypervariable region is a determinant of the different specificity of each one of these toxins; Figure 20.2.

There is evidence that Domain I, specifically block 1 (this region corresponds to the α-helix 5), is involved in pore formation. Mutants of Cry1Ac in α-helix 5 were equally effective as the wild type when binding to the receptor in brush border membrane vesicles (BBMV) obtained from the midgut of three susceptible insects (*Manduca sexta*, *Heliothis virescens* and *Trichoplusia ni*); however, the activity in bioassays of some of these mutants decreased dramatically (between 10 and 1000 times) when proline or other charged residues were introduced. It was proved that these mutants lose the capacity to inhibit the leucine transport dependent on the K^+ gradient in the BBMV; this is an indirect measure that indicates that these proteins are capable of altering the vesicle membrane permeability. Other results that support the hypothesis that the central helix is a structural component of the pores were obtained when peptides corresponding to the α-helix 5 of Cry3A and Cry1Ac were synthesized; both peptides maintain the capacity of the whole molecule to form ionic channels selective to cations in black lipid bilayers. In the literature it is proposed that at least one α-helix of Domain I (α-helix 7) can also participate as an structural component of the pore. This possibility has not been eliminated, although the evidence available suggests that α-helix 7 could participate in addition to the β-1 of Domain II as a bridge of communication between these two domains to allow the molecule to make the conformational change neccesary for the toxin to pass from a soluble conformation to a state that will insert into a hydrophobic environment, which is present in biological membranes.

During experiments of site-directed mutagenesis, where one or more amino acid residues present in the protein structure are substituted by others, it has been confirmed that the regions determinant of the high specificity of each one of the toxins are those corresponding to the loops in the apex of the molecule (Domain II). When this region in proteins Cry1Aa, Cry1Ab, Cry1C and Cry3A was mutated, it was observed that the affinity of the mutant proteins to the receptor are substantially changed.

Three of the five regions of conserved amino acids in the cry family are localized in Domain III (blocks 3, 4 and 5). The evidence obtained from several authors suggests that preservation of the molecule's integrity depends on the maintenance of the globular structure of this domain. This structure depends on the salt bridges between the positively charged residues of block 4 and its negative counterparts in block 5.

The results of experiments in which fragments of Cry1Aa and Cry1Ac were interchanged and of the interactions between these chimeric proteins and the receptor obtained from *Lymantria dispar* suggest that some residues of Domain III also participate in the interaction with the receptor. It is not known if this is a general function of the domain in the *Bt* toxins.

It is proposed that when δ-endotoxins form pores in the membrane they cause the death of the epithelial cells because they disturb the system that mantains the pH gradient. This inactivation is the result of the alteration in membrane permeability because the target membrane is less permeable to cations. When the motive force that mantains the gradient of 1000 times more protons in the cytoplasm than in the lumen is disturbed, the cytoplasm becomes alkalinized and this disturbs the cellular metabolism, causing the destruction of the intestinal tissue. Once this physical barrier is destroyed, the spores obtain access to the haemolymph, where they proliferate because there is plenty of nutrients. The larva dies due to inanition and septicaemia.

20.5 Transgenic Plants Resistant to Insects

Since the begining of agrobiotechnology in 1983, *Bt cry* genes have been identified as key elements in order to increase the agricultural productivity, decreasing the use of agrochemicals (Fraley, 1994). Many important crops and different types of plants have been genetically transformed, and in many cases *Bt* genes have been introduced. The main results are reported below.

During the period 1987–1995, 2261 field trials were done with transgenic plants in the USA, in 7095 different locations. From that total, 23.1 % correspond to insect-resistant plants; 27.8 % to herbicide resistance; 26.8 % to improved characteristics; 11.5 % to virus resistance and 2.9 % fungus resistance.

In the USA from 1993 to 1996, 17 transgenic plants were approved for widespread use and production, six of which were resistant to insects (lepidoptera and coleoptera) due to the introduction of the δ-endotoxin *Bt* genes (Table 20.3).

Table 20.3 Global status of applications for the commercialization of transgenic crops with *Bt* insect resistance

Country	Crop	Year of approval (company)
Argentina	Corn/maize	P (Ciba)
	Corn/maize	P (Monsanto)
	Corn/maize	P (Northrup King)
Australia	Cotton	P (Monsanto)
Canada	Corn/maize	1996 (Mycogen-Ciba)
	Potato	1996 (Monsanto)
European Union	Corn/maize	P (Ciba)
	Corn/maize	P (Monsanto)
Japan	Corn/maize	P (Ciba)
	Corn/maize	P (Northrup King)
	Potato	P (Monsanto)
Mexico	Cotton	1996 (Monsanto)
	Corn/maize	P (CIMMYT)
	Potato	1996 (Monsanto)
USA	Corn/maize	1995 (Ciba)
	Corn/maize	1996 (Northrup King)
	Corn/maize	1996 (Monsanto)
	Cotton	1995 (Monsanto)
	Potato*	1995 (Monsanto)
	Potato	1996 (Monsanto)

P = pending; * resistance to coleopterans by a *Bt* toxin
Source: James and Krattiger, 1996

Now the methodology for plant transformation of dicotyledons is available and there have been important improvements in the transformation of monocotyledons.

In Mexico the International Center for Improvement of Wheat and Corn has been developing, with financial support from the United Nations, a project to obtain corn varieties with resistance to insects. In Feburary 1996, the first transgenic corn was field-tested, containing the *cry*1Ab gene of *Bt*. This field trial was very important because Mexico is the centre of origin of corn, and the possibility exists of genetic flux between transgenic plants and its ancestor, the teocintle; in this case it was neccesary to plan and design a very careful trial and to obtain the approval of the National Agricultural Biosafety Committee of the Secretary of Agriculture. Also it was neccesary to assess the possible implications of introducing transgenic corn in productive areas where teocintle could exchange genetic material with transgenic corn.

20.6 Novel Systems using *Bt*

Recently, the use of *Bt* genes and its proteins has increased, because several research groups have developed novel systems to use and produce them. For example:

- introduction of *Bt* genes in *Pseudomonas fluorescens* (CellCap technology);
- introduction of *cry* genes into algae to control *Anopheles* mosquitoes (malaria);

- introduction of *Bt* genes into endophytic bacteria in order to control sucking insects (aphids);

- production of *Bt* toxins in baculoviruses as an alternative expression system.

All these efforts are based on the fact that cry proteins are recognized as non-toxic to humans, mammals and other commercially important species (Quintero, 1996). It also means that *Bt* will be used against a wide variety of pests that affect not only agriculture but also other areas such as human health (e.g. in destruction of mosquitoes that transmit malaria).

20.7 Concluding Remarks

We believe that the discovery of novel *Bt* activities and strains will continue because there are many insects, mostly in the developing world, that are not effectively controlled with the known cry toxins. Our knowledge about the structure and mode of action of *Bt* toxins will help us to increase the speed of discovery of new cry proteins.

The great challenge of this research field is to avoid or at least delay insect resistance, and that is why it is necessary to improve the understanding of the mechanism of action of cry toxins in order to develop new strategies to face this problem. The increasing use of transgenic plants makes this challenge bigger.

References

BRAVO, A. and QUINTERO, R., 1993, Importancia y potencial de *Bacillus thuringiensis* en el control de plagas. Regional Office of FAO for Latin America and the Caribbean, Santiago, Chile.

BRAVO, A., CERÓN, J., ARANDA, E., LORENCE, A. and QUINTERO, R., 1995a, Screening of *Bacillus thuringiensis* strains with novel insecticidal activities, in Feng T.-Y. *et al.* (eds) *Bacillus thuringiensis Biotechnology and Environmental Benefits*, Vol. 1., Taiwan: Hua Shiang Yuan Publishing Co, pp.87–103.

BRAVO, A., LORENCE, A. and QUINTERO, R., 1995b, Biopesticides compatible with the environment: *Bacillus thuringiensis* a unique model, *Biocontrol*, 1, 41–55.

CRICKMORE, N., ZEIGLER, D.R., FEITELSON, J., SCHNEPF, E., LERECLUS, D., BAUM, J., *et al.*, 1997, *Bacillus thuringiensis* delta-endotoxin nomenclature, www site: http://epunix/biols.susx.ac.uk/Home/Neil_Crickmore/Bt/index.html

FAO-MÉXICO, 1994, *La agricultura en México perspectivas al año 2010*, México, D.F.

FEITELSON, J.S., 1993, The *Bacillus thuringiensis* family tree, in Kim, L. (ed.) *Advanced Engineered Pesticides*, USA: Marcel Dekker, Inc, pp.63–71.

FRALEY, R.T., 1994, The contributions of plant biotechnology to agriculture in the coming decades, in Krattiger A.F. and Rosemari A. (eds), *Biosafety for Sustainable Agriculture*, Cambridge: International Service for the Acquisition of Agri-biotech Applications/Stockholm Environment Institute, pp.3–28.

GROCHULSKI, P., MASSON, L., BORISOVA, S., PUSZTAI-CAREY, M., SCHWARTZ, J.L., BROUSSEAU, R. and CYGLER, M., 1995, *Bacillus thuringiensis* CryIA(a) insecticidal toxin: crystal structure and channel formation, *J. Mol. Biol.*, 254, 447–464.

JAMES, C. and KRATTIGER, A.F., 1996, *Global Review of the Field Testing and Commercialization of Transgenic Plants: 1986 to 1995*, Briefs No.1, Ithaca: The International Service for the Acquisition of Agri-biotech Applications.

LI, J., CARROLL, J. and ELLAR, D.J., 1991, Crystal structure of insecticidal δ-endotoxin from *Bacillus thuringiensis* at 2.5. A resolution, *Nature*, 353, 815–821.

LORENCE, A., DARSZON, A., DÍAZ, C., LIÉVANO, A., QUINTERO, R. and BRAVO, A., 1995, δ-Endotoxins induce cation channels in *Spodoptera frugiperda* brush border membranes in suspension and in planar lipid bilayers, *FEBS Lett.*, **360**, 217–222.

ORTEGA, C.J., ESPINOSA, T.F. and LOPEZ, C.L., 1994, El control de riesgos para la salud generados por los plaguicidas organofosforados er Mexico: retos ante el tratado de libre comercio, *Salud Publica de Mexico*, **36**, 624–632.

QUINTERO, R., 1996, Biotecnología para la agricultura, in Solleiro, J.L., Del Valle M.C. and Moreno E. (eds), *Posibilidades para el Desarrollo Tecnológico del Campo Mexicano*, Vol. I., México, D.F.: Instituto de Investigaciones Económicas/Programa Universitario de Alimentos/Centro para la Innovación Tecnológica and Editorial Cambio XXI, pp.259–275.

YAMAMOTO, T. and POWELL, G.K., 1993, *Bacillus thuringiensis* Crystal Proteins: Recent advances in understanding its insecticidal activity, in Kim, L. (ed.) *Advanced Engineered Pesticides*, USA: Marcel Dekker, Inc, pp.3–42.

Bacillus thuringiensis: Relationship Between *cry* Genes Expression and Process Conditions

REYNOLD R. FARRERA AND MAYRA DE LA TORRE

21.1 Introduction

Integrated pest management (IPM) is becoming widespread in current agricultural practice, due to the ever-increasing resistance to chemical insecticides shown by insect populations. IPM implies the combined use of different strategies, among them the biological control.

Bacillus thuringiensis (*Bt*) is the most important bioinsecticide: in 1990 the world agrochemical market was US$26 000 million. Biological products cut a share of 0.5 % of that market and 90–95 % of those sales corresponded to *Bt* products (Feitelson *et al.*, 1992). Nowadays, *Bt* research is primarily focused on its molecular biology, genetics and isolation of new strains.

21.2 Mode of Action and Specificity of *Bt* δ-Endotoxins

Bt is a pathogen for early larval stages of lepidoptera, diptera and coleoptera as well as some adult coleoptera, although there are some varieties with reported activities against nematodes (Edwards *et al.*, 1992), lice (Gingrich *et al.*, 1974) and mites (Payne *et al.*, 1993). The bacteria produce a parasporal crystal during sporulation and both the spore and parasporal crystal are ingested by the larvae. The proposed infection model is based on research performed on lepidoptera larvae and is as follows: the ingested crystal is dissolved by the strong alkaline medium of the insect intestine (pH 10) and the protein is partially processed by digestive proteases to protease-resistant and highly toxic polypeptides known as δ-endotoxins.

The δ-endotoxins (53 and 78 kDa) bind to specific receptors on the cell membranes of midgut pileous epithelium through a surface-recognition effect. One or two specific gut epithelium receptors are supposed to exist for each *Cry* toxin. The experimental evidence gathered on the lysis mechanism is explained using as a model the tridimensional structure elucidated for *CryIIIA* and more recently for *CryIA(a)* where three domains are identified. Toxins bind to the receptors with a bundle of three antiparallel β-sheets (Domain II).

This binding causes a conformational change in the protein in which Domain I is inserted into the membrane and forms an osmotic pore at the binding site (Li *et al.*, 1991; Grochulski *et al.*, 1995). The subsequent cell death and lysis leads to the formation of extensive tears on the epithelium and contamination of haemolymph with the gut content. This causes an increase in haemolymph pH by which all the nerve cells lacking a myelin sheet and in contact with it are inhibited, resulting in paralysis, first of the intestine and later in general. Descent of gut pH allows spore germination; the bacilli reproduce and emigrate through the epithelial wounds to the haemocoel and then to all the organs, causing septicaemia (Andrews *et al.*, 1987).

After a period, which can extend from 24 h to 7 days depending on the insect species, its physiological state and the toxin dose consumed, the paralysed larvae die of starvation, dehydration or septicaemia. If the ingested dose is sublethal the larvae will show defective development.

Some lepidoptera are insensitive to isolated crystals. The reason for this resistance is an intestinal pH inadequate for the dissolution of the crystal or deficient processing of the protoxin by digestive proteases, which renders the toxin inactive. Therefore, solubility of the crystals is important for the biological activity of *Bt* (Aronson *et al.*, 1986).

It has been reported that the appearance of resistance to *Bt* by populations of pest insects implies a selection of individuals, which present a mutation-mediated deactivation of the cry-specific receptors in their ephitelium membranes. As commercial *Bt* strains tend to have several cry proteins in their crystals, the simultaneous deactivation of the various receptors is very improbable (Van Rie *et al.*, 1990; McGaughey and Whalon, 1992).

Similarly, it has been demonstrated that the range and intensity of toxicity shown by a given *Bt* strain depends on the amino acid source used to grow the microorganism (Dulmage *et al.*, 1990). Our research group has suggested that it could be due to a phenomenon of differential regulation of the *cry* genes, appearing when different substrates are used. This would provoke a variable proportion of crystal-composing proteins which would in its turn modify the solubility and toxicity of the crystals (Aronson, 1994).

21.3   Molecular Biology of *Bt*

In 1989 Höfte and Whiteley proposed a classification of *cry* genes, based on their biological activity, comprising four major groups: *cryI*, *cryII*, *cryIII* y *cryIV*. Later, other groups called *cryV*, *cryVI*, *cytA* y *cytB* were added. Table 21.1 shows this gene classification and some properties of their encoded proteins. *Cry* proteins require partial alkaline proteolysis to be transformed into the active toxins. Protoxins of 130–140 kDa (*CryI* and some of the *CryIV*) are processed into 65–70-kDa toxins, while those of 70–75 kDa (*CryII*, *CryIII*) are processed into 60–65-kDa toxins (Table 21.1) (Aronson, 1993). Recently another more extensive classification based on the gene sequence has been proposed (Crickmore *et al.*, 1998).

Extrachromosomal DNA is common in *Bt*. Approximately 10–20 % of total DNA is found in plasmids and lineal elements of molecular weight 5 mDa to more than 200 mDa (Lecadet *et al.*, 1981; Aronson, 1993). The *cry* genes are regularly located in high-weight plasmids of more than 200 mDa and frequently are multiple copied. The presence of closely related pairs of genes in the same plasmid increases the probability of naturally occurring recombinations. Thus, *CryIA(b)* toxin is thought to be the result of intercrossing of *cryIA(a)* and *cryIA(c)* genes.

Table 21.1 *cry* Genes and their products

Genes	Type of crystal	Protoxin (kDa)	Toxin (kDa)	Specificity
***cryI* group**: *cryIA(a)*, *cryIA(c)*, *cryIA(d)*, *cryIB*, *cryIC(a)*, *cryIC(b)*, *cryID*, *cryIE(a)*, *cryIE(b)*, *cryIF*, *cryIG* *cryIA(b)*	Bipyramidal	130–140	65–70	Lepidoptera
***cryII* group**: *cryIIA*, *cryIIB*, *cryIIC*	Cubic	65–70	60–65	Lepidoptera and Diptera
***cryIII* group**: *cryIIIA*, *cryIIIB*, *cryIIID*, *cryIIIE*	Flat Square	73	60–70	Coleoptera
cryIIIC	Bipyramidal	129	72	
***cryIV* group**: *cryIVA*, *cryIVB*, *cryIVC*, *cryIVD*	Amorphous	130 77 72	72	Diptera
***cryV* group**: *cryVA(a)*, *cryVA(b)*, *cryVB*, *cryVC*				Coleoptera and Lepidoptera
***cryVI* group**: *cryVIA*, *cryVIB*				Nematodes

Protoxin genes are usually flanked by one or more insertion sequences or transposons, which is the probable cause of the observed mobility of these genes (Green *et al.*, 1989; Carlson and Kolsto, 1993). There is evidence that *Bt israelensis* presents cluster transposition of its *cry* genes although these genes are regularly stable (Aronson, 1993).

The strain *Bt kurstaki* HD-1 contains at least 12 classes of extrachromosomal DNA, including a lineal element and a transducer phage. This strain has four out of its five *cry* genes in a 110-mDa plasmid (*cryIA(a)*, *cryIA(c)*, *cryIIA* y *cry IIB*) and the fifth in a 44-mDa unstable plasmid (*cryIA(b)*). Aronson (1993) studied the occurrence of this last plasmid in HD-1 populations and found that around 30 % of the colonies have lost it. As a result the plasmid-lacking cells formed a parasporal crystal with a different composition and specificity than those formed by the plasmid-containing cells (Aronson, 1994).

The *Bt* parasporal crystal can account for up to 25 % of the sporulated cell's dry weight. To produce such a crystal the cell has to synthesize some 1–2 million δ-endotoxin molecules during its stationary growth phase. *Bt* posses several mechanisms to achieve this remarkable cry protein overexpression and accumulation (Agaisse and Lereclus, 1995):

- Expression of *cry* genes from strong promoters known as BtI and BtII, which are expressed sequentially during sporulation. An exception to this is the gene *cryIIIA*, which is expressed from vegetative growth to the eighth hour after the end of the exponential phase.

- Presence of multiple copies of *cry* genes in plasmids, which increases the *Bt* capacity to produce *Cry* proteins.
- Extended mRNA stability, with a half-life of about 10 min, compared with an average of 3 min for other proteins.
- Crystallization of recently synthesized cry proteins, which stands as a mechanism to prevent their degradation by endogenous proteases.

21.4   Production of *Bt*

21.4.1   *Industrial Processes*

Bt have been produced using two fundamental technologies: semisolid and submerged fermentations.

Semisolid fermentation

Mechalas (Rowe and Margaritis, 1987) patented a procedure in which the semisolid mixture is prepared directly in the fermentation vessel. This was a non-hermetic metallic tank with a perforated bottom. The approximate composition of the medium was 54 % wheat bran, 38 % expanded perlite, 5 % soybean paste and 3 % dextrose, mixed with Ca(OH)$_2$, NaCl and CaCl$_2$. The mixture, heated directly with steam for 60 min to reduce the bacterial load, was inoculated and its moisture adjusted to 50–70 % and pH to 6.9. Loads of 250 kg were used and non-sterile air, humidified to 95–100 % saturation at 30–34°C, was injected at a rate of 0.4–0.6 vvm (volume units of air at standard temperature and pressure per volume unit of fermented solid per minute) during the first 3 h and then at a rate of 1–1.2 vvm. The course of the culture was controlled modifying the temperature, moisture and air flow volume. The fermentation time used was 36 h, after which dry air was injected to 4 % residual moisture of the solids. The dried mixture was milled and the resulting powder used to formulate the insecticide. This process yielded 3×10^9 to 17×10^9 spores per gram of dried mix. The semisolid fermentation was the first technology used to produce *Bt*, but has been abandoned in favour of the submerged fermentation.

Submerged fermentation

Several descriptions of this production method have been published, as well as some patents (Dulmage, 1971; Goldberg *et al.*, 1980; Salama *et al.*, 1983; Arcas *et al.*, 1984; De Urquijo, 1987; Dulmage *et al.*, 1990).

In the typical process the total nutrient concentrations are 10 % w/v. Media have been formulated with 1–7 % of carbohydrates (w/w), 0.5–5 % of organic nitrogen sources and minerals, among which the main ones are potassium (up to 0.3 %) and phosphates (up to 0.15 %), plus calcium, magnesium, manganese, cobalt, zinc, iron and copper salts. Organic nitrogen sources such as corn-steep liquid, fish meal, blood meal, corn gluten, brewer's yeast, casein and cottonseed meal have been used (Salama *et al.*, 1983; De Urquijo, 1987).

Medium is sterilized in a steam heat exchanger then cooled and injected to stirred-tank fermentors with a typical volume of 12 000 gallons. Subcultures are minimized to avoid plasmid loss and inoculum rates of 2–5 % v/v are used. Our research group have found that after three to four subcultures in liquid medium at an initial pH of 7.0, or after seven steps at an initial pH of 7.4, the colonies of strains HD-1 and HD-73 of *Bt kurstaki* may show lytic plaques and the bacilli may present atypical sygmoidal, elongated or spiral shapes.

Process temperature is controlled at 28–32°C and pH between 6.8 and 7.4. Fermentation end is set at 85–95 % sporulation. Batch process fermentation time ranges between 24 and 36 h, and in certain cases may extend to 72 h. Spore concentrations of 10^8 and 10^9 spores per ml of broth have been obtained.

Depending on the producer and product presentation, processes differ in the separation and purification steps. In the case of aqueous concentrates, the fermented broth is concentrated and washed by centrifugation, and then the product is formulated, standardized, dosaged and packed in plastic or metal vessels. Our group has demonstrated that filtering with a filter aid is an economical alternative (Villafaña *et al.*, 1996). If the product is an oil-based concentrate, the washed and centrifuged cream is mixed with an oil-based suspension during formulation. If the presentation is a dry wettable powder the broth is concentrated and washed by centrifugation, preformulated with preservatives, humidifiers and dispersants and spray-dried. The powder is used for formulation and packed. Diluents, UV protectors, insect attractors, adhesives, flow enhancers and other additives are used for final formulation.

Submerged fermentation shows several advantages over semisolid fermentation. The handling of an axenic culture, a better control of fermentation conditions, better spore and crystal yields and bioinsecticide formulation flexibility (Rowe and Margaritis, 1987).

According to Dulmage *et al.* (1990) the industrial fermentation method of choice is a batch culture, with the following advantages:

- plasmid losses are minimized
- the appearance of acrystalliferous and asporogenic mutants, and cell lysis due to phages, are reduced (Besaeva *et al.*, 1986)
- it is relatively simple to achieve sporulation synchronization because the lack of nutrients triggers sporulation
- product biological activity is higher than that obtained with other culture systems.

21.4.2 *Experimental Production Alternatives*

At the laboratory and pilot plant scales, three alternatives have been studied for submerged *Bt* production: extended-batch, fed-batch and continuous culture. The efficiency of these systems has been measured as spore production although no relationship between high spore counts and toxicity has been demonstrated. Additionally, the ways in which the culture evolves through the vegetative, transition and sporulation stages have been shown to influence the *Cry* yield (Aronson and Dunn, 1985). A brief summary of each option is presented.

Extended-batch fermentation

Roy and co-workers (1987) studied the plasmid stability of *Bt kurstaki* (HD-1) using extended-batch cultures: the bacterium was batch-grown for 4 h to permit vegetative growth, then half of the working volume was drained and the rest was amended with fresh medium to the original volume and subjected to repeated cycles of growth and draining. The product was spread on nutrient agar plates, left to sporulate and then screened for Spo⁻Cry⁻ phenotypes. The plasmid profile remained constant for 328 h of cultivation and the loss of sporulation and crystal production capability was independent of plasmid loss. They suggested that the crystal production impairment was associated with a defective sporulation. Selinger and co-workers (1988), using a similar culture system and strain, found that from 10 to 25 cycles of 2 h, and an additional 38-h sporulation stage, bacilli with crystal and spore predominated, but after 25 culture cycles the crystal and spore-producing capability was rapidly lost and asporogenic and acrystalliferous strains prevailed. They compared these cells with batch-produced ones and found that the plasmid patterns were similar, even after 25 culture cycles. They concluded that *Bt* does not lose plasmids after prolonged culture but does lose sporulation and crystal production capabilities due to a defective sporulation process, which affects the crystal formation.

Fed-batch fermentations

This culture is a batch in which the specific growth rate is controlled by feeding fresh concentrated media to a vessel, which slowly increases its working volume. Constant, variable or intermittent feedings can be used. The advantages of this method are minimization of substrate or catabolic inhibition effects and achievement of high cell concentrations. An increase in cell concentration could imply an increase in *Cry* concentration. Our team has performed fed-batch experiments with *Bt kurstaki* HD-73, in which concentrations of 2.7×10^{10} spores ml^{-1} have been achieved in 23.5 h. In some of our fed-batch experiments, glucose accumulated and the biomass yield based on substrate indicated that the bacterium used amino acids as another carbon source (Rodríguez and de la Torre, 1996).

Fed-batch *Bt* fermentations do not seem to favour *Cry* protein production. Kang *et al.* (1992), working with *Bt kurstaki* HD-1 under an intermittent feed method, found that high glucose concentration rendered high biomass growth but low sporulation and concluded that in fed-batches, fast growth enhanced sporulation. Avignone-Rossa and Mignone (1993) compared simple batch and fed-batch fermentations of *Bt israelensis* and found that, in spite of similar biomass and spore concentrations, the toxicity of the fed-batch product was much lower. They concluded that fed-batch fermentations produce high spore counts but diminish notably the associated toxicity and assumed that the toxic component synthesis is affected if an exponential growth is not achieved.

Continuous culture

With continuous culture it is possible to maintain a microbial population in a steady state and manipulate the specific growth rate and physiological state of the microorganism. In contrast, prolonged culture periods of *Bt* favour the most energy-efficient mutants, which frequently are not the overproducers, as well as the appearance of acrystalliferous mutants (Dwivedi *et al.*, 1982). Blokhina and co-workers (1984) found that in continuous culture *Bt* dissociates into R ('Rough') and S ('Smooth') variants, with significant differences in their sporulation, toxin synthesis and phage-resistance capabilities. Some authors (Ackermann and Smirnoff, 1978; Besaeva *et al.*, 1986) have stated a high phage lysis susceptibility of *Bt* strains. Khovrychev *et al.* (1986) analysed the phage titre during continuous culture of *Bt galleriae*, and found it to be constant, in the range of 10^6–10^7 ml^{-1}, which does not cause an extent lysis of the culture.

Several studies have been performed on *Bt* continuous culture in staged systems with at least one stage dedicated to vegetative growth and another to sporulation. Khovrychev *et al.* (1986) used a two-stage culture of *Bt galleriae*, with decreasing dilution rates (D_1/D_2 of 4) and a limitation of glucose. They kept the culture for 15 days without asporogenic mutant selection. Nevertheless only 20–30 % of the bacilli sporulated. As the total residence time in both stages was only 20 h and the microorganism required 27 h to complete its cycle, many cells could be eluted out from the system before being able to sporulate. Later they used a three-stage system with dilutions $D_1:D_2:D_3$ of 8:2:1, which was kept up to 11 days, but not in a steady state. With this system they obtained 30 % sporulation in the second stage and 80 % in the third. If dilution at the third stage was reduced to obtain a total residence time of 60 h, sporulation ratio decreased due to spore germination. As an interesting fact, they mentioned the finding of crystal-containing cells without spore. The R and S variant ratio and the phage titre remained constant while increasing the number of stages (Khovrychev *et al.*, 1990).

Our group performed continuous cultures of *Bt kurstaki* HD-73 in a single-stage system with different dilution rates. Steady states were achieved in all cases and three different metabolic states were observed, according to biomass yield based on glucose and microscopic examination. Between D of 0.18 and 0.31 h^{-1} vegetative cell, sporulating bacteria and free spores coexisted while both glucose and amino acids from the medium were consumed. With a D between 0.42 and 0.47 h^{-1} glucose was the main carbon source and only vegetative cells were observed. At a D of 0.5 h^{-1} the biomass yield based on substrate decreased, indicating a change in metabolism. When varying the glucose concentration in the feed from 20 to 30 g l^{-1} the D value interval in which only vegetative cells existed and mainly glucose was consumed broadened to 0.24 h^{-1} and 0.45 h^{-1} (Rodríguez and de la Torre, 1996).

21.4.3 *Factors Affecting Cry Production*

Commercial production of *Bt* is performed with culture media formulated with complex nutrient sources. Nevertheless, spore counts, toxic potency and toxicity range of the product have been reported to depend on the culture medium (Dulmage, 1971; Smith, 1982; Dulmage *et al.*, 1990; Yudina *et al.*, 1992). Insecticidal activity of the crystal produced by the same *Bt* strain can be affected by culture medium and operation conditions. Besides, a high cell growth does not always ensure an elevated *Cry* protein production or an increased insecticidal activity. In a classic work, Dulmage (1971) reported a great variability in *Bt* culture toxicity when using two different varieties of the bacterium and two different media, one based on tryptone and corn meal and the other based on bactopeptone and cottonseed meal. The toxicity of the same strain changed from 206 to 680 international units per millilitre.

Our research group has found that *Cry* protein yield is affected by the carbon:nitrogen ratio and the oxygen availability during the different culture phases. The mechanism through which these effects take place remains to be elucidated but certain phenomena could be part of the overall explanation – for example, competition of the different *cry* genes, among themselves and with other sporulation genes for the transcription and translation elements; the appropriate onset of the metabolic fluxes in the functional routes during sporulation; the availability of energy; the adequate balance between reduced and oxidized coenzymes.

Carbon source

Bacillus thuringiensis can use a variety of carbon sources: glucose, fructose, maltose, ribose, sucrose, lactose, starch, whey, soybean oil, glycerol, organic acids and amino acids. Glucose is by far the most appropriate either for growth or for sporulation (Smith, 1982; Arcas *et al.*, 1984; De Urquijo, 1987). The lack of assimilable carbohydrates produces a defective sporulation in most cases (Nickerson and Bulla, 1974). There are experimental results indicating that the change in the carbon source affects the biological activity and the morphology of the crystals obtained from the same strains, possibly due to alteration on the rate of synthesis of the different *Cry* proteins (Yudina *et al.*, 1992).

A deficiency of glucose in the medium without an alternative carbon source may trigger sporulation. Alternatively, glucose present at the end of fermentation can provoke a dysynchronization of sporulation (De Urquijo, 1987). Rajalakshmi and Shetna (1977) reported that in batch process sporulation is totally inhibited by the presence of cystine along with glucose and other nutrients. Our research group has confirmed that sporulation can be triggered by glucose or amino acid deficiencies.

Nitrogen source

To obtain good sporulation and parasporal crystal formation in *Bt*, free amino acids have to be supplied in the medium (Nickerson and Bulla, 1974; Egorov *et al.*, 1984). Its absence provokes a delayed sporulation and a low yield of protoxin, with consequent low biological activity (Goldberg *et al.*, 1980; Dulmage, 1981). The role of these free amino acids is not clear (Conner and Hansen, 1967). Probably they fulfil the role of rapid sources of nitrogen in early stages of fermentation, besides the fact that they can be used as carbon sources (Egorov *et al.*, 1982; Sakharova *et al.*, 1984, 1985). Amino acids used for crystal synthesis are not taken directly from the medium but come from the turnover of structural proteins of the sporangium (Monro, 1961).

Sakharova *et al.* (1985) reported that *Bt* uses as a carbon source first amino acids, then glucose and later amino acids again and observed a diauxic growth pattern. Anderson (1990) also reported a diauxic growth. Our research team has not found evidence of diauxic growth or sequential assimilation of glucose and amino acids in *Bt kurstaki* HD-1 and HD-73; on the contrary, simultaneous consumption of glucose and amino acids have been observed (Martínez, 1994).

The fact that protein-rich media can inhibit sporulation has been stated. Pearson and Ward (1988) have even recommended protein-rich media for inocula development, where sporulation is undesirable. Egorov *et al.* indicated that protease production and spore production in several strains seem to be regulated by nitrogen (Egorov *et al.*, 1982, 1984).

Carbon:nitrogen ratio

Anderson (1990) suggested a relationship between carbon:nitrogen ratio and *Cry* production. He found that in a media with low C:N ratio, the biomass yield based on glucose (Yx/s) was higher than 1, suggesting an extensive use of amino acids as carbon and energy sources. At high C:N ratios, glucose was not depleted at the end of the fermentation and biomass yield decreased. Finally, in balanced media (C:N of 7.5:1) Yx/s and glucose consumption achieved maximal values. Additionally he found that, when using oxygen transfer coefficients (k_La) close to $120\,h^{-1}$ the above-described influence of the C:N ratio was negligible and suggested that the relative importance of balancing the medium to obtain high *Cry* production depended on the existence or not of good oxygenation. Similarly, Dulmage *et al.* (1990) insisted on the need for balancing carbon and nitrogen sources to avoid pH lower than 5.8 in the middle fermentation or higher than 8 at its end, which can inhibit growth or dissolve crystals.

Specific growth rate

Maximum specific growth rates reported for several *Bt* strains vary from 0.29 to 1.9 h^{-1} with the most common value close to 0.8 h^{-1} (Rodríguez *et al.*, 1991). Our research group has found that a high duplication rate of vegetative cells is counterproductive for *Cry* production, possibly due to accentuated plasmid loss. Alternatively, it might be that under high growth rates the cell reaches a physiological condition in which the metabolic effort is directed to reproduction and not to accumulation of reserve materials. These materials, like polyalcanoates and, more specifically, poly-β hydroxybutyrate (PHB), apparently act as the source of energy for sporulation and *Cry* synthesis. With this scheme, high specific growth rates would lead to cultures with high spore count but low toxicity.

Oxygen

Oxygen uptake by *Bt* can be very complex (Flores *et al.*, 1997). It has been repeatedly reported that high aeration rates are essential for spore and toxin formation (Dulmage, 1981; Ignatenko *et al.*, 1983; Pearson and Ward, 1988; Anderson, 1990).

Rowe and Margaritis (1987) mentioned that during transition and sporulation phases the oxygen demand is only 30 % of that used in the exponential phase. In contrast, Lüthy *et al.* (1982) suggested that an increase of oxygen consumption could be expected during the transition from vegetative to sporulation phases because the trycarboxylic acid cycle becomes fully active. Arcas (1985) stated a proportionality between sporulation and oxygen transfer coefficient ($k_L a$).

Flores *et al.* (1997), in our laboratory, found a linear correlation between *Bt* spore production and $k_L a$ and scaled up the process from 1 to 1000 litres, with constant $k_L a$ as scale-up criterion. In the biggest fermenter, the specific growth rate of the culture decreased but productivity and biomass yield (based on substrate) remained in the same magnitudes (2.2×10^{11} spores l^{-1} h^{-1} and 0.406 g cell/g glucose respectively), due to a decrease in the fermentation time.

Anderson (1990) found that in vegetative phase, the bacterium grew independently from $k_L a$ in a range of 5–120 h^{-1} and the glucose metabolism was directed to lactate and acetate accumulation. Nevertheless, the effect of oxygen demand was evident in the transition and sporulation phases in which oxygen demand increased compared to the vegetative phase. Crystal protein production and biomass increased with $k_L a$, with a maximum at the highest tested value (120 h^{-1}).

Arcas *et al.* (1984) found a correlation between initial glucose concentration, aeration and spore yield. If initial glucose was elevated to 56 g l^{-1} and the medium was vigorously shaken to maintain a dissolved oxygen of 20 % of saturation level or more, spore counts of 7.36×10^9 spores ml^{-1} for batch and up to 1.2×10^{10} spores ml^{-1} for fed-batch, could be obtained (Arcas *et al.*, 1987).

It is probable that the high oxygen transfer prevents excessive accumulation of organic acids produced by glucose catabolism, increasing the fluxes of the first part of the trycarboxylic acid cycle, and therefore pH does not decrease drastically. Nevertheless, the correlation can be more complex. Results obtained in our laboratory suggested a relation-ship between oxygen availability, storage of PHB and *Cry* protein production (Martínez, 1994). Wakisaka *et al.* (1982) reported a correlation between PHB granule appearance and δ-endotoxin formation in *Bt*. They found that at low potassium concentrations ($11 \mu g \ ml^{-1}$) no δ-endotoxin was formed but large granules of PHB accumulated. At medium concentrations ($60-100 \mu g \ ml^{-1}$) either PHB granules or δ-endotoxin could be found in sporulated cells, but total biomass decreased. Finally, at high potassium values (over $100 \mu g \ ml^{-1}$) PHB was totally consumed and δ-endotoxin production was normal.

Potential of hydrogen (pH)

Optimum pH for *Bt* growth is found 6.8–7.2. Nevertheless, the use of substrates during fermentation and metabolite accumulation provokes a remarkable pH variation in the medium. In the early stages and during the vegetative growth, the predomination of glycolysis and the accumulation of lactate, pyruvate and acetate causes a descent in pH. If 5.8 is passed, a set of pH-mediated problems occur, such as growth inhibition, bacilli malformations and defective sporulation. During transition, fatty acids are consumed and pH goes up. During sporulation these acids are depleted, PHB is metabolized and amino acids are catabolized through the trycarboxylic acid cycle. The amino compounds excreted by the microorganism into the medium raise the pH even more. At the end of fermentation, it can go to 8.3 or higher.

These spontaneous pH variations have to be carefully controlled during commercial fermentations, because crystal productivity can be negatively affected either by low-pH-mediated growth inhibition, which reduces cell multiplication or impairs sporulation, or by high pH, which can cause the dissolution of the crystal if pH approaches 9. Smith (1982) found that a final pH in the range of 5.3 to 8.1 does not affect the level of protoxin produced.

Rowe and Margaritis (1987) mentioned results of several researchers, asserting that an increase of glucose in the medium without buffering or any other pH control lowers the pH during vegetative growth and can provoke a growth inhibition when glucose exceeds $8 \ g \ l^{-1}$. On the other hand, if buffering is employed an increment of glucose, accompanied by increased agitation, would lead to an increased spore yield and toxicity. Additionally, it has been suggested that elevated aeration enhances the acid metabolite consumption and avoids the excessive pH descent (Foda *et al.*, 1985).

Temperature

Bt grows between 20°C and 42°C, with an optimum for endotoxin production at 28–32°C, and an optimum for growth around 30°C. High temperatures prevent endotoxin synthesis and favour selection of acrystalliferous and asporogenic mutants by plasmid loss (Ignatenko *et al.*, 1983; Rowe and Margaritis, 1987).

Limiting nutrient

Several works suggest that the nutrient which limits the growth can have an effect on *cry* gene expression. Liu and Bajpai (1995) modified the Arcas medium from its yeast extract limitation to a medium where carbon or both carbon and organic nitrogen were limiting. They obtained an increase in the final cell density, total amount of protein and insecticidal potency of *Bt* var. *kurstaki* HD-1 products. Previously, Sakharova *et al.* (1984) reported that the nature of the limiting nutrient influenced mainly the spore quality and yield of *Bt* var. *galleriae*.

21.5 Conclusions

The different *Bacillus thuringiensis* varieties produce a wide range of *Cry* insecticidal proteins, which exemplifies the progressive adaptation of the microbe to the ecological niche constituted by insects. The bacteria have developed several strategies to massively overproduce the endotoxins: for example, multiple copies of *cry* genes, strong promoters, extended half-life mRNAs and the crystallization process to resist proteases. The insect-icidal *Bt* preparations have to include viable spores and sufficiently toxic crystals to ensure the biological activity against the selected target insects. To achieve this, from the bioprocess point of view, it is necessary to bear in mind all the problems arising from the very varied strains and those arising from the process conditions including media formu-lation. Phenomena which depend on the strain include the instability of *cry*-genes-carrying plasmids, the possibility of recombination among those genes, the optimum copy number of each gene, the regulation of *cry* gene expression, the assembly of the parasporal crystal and its influence on the crystal solubility, the origin of the Spo⁻Cry⁻ phenotype appearing in the cultures, and the prioritization of sporulation over parasporal crystal formation.

Medium composition and process conditions should direct all of those phenomena to the overproduction of *Cry* and good cell growth. It is paramount to obtain an acceptable understanding of the genetics and molecular biology of *Bt* and correlate those findings with fermentation parameters. In such a case it would be possible to optimize the process to desired objectives, such as production of a parasporal crystal with special character-istics, like a constant proportion of cry proteins or specific solubility.

References

ACKERMANN, H.W. and SMIRNOFF, W.A., 1978, Recherches sur la lysiogenie chez *Bacillus thuringiensis* et. *B. cereus*, *Can. J. Microbiol.*, **24**, 818–826.

AGAISSE, H. and LERECLUS, D., 1995, How does *Bacillus thuringiensis* produce so much insect-icidal crystal protein? *J. Bacteriol.*, **177**, 6027–6032.

ANDERSON, T.B., 1990, Effects of carbon:nitrogen ratio and oxygen on the growth kinetics of *Bacillus thuringiensis* and yield of bioinsecticidal crystal protein, unpublished MSc thesis, The University of Western Ontario, Canada.

ANDREWS, R.E. JR., FAUST, R.M., WABIKO, H., RAYMOND, K.C. and BULLA, L.A. JR., 1987, The biotechnology of *Bacillus Thuringiensis*, *Crit. Rev. Biotechnol.*, **6**, 163–232.

ARCAS, J., 1985, Producción de bioinsecticidas, unpublished PhD thesis, Universidad Nacional de la Plata, Argentina.

ARCAS, J., YANTORNO, O., ARRARÁS, E. and ERTOLA, R., 1984, A new medium for growth and delta-endotoxin production by *Bacillus thuringiensis* var. *kurstaki*, *Biotechnol. Lett.*, **6**, 495–500.

ARCAS, J., YANTORNO, O. and ERTOLA, R., 1987, Effect of high concentration of nutrients on *Bacillus thuringiensis* cultures, *Biotechnol. Lett.*, **9**, 105–110.

ARONSON, A.I., 1993, The two faces of *Bacillus thuringiensis*: insecticidal proteins and post-exponential survival, *Mol. Microbiol.*, **7**, 489–496.

ARONSON, A.I., 1994, Flexibility in the protoxin composition of *Bacillus thuringiensis*, *FEMS Microbiol. Lett.*, **117**, 21–28.

ARONSON, A.I. and DUNN, P.E., 1985, Regulation of protoxin synthesis in *Bacillus thuringiensis*: Conditional synthesis in a variant is suppressed by D-cycloserine, *FEMS Microbiol. Lett.*, **27**, 237–241.

ARONSON, A.I., BECKMAN, W. and DUNN, P., 1986, *Bacillus thuringiensis* and related insect pathogens, *Microbiol. Rev.*, **50**, 1–24.

AVIGNONE-ROSSA, C. and MIGNONE, C., 1993, Delta-endotoxin activity and spore production in Batch and Fed-batch cultures of *Bacillus thuringiensis, Biotechnol. Lett.*, **15**, 295–300.

BESAEVA, S.G., MIKHAILOV, A.A., PETROVA, T.M., TUR, A.I. and BYSTROVA, E.V., 1986, Spontaneous induction of bacteriophage in *Bacillus thuringiensis*, *Microbiology (Russia)*, **56**, 816–818.

BLOKHINA, T.P., SAKHAROVA, Z.V., IGNATENKO, N., RABOTNOVA, I.L. and RAUTENSHTEIN, Y.I., 1984, Variability in *Bacillus thuringiensis* under various growth conditions, *Microbiology (Russia)*, **53**, 427–431.

CARLSON, C.R. and KOLSTO, A.B., 1993, A complete physical map of a *Bacillus thuringiensis* chromosome, *J. Bacteriol.*, **175**, 1053–1060.

CONNER, R.M. and HANSEN, P.A., 1967, Effects of valine, leucine and isoleucine on the growth of *Bacillus thuringiensis* and related bacteria, *J. Invert. Pathol.*, **9**, 12–18.

CRICKMORE, N., ZIEGLER, D.R., FEITELSON, J., SCHEPF, E., VAN RIE, J., LERECLEUS, J. et al., 1998, Revision of the nomenclature for the *Bacillus thuringiensis* Pesticidal crystal proteins, *Microb. Mol. Biol. Rev.*, **62**, 807–813.

DE URQUIJO, N.E., 1987, Producción de *Bacillus thuringiensis* para el control de ciertas plagas de importancia agrícola y médica en México, unpublished MSc thesis. CINVESTAV-IPN, México.

DULMAGE, H.T., 1971, Production of the spore delta endotoxin complex by variants of *Bacillus thuringiensis* in two fermentation media, *J. Invert. Pathol.*, **18**, 385–389.

DULMAGE, H.T., 1981, Production of bacteria for biological control of insects, in Papavizas, G.C. (ed.) *Biological Control of Crop Production*, vol. 5, Beltsville symposia in agricultural research, Totowa, NJ: Allanheld, Osman and Co., pp.129–141.

DULMAGE, H., YOUSTEN, A.A., SINGER, S. and LACEY, L.A., 1990, Guidelines for production of *Bacillus thuringiensis* H-14 and *Bacillus sphaericus*. Geneva: UNDP/WORLD BANK/WHO Special Programme for Research and Training in Tropical Diseases.

DWIVEDI, C.P., IMANAKA, T. and AIBA, S., 1982, Instability of plasmid-harbouring strain of *E. coli, Biotechnol. Bioeng.*, **24**, 1465–1468.

EDWARDS, D.L., PAYNE, J. and SOARES, G.G., 1992, Isolates of *Bacillus thuringiensis* having activity against nematodes, US Patent number 5,093,120.

EGOROV, N.S. and LORIYA, ZH.K. and YUDINA, T.G., 1982, Effect of various carbon sources and purine nucleotides on the synthesis of exoprotease by *Bacillus thuringiensis, Microbiology (Russia)*, **51**, 43–47.

EGOROV, N.J., LORIYA, ZH.K. and YUDINA, T.G., 1984, Influence of aminoacids on the synthesis of exoprotease by *Bacillus thuringiensis, Appl. Biochem. Microbiol.*, **19**, 481.

FLORES, E.R., PÉREZ, F. and DE LA TORRE, M., 1997, Scale-up of *Bacillus thuringiensis* fermentation based on oxygen transfer, *J. Ferment. Bioeng.*, **83**, 561–564.

FODA, M.S., SALAMA, H.S. and SELIM, M., 1985, Factors affecting growth physiology of *Bacillus thuringiensis, Appl. Microbiol. Biotechnol.*, **22**, 50–52.

GINGRICH, R.E., ALLAN, N. and HOPKINS, D.E., 1974, *Bacillus thuringiensis*: Laboratory test against four species of biting lice (*Mallophaga*: Trichodectidae), *J. Invert. Pathol.*, **23**, 232–236.

GOLDBERG, I., SNEH, B., BATTAT, E. and KLEIN, D., 1980, Optimisation of a medium for high yield production of spore-crystal preparation of *Bacillus thuringiensis* effective against the Egyptian Cotton Leaf Worm *Spodoptera litoralis* Boisd, *Biotechnol. Lett.*, **2**, 419–426.

GREEN, B.D., BATTISTI, L. and THORNE, C.B., 1989, Involvement of Tn4430 in transfer *of Bacillus anthracis* plasmids mediated by *Bacillus thuringiensis* plasmid pX012, *J. Bacteriol.*, **171**, 104–113.

GROCHULSKI, P., MASSON, L., BORISOVA, S., PUSZTAI-CAREY, M., SCHWARTZ, J.-L., BROUSSEAU, R. and CYGIER, M., 1995, *Bacillus thuringiensis CryIA(a)* insecticidal Toxin: Crystal structure and channel formation, *J. Mol. Biol.*, **254**, 447–464.

HÖFTE, H. and WHITELEY, H.R., 1989, Insecticidal crystal proteins of *Bacillus thuringiensis*, *Microbiol. Rev.*, **53**, 242–255.

IGNATENKO, Y.N., SAKHAROVA, Z.V., KHOVRYCHEV, M.P. and SHEVTSOV, V.V., 1983, Effect of temperature and aeration on growth and spore formation in *Bacillus thuringiensis*, *Microbiology (Russia)*, **52**, 716–718.

KANG, B.C., LEE, S.Y. and CHANG, H.N., 1992, Enhanced spore production of *Bacillus thuringiensis* by fed-batch culture, *Biotechnol. Lett.*, **14**, 721–726.

KHOVRYCHEV, M.P., SAKHAROVA, Z.V., IGNATENKO, YU.N., BLOKHINA, T.P. and RABOTNOVA, I.L., 1986, Spore formation and biosynthesis of protein crystals during continuous culturing of *Bacillus thuringiensis*, *Microbiology (Russia)*, **55**, 983–988.

KHOVRYCHEV, M.P., SLOBODKIN, A.N., SAKHAROVA, Z.V. and BLOKHINA, T.P., 1990, Growth and development of *Bacillus thuringiensis* in multiple stage continuous cultivation. *Microbiology (Russia)*, **59**, 903–1003.

LECADET, M.M., LERECLUS, D., BLONDEL, M.O. and RIBIER, J., 1981, *Bacillus thuringiensis*: Studies on chromosomal and extrachromosomal DNA, in Levinson, H.S., Sonenshein, A.L. and Tipper, D.J. (eds) *Sporulation and Germination*. Washington: American Society of Microbiology, pp.88–92.

LI, J., CARROL, D., and ELLAR, J., 1991, Crystal structure of insecticidal delta-endotoxin from *Bacillus thuringiensis* at 2.5 Angstrom resolution, *Nature*, **353**, 815–821.

LIU, W.-M. and BAJPAI, R.K., 1995, A modified growth medium for *Bacillus thuringiensis*. *Biotechnol. Progr.*, **11**, 589–591.

MARTÍNEZ, M.M., 1994, 'Efectos de la limitación de oxígeno a partir de la fase de transición en el metabolismo y producción de δ-endotoxina de *Bacillus thuringiensis*', unpublished MSc thesis, CINVESTAV-IPN, México.

MCGAUGHEY, W.H. and WHALON, M.E., 1992, Managing insect resistance to *Bacillus thuringiensis* toxins, *Science*, **258**, 1451–1455.

MONRO, R.E., 1961, Protein turnover and the formation of protein inclusions during sporulation of *Bacillus thuringiensis*, *Biochem. J.*, **81**, 225–232.

NICKERSON, K.W. and BULLA, L.A. JR., 1974, Physiology of sporeforming bacteria associated with insects: minimal nutritional requirements for growth, sporulation and parasporal crystal formation of *B. thuringiensis*, *App. Microbiol.*, **28**, 124–128.

PAYNE, J.M., CANNON, R.J.C. and BAGLEY, A.L., 1993, *Bacillus thuringiensis* isolates for controlling acarides. US Patent number 5,211,946.

PEARSON, D. and WARD, O.P., 1988, Effect of culture conditions on growth and sporulation of *B.t.i.* and development of media for production of the protein crystal endotoxin, *Biotechnol. Lett.*, **10**, 451–456.

RAJALAKSHMI, S. and SHETNA, Y.I., 1977, The effects of aminoacids on growth, sporulation and crystal formation in *Bacillus thuringiensis*. *J. Indian Inst. Sci.*, **59**, 169–176.

RODRÍGUEZ, M.M., DE LA TORRE, M.M. and DE URQUIJO, N.E., 1991, *Bacillus Thuringiensis*: Características biológicas y perpectivas de producción, *Rev. Latinoam. Microbiol.*, **33**, 279–292.

RODRÍGUEZ, M. and DE LA TORRE, M., 1996, Effect of the dilution rate on the biomass yield of *Bacillus thuringiensis* and determination of its rate coefficients under steady state conditions, *Appl. Microbiol. Biotechnol.*, **45**, 546–550.

ROWE, G.E. and MARGARITIS, A. 1987. Bioprocess developments in the production of bioinsecticides by *Bacillus thuringiensis*, *Crit. Rev. Biotechnol.*, **6**, 87–127.

ROY, B.P., SELINGER, L.B. and KHACHATOURIANS, G.G., 1987, Plasmid stability of *Bacillus thuringiensis var. kurstaki* (HD-1) during continuous phased cultivation, *Biotechnol. Lett.*, **9**, 483–488.

SAKHAROVA, Z.V., IGNATENKO, YU.N., KHOVRYCHEV, M.P., LYKOV, V.P., RABOTNOVA, I.L. and SHEVTSOV, V.V., 1984, Sporulation and crystal formation in *Bacillus thuringiensis* with growth limitation via the nutrient sources, *Microbiology (Russia)*, **53**, 279–284.

SAKHAROVA, Z.V., IGNATENKO, YU.N., SHCHUL'TS, F., KHOVRYCHEV, M.P., RABOTNOVA, I.L., 1985, Kinetics of the growth and development of *Bacillus thuringiensis* during batch culturing, *Microbiology (Russia)*, **54**, 604–609.

SALAMA, H.S., FODA, M.S., DULMAGE, H.T. and EL-SHARABY, A., 1983, Novel fermentation media for production of delta-endotoxins from *B. thuringiensis*, *J. Invert. Pathol.*, **41**, 8–19.

SELINGER, L.B., DAWSON, P.S.S. and KHACHATOURIANS, G.G., 1988, Behaviour of *Bacillus thuringiensis var. kurstaki* under continuous phased cultivation in a cyclone fermentor, *Appl. Microbiol. Biotechnol.*, **28**, 247–253.

SMITH, R.A., 1982, Effect of strain and medium variation on mosquito toxin production by *Bacillus thuringiensis var. israelensis*, *Can. J. Microbiol.*, **28**, 1089–1092.

VAN RIE, J., MACGAUGHEY, H., JOHNSON, D.E., BARNETT, B.D. and VAN MELLART, H., 1990, Mechanism of insect resistance to the microbial insecticide *Bacillus thuringiensis*, *Science*, **247**, 72–74.

VILLAFAÑA, J.R., GUTIÉRREZ, E. and DE LA TORRE, M., 1996, Primary separation of the enthomopathogenic products of *Bacillus thuringiensis*, *Biotechnol. Progr.*, **2**, 564–566.

WAKISAKA, Y., MASAKI, E. and NISHIMOTO, Y., 1982, Formation of crystalline delta-endotoxin or poly-beta-hydroxy-butyric acid granules by asporogenous mutants of *B. thuringiensis*, *Appl. Envir. Microbiol.*, **43**, 1473–1480.

YUDINA, T.G., SALAMAKHA, O.V., OLEKHNOVICH, E.V., ROGATYKH, N.P. and EGOROV, N.S., 1992, Effect of carbon source on the biological activity and morphology of paraspore crystals from *Bacillus thuringiensis*, *Microbiology (Russia)*, **61**, 577–584.

Cleaner Production Activities in Zimbabwe

RICHARD T. TAWAMBA, R. GURAJENA AND CHRISTOPHER J. CHETSANGA

22.1   Background

The United Nations Conference for Environment and Development (UNCED) held in Rio De Janeiro in June 1992 put on the agenda the need to address issues of industrial development and the environment (UN, 1992; OECD, 1995). In order to attain these objectives and fulfil the requirements of Agenda 21, the Government of Zimbabwe signed a series of conventions, amongst which is the promotion of Cleaner Production (CP) by Zimbabwe industry. This set the stage for the execution of measures to safeguard the environment from human activities. It also paved the way for different stakeholders to play a part in contributing to a cleaner environment. The United Nations Environment Programme (UNEP), through the United Nations Industrial Organization (UNIDO), promotes cleaner production on the grounds that it is easier to prevent pollution at source than it is to clean up afterwards. In essence, cleaner production represents a preventive environmental strategy for industrial production that reduces risks to human health and the environment. The concept emerged from efforts in Europe in the 1970s to promote 'zero/low' waste technologies (UNEP, 1994). In 1989, UNEP established a cleaner production programme to promote cleaner production in industrialized countries and to assist currently industrializing countries in adopting cleaner industrial practices (UNIDO/UNEP, 1995).

Realizing that there could be both financial and human resource constraints to implement the programme, UNIDO/UNEP appealed to donor agencies for financial support. Funding was thus made available through the Dutch, Danish and Austrian governments, which donated funds for launching cleaner production projects in a few developing countries, namely Tanzania, Czech Republic, India, China, Mexico, Brazil and Tunisia.

In 1993, UNIDO approached the Government of Zimbabwe to nominate a host institute for the Cleaner Production Centre of Zimbabwe (CPCZ). The Ministry of Industry and Commerce asked the Scientific and Industrial Research and Development Centre (SIRDC) to take on this project. At that stage, SIRDC was under formation, did not have the necessary technical staff and was thus not ready to host the centre. The host requirement was solved when the Environment Forum of Zimbabwe (EFZ) offered to host the

Table 22.1 Summary of Budget Provisions for Phase 1

Inputs	UNIDO (US$)	EFZ
International expert	270 000	–
Adminstrative assistant	–	0*
Project travel	90 000	–
Cleaner production promoter	75 000	–
Deputy Cleaner Production promoters ($\times$ 3)	–	0
Expendable equipment	36 000	–
Non-expendable equipment	27 500	–
Miscellaneous	15 500	–
Office space	–	0

0 indicates that the expected contribution is still outstanding

Cleaner Production Centre. The EFZ is an environmental organization composed of industrial enterprises in Zimbabwe. The EFZ agreed to provide accommodation for the CPCZ. This arrangement led to the appointment of the Director of CPCZ, and the Deputy Director on secondment from SIRDC.

The Director and the Deputy went to Europe for training. A programme of activities and funding provisions were prepared for the centre. An Advisory Board was put in place to oversee and guide the activities of the centre. UNIDO exercised overall oversight for the CPCZ.

22.2 Project Inputs

UNIDO prepared a three year budget, which is renewable subject to successful performance. Phase 1 is being funded by Austria, Denmark and the Netherlands. The budget items are shown in Table 22.1. EFZ, as the host institute, was to provide infrastructure and staff to ensure sustainability of the centre after donor funds were exhausted. Arrangements are under way for EFZ to meet its commitments to the project. The project is currently housed in UNIDO offices.

22.3 Institutional Arrangement

Sustainable host institution arrangements are not yet in place. Questions usually raised include 'What is CPCZ?' 'Is it an NGO, a parastatal or an agency of the Government of Zimbabwe?' 'Who finances the long-term activities of CPCZ?' At inception, it was envisaged that funds generated from demonstration projects would sustain the activities of the centre. This has not yet materialized because most of the client companies have not yet paid for the services rendered. The small and medium enterprises could not afford the fees.

UNIDO/UNEP are currently supporting the centres listed in Table 22.2. To ensure the sustainability of the project it is essential to set it up an organization with the requisite infrastructure and backup staff.

Table 22.2 CPCZ organizational hosting in other countries

Country	Host Institute
Tanzania	Tanzania Research and Development Organization
India	National Productivity Council
Asian Region	Asian Institute of Technology Transfer
China	National Environmental Protection Agency
Czech Republic	Czech Institute of the Environment
Mexico	Instituto Tecnologico de Monterrey

22.4 Demonstration Projects

The CPCZ has been involved in a number of demonstration projects to show the value of cleaner production technology to industry in general.

More specifically, demonstration projects are aimed at:

- demonstrating to company management and employees the importance and benefits of continuing a waste minimization initiative beyond the pre-assessment phase;
- involving all employees and operators in the identification of waste resources and the possibilities for prevention or reduction of waste at source;
- raising awareness of all employees to the cleaner production technology in a constructive and quantitative way by means of training and involvement;
- using external resources available to provide a different perspective/viewpoint to existing waste sources and their potential for minimization;
- training of company personnel in environmental auditing.

A number of demonstration projects were carried out in companies, some of which are members of the EFZ. These include the brewing, food and timber processing industries. The outcome of the CP programme among some of these companies is described below.

22.4.1 *Sugar Refineries*

Demonstration projects were carried out at two branch sugar refineries located in Harare (capital city) and Bulawayo.

Harare

The company embarked on CP projects so as to be at the forefront with new developments and also as a directive from top management. The concept and principles were clearly understood by both senior management and technical staff. However, the programme lacked continuity due to a high staff turnover. Of the options identified, very few were implemented because of lack of capital.

CP achievements included reduction in loss of sugar, reduction in airbone dust, successes with mills, commissioning of two mud clarifiers and reduction in water leaks, resulting in savings on water consumption.

Table 22.3　Overview of cleaner production options for Harare sugar refineries

Option	Environmental benefits	Economic benefits
1. Water meter installed in clarifier	Water saving, wastewater reduction	Costs: low Benefits: not computed
2. Dewatering systems for muds	Water saving, wastewater reduction	Costs: US$17 000 Benefits: US$21 500
3. Reuse excess cooling on site	Water savings	Costs: not computed Benefits: US$80 000

Source: IVAM, 1996

Some of the environmental and economic benefits, which were highlighted in the demonstration projects, are shown in Table 22.3.

Bulawayo

The programme was started in Bulawayo after the company learnt of the success at their Harare plant. The programme was launched at a time when the company was experiencing high process losses. Initially, there was very little understanding and appreciation of the programme by management and personnel, but the CP project managed to bring awareness of the options that were available to the existing problems.

However, due to lack of necessary measuring instruments, none of the options identified were implemented. As with their Harare plant, funding was also a major constraint in implementing some of the options, but this is to be allowed in the 1998–99 budget. The company implemented methods, which reduced high process losses and energy consumption after CP methodology.

22.4.2　Breweries – Harare

As a requirement of the group of companies of which the breweries are a part, an environmental policy is in place, and environmental issues are always covered in monthly reports. The company participated in the CP programme as a result of the potential benefits, which were realized by the Group management. Most of the options generated involved good housekeeping measures, which could be achieved at low costs (Table 22.4).

Of the options identified, 30 % were implemented. The options not implemented were expensive and could not be covered in the annual budget. Other drawbacks were that some of the options chosen could bring benefits only in the long-term, which are hardly appreciated over short-term benefits. The other constraint was that existing equipment was too old and new equipment expensive.

22.4.3　Food Industry – Mutare

The company embarked on its CP project after a feasibility study had been done by CPCZ staff. They expected to find a solution to get rid of tomato waste. The project team

Table 22.4 Overview of cleaner production options for national breweries (Harare)

Option	Environmental benefits	Economic benefits
1. Install filter press in clarifier	Reduction of beer losses	Costs: US$17 000 Benefits: US$63 000
2. Reuse sterilant water from filter unit	Water savings Reduction in wastewater	Costs: US$2000 Benefits: US$2000
3. Use air for cake removal in filter press	Water savings	Costs: 0 Benefits: US$700

Source: IVAM, 1996

appreciated CP although they felt that this did not work all the time as the CPCZ staff did not find a solution to their tomato waste problem. The team also felt that implementation of the options required money. Of the options identified, 25 % were implemented and cost savings were achieved in areas where CP was focused. The project team felt that the time given for the demonstration projects was too short for the entire CP project to be understood and appreciated by all. It was also felt that CP demonstrations added awareness. The company, however, stated that CPCZ provided all sorts of suggestions but no immediate solutions were realized. In particular, the problem with tomato waste was not solved.

22.4.4 *Food Industry – Harare*

The company embarked on the CP programme because it views environmental issues seriously. They saw this as an opportunity to be proactive and start the environmental management process.

The CP concept was well understood at the technical level, but top management lacked commitment to the programme as they did not appreciate the importance of making it a priority. As a result, insufficient training was given to the rest of the workforce. The other reason for this was that too many things were being introduced into the organization, for example, ISO 9000 and Total Quality Management training.

A total of 39 options were identified. Of these, 30 % were implemented, resulting in a reduction in water and energy consumption. The company realized the benefits of the programme; hence it felt that the programme should be continued.

However, not all the savings could be quantified due to problems in data capture and analysis. The company felt that more staff was required in order to derive maximum benefit from cleaner production technology.

22.5 Information Dissemination

The CP programme has identified workshops as an effective means of disseminating information about CP technologies. A series of workshops were held in all the major cities (Gweru, Bulawayo, Harare and Mutare). These were organized for companies that intended to participate in demonstration projects.

The current legislative framework does not make it mandatory for industries to adopt CP policies. In order to sensitize decision makers about the need to promote such methods through legislation and regulation, workshops were held in Harare and Bulawayo. New types of policies and economic instruments were reviewed within the Zimbabwe context. One of the weaknesses of the workshops was that they were attended by members of local authorities, who do not formulate policy or legislation. Members of parliament, who are the legislators, did not attend the workshops. As part of its information dissemination, the CPCZ also produces newsletters which are distributed to various stakeholders.

22.6 Cleaner Production Manual

As with CP programmes in other countries, the CPCZ is required to produce a manual in vernacular or other suitable languages. The need for this became more evident when the evaluation exercise revealed that training did not take place in most companies (Geisser, 1994). This could have been due to unsuitable training modules or the unsuitability of the language used during training. In very few cases could the CP methodology be used to quantify the benefits that accrue to companies. In times when resources are scarce, managers are under pressure to justify environmental expenditures. Cost/benefit analyses are necessary to ensure that options are properly assessed with optimum payback periods, and non-use benefits like improved company image might prove difficult to quantify.

22.7 Barriers Encountered During Demonstration Projects

Several barriers were encountered during the process of carrying out demonstration projects and these can generally be presented and explored in detail under the following groups.

22.7.1 Technical Barriers

In most cases there was no adequate process monitoring equipment, which made quantitative material balances impossible. Most of the equipment and machinery was old and no longer reliable.

22.7.2 Lack of Local Role Models

There being no local case studies, continuous reference was made to success stories carried out in other developing and in developed countries. Management assumed a sceptical attitude about the whole exercise.

22.7.3 Financial Barriers

Prevailing high bank interest rates, at 30 % in 1997, have discouraged any meaningful investment into cleaner production technologies. This situation is exacerbated by the fact that the capital expenditure has a long payback period.

22.7.4 **Institutional Barriers**

Cleaner production technology is still new in Zimbabwe. There are not many institutions with the relevant expertise. There is a need to popularize this concept by networking with the increasing number of institutions that are now promoting cleaner production.

22.7.5 **Policy Barriers**

At present there is neither policy nor legislation at the central government level to give impetus to the promotion of the cleaner production concept in industry. In such an environment, industry is reluctant to invest in new technologies.

22.7.6 **Attitudinal Barriers**

Some organizations were defensive as they felt that they had qualified employees who could do the job. In some cases the qualifications of CPCZ demonstrators were questioned.

22.8 **Conclusions**

The potential for cleaner production intervention in Zimbabwean industry is quite high. The main interest in cleaner production stems from improved business operations while environmental benefits are of secondary concern. There still exists the need to introduce regulatory programmes that promote, rather than inhibit, the adoption of cleaner production. The present regulatory systems focus on performance standards that do not encourage process changes.

Demonstration projects were reasonably successful and are important local case studies, which can be published or used in workshops and seminars. The assessments proved that cleaner production can be implemented for an acceptable price.

References

GEISSER, K., UNEP, 1994, What's Happening in Industry? Presented at the Third High Level Advisory Seminar on Cleaner Production – Poland.

IVAM, 1996, Support to the NCPC in the Republic of Zimbabwe, Third International Mission, 8–20 March.

MAROVATSANGA, L.T., 1996, Report on the Evaluation and Review of the In Plant Cleaner Production Demonstration Projects for the Period 1995–1996 of the CPC-Zimbabwe.

OECD, 1995, *Promoting Cleaner Production in Developing Countries*.

UN, 1992, *Report of the United Nations Conference on Environment and Development*, Vol 1, 3–14 June.

UNEP, 1994, *Industry and Environment*, **17**.

UNIDO/UNEP, 1995, *Cleaner Production Manual* – Part Four.

APPENDIX A

Sample Calculation

CMA ash efficiency calculation, i.e. $\eta_m = (M_{in} - M_{out})/M_{in}$. To the first run of CMA ash (0.6525 g) the total ferrous sulphate solution volume (V_f) fed into the system was 56 ml. The concentration of ferrous sulphate, C_f, was 0.001 g ml^{-1}. The total mass of ferrous sulphate entering the system was:

$$M_{in} = C_f V_f = 0.001(g/ml)56(ml) = 0.056(g)$$

The M_{out} can be calculated from the weight difference of the filter papers plus CMA ash before and after the run,

$$M_{out} = M_{in}[(W_{a1} - W_{b1}) + (W_{a2} - W_{b2})] = 0.0157 \text{ g}$$

Where: $W_{a1} - W_{b1}$ is the weight difference of first filter paper and ash and $W_{a2} - W_{b2}$ is the weight difference of second filter paper and ash.

The efficiency for this run was:

$$\eta = (M_{in} - M_{out})/M_{out} = 72\%$$

Calculation of the Reynolds and Peclet Numbers

The Reynolds number based on particle diameter is: $Re = D_f \rho u / \mu$. Here, D_f was substituted with average diameter of CMA ash particle; $D_f = 2.03E - 05m$; $\mu = 2.10E - 05 PaCDOT$; $\rho = 1.3kg/m^3$; $A = \pi R^2 = 0.00363m^2$; $u = q/A = *60(s/min) = 9.18 \times 10^{-3}m/s$. $Re = D_f \rho u / \mu = 2.03 \times 10^{-0.5} CDOT1.3CDOT9.18 - 03/2.10 \times 10^{-5} = 1.15 \times 10^{-2}$.

For the Peclet number $N_{Pe} = D_f u / D$; $D_f = 2.03 \times 10^{-5}m$; where k is the Boltzman constant.

Under the condition of 1 atm and 298 K, the relationship between the slip correction factor C_c and the particle diameter D_p can be described by the following equation:

$$C_c = 1.647 D_p^{-0.4211}.$$

When the fluid temperature is 381 K and the average particle diameter is $0.066\mu m$, the slip correction coefficient C_c is 6.615. $k = 1.38 \times 10^{-23} J/(molCDOTK)$; $N_{Pe} = 6.6 \times 10^{-8} mCDOT0.0918/3.35 \times 10^{-5} = 5.57 \times 10^{-2}$.

Index

Abbott 277
absorbable organic halogens (AOX) 214
acetyl coenzyme A 158, 193, 197
acid discharges 124
Acinetobacter spp.
  cometabolic reactions 183
  A. calcoaceticus 199, 202
  A. eutrophus 197
acrylics 148
actinomycetes 124
activation by insertion 202
adsorption 69, 111, 113, 139–42
aerobic processes 157–8, 182, 191
aerosols 69, 71
agar and agarose 143
Agave tequilana 101
AgrEvo 269
agricultural wastes, re-use and treatment 88, 91,
      127, 130, 131
agriculture 45, 47, 50, 51, 185
agro-industrial sector, wastewaters 90
agrochemicals 277
Agrocybe praecox 219
air pollutants/pollution
  atmospheric distribution 185
  biofiltration 11
  coal combustion 68
  collection models 66
  control 65–6
  experimental study 69–71
  and fertilizer use 45
  human health effects 65
  major 64
  palm oil industry 250, 251
  particulate 63–80
  source 65
airlift reactors 220
Akao 266
Alcaligenes 183
alcohol ethoxylate, removal 91
alcoholic beverages, feedstocks 101

algae, metal biosorption 148
algal
  biomass *see* biomass
  culture effluent 91
  ponds 109
AlgaSORB 147, 148
alginates 90, 113, 147, 148
n-alkanes 169–70, 171, 192–3
aluminium, adsorption 140
amended soil liners 57
amine groups 139
amino acids, removal 91
ammonia 47, 51, 64
amylase 233
anaerobic
  digestion
    animal waste 233, 234–5
    and chemical oxygen demand reduction 248
    palm oil mill effluent 254
    tequila vinasse 101–6
  hydrocarbon biodegradation 191
  microbial metabolism 157–8
  respiration 157–8, 182
  technology 89, 90, 92–3, 95
analytical determinations 148
animal
  fats 266, 269
  feed
    enzyme applications 5–6, 7
    fish 118
    Lemnaceae biomass 86, 94, 233, 238
    oil 266
    plant biomass 85
    protein hydrolysates 257
    replacement of components 233
    seafood processing waste 255
  production
    cleaner production 227–43
    economic factors 232, 234
    environmental damage 227
    and nitrogenous pollution 107

waste 91, 233–9
  see also agricultural waste
anionic metal complexes 141
anthracene 195
  see also polyaromatic hydrocarbons
aphid control 283
aquaculture 88, 248, 254, 257
aquatic plants 84–5
aquifers
  bioremediation 155–66
  contaminated 160–1, 163–4
arachidonic acid 269
Argentina
  soybean oil 266
  transgenic crops 282
aromatic hydrocarbons
  biodegradation 195–200
  ring cleavage 195–6, 197
arsenic compounds 68
Arthrobacter 171, 183
Arthrospira see Spirulina
Ascophyllum nodosum 148
ash, coal 64
Asia
  integrated systems 92
  pig production 227
Aspergillus terreus 269
ASTM tests, barrier liners 56, 57, 58, 60
Australia
  sugarcane and sugar processing industry 245,
    246
  transgenic crops 282
  waste burning prohibited 248
Austria 299, 300
automobile pollution 185
Azolla pinnata 89
  see also Lemnaceae

Bacillus spp.
  cometabolic reactions 183
  metal biosorption 146, 148
  B. subtilis 257, 258
  B. thermoproteolyticus 257
  B. thuringensis
    carbon:nitrogen ratio 291, 292
    Cry gene expression 285–98
    galleriae 290–1, 295
    genes, introduction into other species 282–3
    insecticidal use 277–8, 285–98
    israelensis 287, 290
    kurstaki 287, 289–90, 291, 295
    production 288–95
bacteria
  denitrifying 47
  nitrogen metabolism 49–51
  see also microbial; microorganisms
bacteriochlorophyll 108
bagasse 246, 248
baghouse filtration 63, 64, 66–7, 69–71
Bangladesh 94
barrier liners 55–61
batch systems 148
beer production 6, 8, 302
bentonite 57, 58
benz[*a*]anthracene 167

benzene 195, 199
  see also industrial solvents
benzo[*a*]pyrene 195
  see also PAHs
Berijerinckia 183
beryllium 140
BIO-FIX 147–9
Bio-Spirulemna system 234, 236, 237–40
bioaugmentation 181
bioavailability 162, 167–77, 186
biocell 162
biochemical oxygen demand (BOD)
  palm oil mill effluent 249
  pig waste 231
  polluting strength 123
  removal 88–9, 125, 130–1, 235
  seafood processing industry wastewater 255
BIOCLAIM 146–7, 148
biodegradation
  agrochemicals 277
  definition 167
  hydrocarbons 191
  PAHs 168
  xenobiotics 188
biodiversity 20, 22, 275, 276
biofiltration 11, 181
bioflocculation 110
biofuels 272
biogas 92, 103, 233–9, 248, 251, 253
bioinsecticides 6, 275–84
  see also Bacillus thuringiensis
biomass
  animal feed 233
  aquatic plants 84–5
  commercial exploitation 108
  fish food 134, 254
  harvesting 110
  immobilization 136, 142–3
  Lemnaceae 93, 233, 235, 238
  living or non-viable 136–7
  metal accumulation 136
  in pig production 233–9
  Spirulina 235, 238
biopile 162
bioreactors 114–15, 162–3, 181
bioremediation
  advantages and disadvantages 159, 168, 179
  clean technology 164–5
  contaminated soil 11, 155–66, 179–89
  definition 167
  economic aspects 159, 165
  management technology 165–6
  microorganisms 183
  monitoring 164
  multidisciplinary 180–1
  off-site 159, 161
  scaling-up 161–4
  in situ 164
  technologies 180–1
  USA 179, 180
biosorbents, commercially available 146–9
biosorption 139, 141–2, 145–50
biostimulation 159, 181
biosurfactants 171
biotransformation, definition 167

biotreatability studies 160–1
bioventing 181
Bjerkandera sp. 216, 218, 219
bleaching processes, pulp and paper industry
    211–26
Border Development Cooperation Commission
    (BECC) 10
Brazil
    aid funding 299
    animal production 227
    ethanol production 248
    soybean oil 266
    sugarcane and sugar processing industry 245
    wastewater treatment 93
brewery, effluents 90
brewing *see* beer production
Brownian motion 67, 68
Brundtland Report 20
Bryoria spp. 148
bubble columns 220
buffer capacity of soil 156
Bulawayo (Zimbabwe) 302, 303, 304
Business Council for Sustainable Development 4
butyric acid 269

cadmium 140, 185
calcium alginate 143
calcium magnesium acetate 64, 66, 69–71, 75–8,
    307
calcium magnesium carbonate 64, 69–71, 78
calcium oxalate 93
calcium salts 65–6, 141
Calgene 268, 269, 270, 272
Canada
    International Development Centre 23
    rapeseed production 266
    transgenic crops 267, 282
Candida spp. 192
    C. tropicalis 258
canola 266, 268, 269
Caprenin 270
capsule, microorganisms 138, 140
carbon:nitrogen ratio, *Bacillus thuringensis* 291,
    292
carbon
    adsorption 135
    as pollutant 123, 124
    recycling 234
    source 157, 163, 184, 292
carbon dioxide 157, 161
    see also biogas
carbon monoxide 64
carboxylate groups 139, 140
carcinogenic pollutants 167, 191, 192, 202
β-carotene pigment, from biomass 238
carrageenan 112, 143
catalase 218
catechol 159, 195
cattle 227, 231
cause evaluation 9
cell
    immobilization 110–11
    leakage 117
    membrane 138
    surface 139–42

CellCap 282
cellulose 211
ceramics 143
chemical flocculation 110
chemical industry, wastes 64, 91
chemical oxygen demand (COD)
    components 124
    pig production 234
    pulp wastewater 215
    reduction/removal 90, 103, 105, 106, 248
    seafood processing waste 255, 257–8
chemoautotrophs and chemoheterotrophs
    181
chickens *see* poultry
China
    aid funding 299
    integrated systems 92
    pig production 227
chitin and chitosan 255, 256, 257
Chlamydomonas reinhardii 148
Chlorella spp. 112–13, 258–9
    C. homosphaera 148
    C. pyrenoidosa 108
    C. vulgaris 139, 148
chlorinated industrial solvents 156
chlorine bleaching 214
chlorodibromomethane 91
chlorofluorocarbons 19
chocolate 267, 270
chromium 139, 140
Ciba 282
citric acid cycle 158, 193
Citrobacter spp. 137
clean-up biotechnology 155–66
cleaner pig production units (CPPU) 231–3
cleaner production
    animal production 227–43
    bioremediation 164–5
    concept 4, 5
    difficulties 304–5
    economic aspects 23, 26, 245–64
    microbial processes 245–64
    oil agroindustry 265–74
    options 9
    pulp and paper industry 211–26
    seafood processing 256–9
    training 304
    Zimbabwe 299–305
Cleaner Production Centre of Zimbabwe (CPCZ)
    299–305
climate change convention 20
club-rush (*Schoenoplectus lacustris*) 124
coal
    ash 64
    cleaning 63, 65
    combustion 63–80
    conversion 107
cobalt 140
cocoa butter 270, 272
coconut oil industry 267, 270–1, 272
Coelastrum spp. 108
coffee production 5
coleoptera 277, 281, 285, 287
coliforms, removal 88, 90
collection efficiency 68

Colombia
 earthworm production 254
 wastewater treatment 93
coloured effluents 251, 255
cometabolism 182–3
command-and-control strategy 19–27
compacted clay liners 56–7
compatibility testing, barrier liners 58–60
composting 181, 248, 249, 253
confined vortex scrubber 66
contaminants
 bioavailability 162, 186
 characterization 184–6
 degradative pathways 157–8
 detection 160–1
 free 160
 physicochemical properties 185–6
continuous culture, *Bacillus thuringensis*
  production 290–1
continuous flow reactors 114–16, 148
continuous stirred tank reactors 136
cooperative system, sugarcane and sugar
  processing industry 246, 247
copper 137, 140, 185
Corynebacterium spp. 192
coupling, covalent 112–18
credit facilities 25–6
creosote 185
criteria, selection and ranking 35–7, 38, 39–43
critical micelle concentration 172, 173
crop residues, mineralization 47, 51–2
crude oil *see* oil
Cry genes 277, 279, 281, 282–3, 286–8, 295
cry proteins (δ-endotoxins) 277, 278–81, 283,
  285–6, 291–5
Cuba 245
Cunninghamella elegans 167
cyanide contamination 185
cyanobacteria 108, 109–10, 135–53, 233
cycloalkanes, oxidation 194
cyclone separation 65
cytochrome P-450 192–3, 195
Czech Republic 299

dairy effluents 87, 90, 127
Damköller number 170
Darcy's law 126
data presentation, Vital Issues process 38–9
decision support systems (DSS) 29, 31, 33
degradative pathways, contaminants 157–8
dehalogenation, reductive 182
delignification 212, 213, 215, 216
denitrifying bacteria 47
Denmark 299, 300
denser than water non-aqueous-phase liquids
  (DNAPLs) 156
Department of Energy, US 30
detergents 86, 107
development *see* sustainable growth and
  development
diatomaceous earth 143
diesel *see* petroleum hydrocarbons
diffusion *see* Brownian motion
dioxins 168
dioxygenases 199–200, 202
diptera 277, 285, 287

dispersed bed reactors 145, 146
dissolved air flotation 255
distillery wastes 101–6, 248, 257
DNA 202, 286–8
 see also plasmids
downflow reed beds 124, 128–9
drip irrigation 95
duckweed *see* Lemnaceae
Dunaliella spp. 108
Dupont 268, 269, 272, 277

earthworms 248, 253, 254
eco-efficiency 4
economic aspects
 animal production 232, 234
 bioremediation 159, 165
 cleaner production 23–4, 26, 245–64
 coconut industry 270, 272
 oil agroindustry 266–7
 palm oil industry 251
 pig production 239
 sugarcane and sugar processing industry 246,
  249
education, as vital issue 43
efficiency calculation, calcium magnesium acetate
  ash 307
effluents
 re-use 95
 see also wastewater
Eichhornia crassipes 84–5, 91, 124
Eisenia foetida see earthworms
electrodialysis 135
electron transfer 157, 158, 182, 185–6
electrostatic precipitator 63, 65
elemental chlorine free (ECF) bleaching processes
  214–16
elution 142, 146, 147
Emden-Meyerhof-Parnas pathway 158
end of pipe requirements 3, 23
δ-endotoxins, cry proteins 277, 278–81, 283,
  285–6, 291–5
energy-yielding processes 157
enthalpy 141
entrapment 111, 143, 144
Environment Forum of Zimbabwe (EFZ) 299, 300
environmental biocatalysis 191–207
environmental damage, animal production 227,
  231
Environmental Management Systems 11
environmental market, growth rates 10–11
environmental policy
 European Union 19, 83
 international 10–11
 waste disposal 19–27
Environmental Protection Agency, US 10, 64
enzyme applications 5–8, 216–19, 233, 269–70
enzymes, white rot fungi 216
epoxy resin 143, 145, 148, 149, 150
erucic acid 272
Escherichia coli see coliforms
ethanol production 248, 249
European Union
 discharge standards 235
 environmental policies 19, 83
 transgenic crops 282
eutrophication 83, 107, 109, 131, 165, 231, 235

evaporation/transpiration, water loss 86
exopolymers 138, 171
explosives manufacture 107
extended-batch fermentation, *Bacillus thuringensis*
    production 289–90
extracellular complexing 137–8
extracellular precipitation 137
Exxon Valdez 179

fabric filters 65, 66–7
fat and oil industry *see* oil agroindustry
fats *see* oils and fats
fatty acids 105, 106, 267–9, 272
fed-batch fermatation, *Bacillus thuringensis*
    production 290
feedstocks, alcoholic beverages 101
fees for water and waste discharge 24
fermentation 157, 182, 269–70
ferredoxin 199, 200
ferrous sulphate 69–75
fertilizers
    consumption, Mexico 48, 49
    and nitrate pollution 107
    nitrogen 45–54
    as pollutants 185
    efficiency 45
    production 107
    see also nitrogen/nitrates
Fiji 245
filters/filtration 65, 66, 67, 72–5
fish
    culture 86, 91, 94
        see also aquaculture
    food
        from seafood processing industry waste 259
        plant/algal biomass 118, 134, 238, 254
        protein hydrolysates 257
    oils, human nutrition 269
    sauce 256, 257
    silage 256
    see also seafood processing industry
five R policies 7–9
fixed bed reactors 146, 147, 148
Flavobacterium 183
fluidized bed reactors 114, 117, 145, 146, 148,
    149, 220
food, future demand 45, 275–7
food industry
    effluents 90–1
    enzyme applications 5, 6, 8
    Mexico 275
    and nitrogenous pollution 107
    Zimbabwe 302–3
formaldehyde 148
formalin 148
fossil fuels 20, 167, 238
front of pipe 23
fructose 265
fuel spills 155
fungal biomass, metal biosorption 148
fungi, hydrolytic activity 124

gallium 137
gas
    adsorption 69
    chromatography 186, 187

gasoline *see* petroleum hydrocarbons
genetic adaptation, microorganisms 200–1
genetic engineering 165, 267–9, 270
genetics
    aromatic hydrocarbon degradation 197–202
    pig production 232, 233
geohydrochemical characterization 160
geohydrological characterization 162
geological considerations 187
geomembranes 57–8
geosynthetic clay liners 57–8
Germany 227
glass beads 143
Global Environment Fund, World Bank 19
globalization 9–10, 270
glucans 270
glutaraldehyde 112, 148
glycolysis 158, 159
gold 139, 140, 148
grass carp 94
gravity separation 65
Green Revolution 276
greenhouse gases 20, 45, 46, 231
groundwater
    chemical transport 185
    contamination 21, 83
    microbial treatment 183
growth
    consumer demand 3, 9, 45
    environmental market 10–11
    food production 275
    population 20, 45, 275
guar gum 147
Gweru (Zimbabwe) 303

halogenated hydrocarbons 183
Harare (Zimbabwe) 301–2, 303, 304
hazardous wastes 21, 167, 179
heavy metals
    adsorption by cyanobacteria 135–53
    in biomass 86, 93
    pollutant 123, 124
    recovery/removal 89, 118, 125, 147–50
    see also individual elements; metals
Heliothis virescens 279, 280
helminth eggs, removal 88, 90, 235
hemicellulose 211
Henkel 266
herbicide resistance 269
heterotrophs 184
high-rate oxygen ponds 233, 235, 248
holistic management, as vital issue 43
horizontal flow reed beds 124, 126–7, 129
human
    activities
        and metal concentrations 135
        and nitrate pollution 107
        and organic contaminants 185
    health
        and air pollution 65
        immune system enhancement 238, 269
        and industrial waste 22
    nutrition
        fish oils 269
        mushroom nutraceuticals 252–3
        plant biomass 85, 238

humic acid 254
hydraulic connections 187
hydrocarbons, biodegradation 185, 191, 192–7
Hydrocotyle umbellata 84
hydrogen sulphide production 137
hydrogeological characteristics 188
hydrolytic activity 124
hymenoptera 277, 285

imidazole groups 139
immobilization 107–21, 136, 143–5
immune system enhancement 238, 269
incinerators 185, 250, 251
India
    aid funding 299
    ethanol production 248
    sugarcane and sugar processing industry 245
    wastewater treatment 93
Indonesia
    coconut industry 270
    palm oil industry 249, 266
industrial
    pollution, Mexico 22–3
    solvents 156, 159, 185
    waste
        and health 22
        reed bed treatment 130
infrastructure
    management 29–44
    wastewater treatment 83
inorganic contaminants, microbial metabolism 157
insect-resistant transgenic plants 281–2
insecticidal use, *Bacillus thuringenis* 277–8,
        285–98
integrated systems
    pest management 285
    waste management 91–3, 245–64
Inter-American Development Bank 19
Intergovernmental Meeting of Technical Experts 10
International Association for Clean Technologies 26
International Centre for Cleaner Technologies and
        Sustainable Development (CITELDES) 16
International Development Centre, Canada 23
international environmental legislation 10–11
international standards 11–15
intracellular accumulation 138
investment portfolio (IP) 29, 32
ion-exchange reactions 135, 136, 139
iron 137, 182, 196
iron-oxidising bacteria 138–9
iron-sulphur protein 199
irrigation 276
ISO 9000 15, 303
ISO 14000 11, 15, 259
isoprenoids 194
IZU-154 216

Japan
    protein hydrolysate production 257
    transgenic crops 282

kappa number 214, 218
ketones 193
Kickuth equations 126
kitchen waste 91

Kraft process 212, 213, 216–19
Kuwambara hydrodynamic factor 67, 68

laccase 216–18, 220
land farming 181
landfills
    containment type 55–61
    leachate, reed bed treatment 130
    pollution 155, 185
latrine wastes 91
lauric acid 267, 272
lead 64, 140
leather and tanning industry, enzyme applications
        5–6
Lemna spp.
    L. gibba 87
    L. minor 87, 89, 91
    L. perpusilla 85, 87
    L. valdiviana 89
    see also Lemnaceae
Lemnaceae
    animal feed 86, 94, 233
    in BIO-FIX 147
    biomass 93, 233, 235, 238
    calcium oxalate 93
    characteristics 85–6
    feed conversion ratio 93
    growth rate 86
    harvesting 86
    heavy metal tolerance 89
    nutrient uptake 87–8
    production and use 93–4
    productivity and chemical composition 237
    protein content 86–7
    waste treatment 83–100, 233–4, 236
lepidoptera 277, 281, 285, 287
Leptothrix spp. 139
Letharia spp. 148
lice 285
light
    intensity 236
    penetration 117
lighter than water non-aqueous-phase liquids
        (LNAPLs) 156
lignin 211
lime, in metal ion removal 135
lipases 253, 269, 270
lipid A 140
lipids
    biochemistry 267–8
    see also oils and fats
lipochemicals 266–7
Lipomyces starkei 269
lipopolysaccharide 140
lipoprotein 140
Lymantria disparis 279, 281

magnesium 141
Malawi
    ethanol production 248
    sugarcane and sugar processing industry 245
Malaysia
    coconut industry 270
    composting 214
    palm oil industry 249, 253, 266

management technology, for bioremediation
165–6
Manduca sexta 279, 280
manganese 139, 182, 217, 218, 219
-dependent peroxidase 220–1
manure *see* animal waste
Maquiladora Programme 9, 21
margarine production 268–9, 272
marine environment, oil 191–2
market wastes 91
mass transfer 163–4, 168
mathematical modelling 170
mercury 64, 65, 66, 139, 140
meta cleavage genes 197
metals
    accumulation 136, 138
    aggregate formation 138
    biosorption 145–50
    chelation 138, 141
    desorption 141, 142
    elution 146, 147
    'hard' and 'soft' 140
    methylation 137
    oxidation 138–9
    as pollutants 185
    potentially toxic 135
    reduction 139
    removal 124, 129–30, 135–42
    selective recovery 148
methaemoglobinaemia 107
methane 231
    see also biogas
methylation 137
Methylosinus 183
Mexico
    aid funding 299
    coconut industry 270
    economic incentives 24
    fertilizer consumption 48, 49
    food production 275
    industrial pollution 22–3
    International Centre for Improvement of Wheat
        and Corn (CIMMYT) 282
    National Science and Technology Council 23
    nitrogen fertilizers 45–54
    population growth 20
    potato production 265
    sugarcane and sugar processing industry 245,
        265
    sustainable development 20–2
    transgenic crops 282
    US border 9–10, 20
    waste treatment 21
    water pollution 21
micelles 173
microalgae 107–21, 235
microbial
    biomass protein (MBP) 256, 257, 258
    consortium 183
    populations, rhizosphere 124
    processes
        clean technologies 245–64
        metal removal 135–42
        palm oil industry 251–4
microenterprises *see* small and medium enterprises

microorganisms
    adaptation 183–4, 200–1
    in bioremediation 167, 183
    characterization 160, 169
    classification 181
    exogenous 161, 165, 168
    exopolymers 138
    genetically engineered 165
    metabolism, aerobic and anaerobic 157–8
    nutrition 184
    thermophilic 191
        see also bacteria
    photosynthetic 108
    survival in adverse conditions 156–9
mine wastewater, reed bed treatment 130
mineral production, particulate emission 64
mineralization, crop residues 51–2
mites *see* hymenoptera
moisture content, and microbial metabolism 186
mono-oxygenase 200, 201
Monsanto 282
Mortiella vinacea 269
mosquitos 86, 282, 283
mud cake (sugar industry) 246
mushroom industry 248, 249, 251, 252–3
Mutare (Zimbabwe) 302–3
mutation, random 201
Mycobacterium sp. 172, 183
Mycococcus sp. 183
Mycogen 282

Nabisco Foods 269
NAD/NADH 193, 194, 199
NAH plasmids 201, 202
naphthalene 91, 159, 169, 171, 172, 197
National Science and Technology Council, Mexico
    23
natural resources *see* biodiversity
Natural Resources Defence Council 65
nebulizers 71
nematodes 278, 285, 287
Netherlands
    aid funding 299, 300
    water treatment 12–14
New Mexico 30
New Zealand 253
nickel 137, 140, 185
nitrification 48
nitrogen oxides 45, 52, 64, 65
nitrogen/nitrates
    balance 48–9, 50
    in biodegradation 182
    chemical production 91
    cycle 46, 84–5, 88, 109–10
    fertilizers 45–54
    fixation 109, 184
    as macronutrient 87–8
    microbial metabolism 49–51, 292
    in pig waste 231, 235
    as pollutant 107, 123, 124
    production 52, 53
    recovery 235
    removal 83, 107–21, 125, 235
    uptake 84, 116
    in water 48–9, 107

nitrosamines, carcinogenic 107
Nitrosomonas 183
Nitzschia palea 108
Nocardia sp. 183, 194
non-aqueous-phase liquids (NAPLs) 156
non-ionic organic contaminants (NOCs) 167, 170, 171
North American Development Bank (NADBank) 10, 19–20
North American Free Trade Agreement (NAFTA) 9, 19
Northrup King 282
Norway 257
Novo Nordisk 277
NP-Alk 169
nutrients 87–8, 131, 161, 294–5

odours 86, 96, 231, 232, 255
oil
    biodegradation 168, 170, 191–207
    marine environment 191–2
    pollution 185, 191
    spills 179, 180
oil agroindustry 265–74
oil flavour compounds 256
oils and fats, polyunsaturated 237, 238, 267
Oleofina 266
Olestra 270
optimization techniques 33
organic compounds 88–9, 157, 185
    see also xenobiotics
organohalogens 185, 214
organometallic compounds 137
ortho cleavage genes 197–9, 202
overland flow, reed bed 124, 129–30
β-oxidation 193, 195
oxidation-reduction *see* redox reactions
oxidative phosphorylation 158
oxoanions 141
oxygen 161, 163, 292–4
ozone 64, 65

packed bed reactors 114, 145, 149, 220, 221
paint manufacturing 185
pairwise comparisons 30, 31, 37–9
palladium 140
palm oil industry 249–54, 266
palm oil mill effluent 249, 250, 253
panel meetings 29, 30, 34–7
paper *see* pulp and paper
particle size distribution 71, 72
particulates 64, 65
partition coefficient 170, 175
Partnership for Pollution Prevention 10
patent protection 272
pathogens 88, 90, 93, 123, 124, 235
Peclet number 68, 73, 76, 307
Penicillium 183
pennywort *see Hydrocotyle umbellata*
peptidoglycan 140
peripheral enzymes, genes 199–200
peroxidases 216, 220–1
peroxides 163, 214–15
persimmon tannin 148
pesticides 185, 276

Pesudomonas sp. 194
pet food *see* animal feed
petrol *see* petroleum products
petroleum products 156, 159, 162, 168, 185
    contamination 180
    microbial degradation 186
    refining 107
    substituted by ethanol 248
Pfizer 269
pH
    and *Bacillus thuringensis* growth 294
    and metal ion adsorption 140–1
    microbial metabolism 186
    in reed beds 124, 129–30
Phanerochaete spp. 183
    P. chrysosporium 216, 217, 220, 221
    P. sordida 216, 219
phenanthrene 172, 195
    see also polyaromatic hydrocarbons
phenol 91
Phillipines 270
Phlebia radiata 219
Phormidium spp. 112, 113, 116, 117
    biomass immobilization 143–4
    P. laminosum, heavy metal recovery 149–50
phosphate/phosphorus
    cell surface ligands 139, 140
    chemical production 91
    human sources 107
    as macronutrient 87–8
    metabolism 110, 184
    pollutant 123, 124
    recovery 235
    removal 83, 107–21, 125, 235
    uptake by macrophytes 84
photoautotrophs and photoheterotrophs 181
photoinhibition 236
photosynthesis 108, 235
photosynthetic bacteria 258–9
Phragmites australis 124, 126, 128
phycobilin pigment 118
phycocyanin pigment 238
Pichia kudriavzenii 257
pig production 127, 227–40
pigments 118, 237, 238
pilot projects 160
plant biomass 85
    see also biomass
plasmids 197–200, 201, 286–8, 289
platyhelminths 277
policy portfolios (PP) 29, 31–2
pollutants/pollution
    accumulation 168
    and agriculture 185
    atmospheric distribution 185
    characterization 160, 169
    major 124
    oil production 191
    strength, biochemical oxygen demand (BOD) 123
    sugarcane and sugar processing industry 245, 246
polyacrylamide 143, 148
polychlorinated biphenyls (PCBs) 167

polycyclic aromatic hydrocarbons (PAHs)
  biodegradation 159, 168, 186, 195, 197
  biotransformation 167
  combustion 68
  cometabolism 183
  health risk 191
polyester foam 148
polyethyleneimine 148
poly-β-hydroxybutyrate 293, 294
Polyporus ciliatus 219
polysaccharides 138
polysulfone 145, 148, 149, 150
polyunsaturated oils and fats 237, 238, 267
polyurethane foam 111, 143
polyvinyl foam 111, 143
polyvinylformal 148
portfolio selection, Vital Issues process 35
potash, recycling 250
potassium, as macronutrient 87–8
potato production, Mexico 265
poultry
  food 233, 238, 255
  world production figures 227, 231
poverty 3, 4, 245
power plants, pollution 185
Procter & Gamble 270
Proctor compaction energy 56, 58
production *see* cleaner production
proteases 233, 257
protein hydrolysates 256, 257
protocatechuic acid 195
protoxins 278
protozoans 277
Pseudomonas spp.
  cometabolic reactions 183
  genetics 197–200
  metabolism 192
  metal biosorption 148
  P. aeruginosa 171, 199
  P. cepacia 199
  P. citronellolis 194
  P. fluorescens 282–3
  P. mendocina 200
  P. paucimobilis 197, 199
  P. pseudoalcaligenes 198, 199
  P. putida 139, 169, 197, 199, 202
Puerto Rico, Water Resources Management
    Initiative 30, 39–43
pulp and paper industry 5, 6, 64, 211–26
pulse jet filters 66
pumping costs 163

quasi-elastic light scattering (QELS) 68

radionuclides, recovery 148
rapeseed oil 266
reuse 7–9, 83–100
recombination 201–2
recovery 6, 7–9, 83–100, 233–9
recreational use, surface water 95–6
recycling 233–9
redox reactions 138–9, 182, 185
reduction of waste 7–9
reductive dehalogenation 182
reed *see Phragmites australis*

reed bed systems, water treatment 123–32
reedmace (*Typha latifolia*) 124
refrigeration plants 107
regional strategies 26–7, 265
regulatory policies, transgenic plants 272
regulatory proteins 200
regulatory systems, new 5
remediation
  soil contamination 3
  see also bioremediation
replacement of toxic materials 7–9
research and technological organizations
    15–16
resource recovery *see* recovery
respiration
  aerobic 157–8, 182
  anaerobic 157–8, 182
respiratory diseases 63, 65, 246
reverse air cleaned filters 66
reverse osmosis 135
Reynolds number 73, 76, 307
rhamnolipids 171
Rhizoctonia 183
Rhizopus arrhizus 148
rhizosphere 124
Rhodocyclus gelatinosus 258
Rhodotorula gracilis 269
Rio, Earth Summit 3, 10, 19, 23, 265
Riotech, New Mexico 30
root zone, rhizosphere 124
rubredoxin system 192–3
run-off from vehicles etc. 130

Saccharomyces cerevisiae 147, 148
SAL plasmids 201
Salatrim 269, 270
Sandoz 277
Sargassum spp. 148
saturated oils 267
Scendesmus spp. 108, 112, 116
  S. quadricauda 148
Schoenoplectus lacustris (club-rush) 124
scrubbers 63
seafood processing industry 254–9
secondary effluent 86
sectoral level 26
seepage 156
Selenastrum capricornutum 148
selenium 137
semisolid fermentation 288
septic systems 185
Serratia spp. 183
  S. marcescens 257
sewage
  domestic 83, 86, 95
  and nitrogenous pollution 107
  sludge
    as pollutant 185
    as raw material 64, 69
  treatment
    microbial 183
    reed bed 130
    tertiary 124
  urban 83
shaker filters 66

shrimp
  cultivation 94
  see also seafood
siderophores 137
silica 143, 148
site characterization 159–61, 186–8
slaughterhouse effluents 90, 91
slime exopolymers 111, 138, 140
sludge
  activated 138, 255
  reed bed treatment 130
  stabilizing and drying 124
small and medium enterprises (SMEs) 19–27
socio-economic biotechnology systems 245–64
sodium carbonate manufacture 107
sodium dodecyl sulphate 172–3
soil
  bioavailability 167–77
  bioremediation 3, 11, 155–66, 179–89
  buffer capacity 156
  characterization 169, 170
  chemical transport 185
  contaminated 3, 11, 155–66, 167–77, 179–89
  function 155–6
  persistence 167, 186
  sorption kinetics 168, 174
source identification 9
soybean
  oil 266
  transgenic 267
sphagnum moss 147
spiked samples 169, 175
spills 155, 156, 167, 179, 180
Spirodela spp. 93, 94
  S. polyrrhiza 84, 85, 87, 89
  see also Lemnaceae
Spirulina spp.
  in animal waste recycling 233
  in BIO-FIX 147
  in Bio-Spirulemna 233–4
  biomass 134, 235, 238
  productivity and chemical composition 236–7
  in sugar industry 248
  widely used 108
stakeholders 29, 32–4
starch processing 6, 8, 257
sterile conditions 163
stirred tank reactors 145, 148, 220
Stokes number 68, 74
Streptomyces albus 148
submerged fermentation 288–9
sugarcane and sugar processing industry 245–9, 265, 301–2
sulphate 182, 185
sulphate process *see* Kraft process
sulphite process 213
sulphur dioxide 64, 65
Summit of the Americas 10
Superfund Programme Rremedial Actions 180
surface water 95–6, 185
surfactants 162, 165, 171–4, 186
suspended solids 86, 89, 124, 125
sustainable growth and development 3–17, 19–22, 96, 245–64, 265
synthetic (composite) liners 57–8

Tanzania 299
tax allowances 24–5
temperature effects 186, 236, 294
teocintle 158
tequila distillery waste 101–6
terpenes 194
textile industry, enzyme applications 5, 6
Thailand
  palm oil industry 249, 250–1
  seafood processing industry 254–9
Thiele module 170
Thiobacillus ferrooxidans 138
thorium 137
TOL plasmids 197, 200, 201, 202
toluene 197, 199
  see also industrial solvents
tomato waste 302–3
Total Quality Management 303
total suspended solids (TSS) 86, 89
totally chlorine-free bleaching 214
toxicity
  metals 135
  methylated metals 137
  pollutants 168, 173
  surfactant 173, 175
trace metals, preconcentration 148
Trametes sp. 183, 219
  T. versicolor 216, 218
transgenic plant crops
  Argentina 282
  commercialization 282
  Cry genes 277
  insect-resistant 281–2
  regulation 272
  target enzymes 268
  USA 267, 281, 282
transposition 201–2
Trichoplusia ni 280
triglycerides 267–8
Triton X-100 171, 172
tubular photobioreactors 235
tuna *see* seafood procesing industry
Tunisia, aid funding 299
Typha latifolia (reedmace) 124

Ulva spp. 147
underground storage tanks 180
Unichema 266
UNIDO 300
Unilever 266, 270
United Nations Conference for Environment and Development (UNCED) 3, 299
United Nations Environment Programme (UNEP) 4, 299, 300
United Nations Industrial Organization (UNIDO) 299
United Nations Programme for Cleaner Production 3
unsaturated oils 267
upflow anaerobic sludge blanket (UASB) 92, 101–6
upward flow columns 148
uranium 137, 140, 148
urban waste 92–3
urbanization 83

USA
    animal production 227
    bioremediation 179, 180
    Department of Energy 30
    Environmental Protection Agency 10, 64
    Mexican border 9–10, 20
    soybean oil 266
    transgenic crops 267, 281, 282

veratryl alcohol 216, 220
Vietnam, earthworm production 254
vinasse 101–6
Vital Issues process 29–44
volatile organic compounds 64, 65
volatilization 137

waste/wastewater
    agro-industrial sector 90
    disposal
        environmental policy 19–27
        fees 24
        inappropriate 155
        sites, microorganisms 183
    pig production 232
    polluting strength 123
    pulp mills 213, 215
    recycling 6, 233–9
    seafood processing industry 255
    stabilization pond effluents 86
    treatment 11
        biological 108–9
        chemical transport 185
        developing countries 83
        infrastructure 83
        integrated systems 91–3
        Lemnaceae-based 83–100
        low cost 83–4
        Mexico 21
        palm oil industry 249–51
        sustainable development 96
        technologies 3
    urban 107, 109
water
    fees for use 24
    hard 141

loss, evaporation/transpiration 86
nitrates in 107
in pig production 232, 234, 239, 240
quality, international collaboration 10
saving 232–3, 239
supplies, Mexico 21
treatment
    microalgae 107–21
    Netherlands 12–14
    reed bed systems 123–32
water hyacinth *see Eichhornia crassipes*
water lettuce *see Pistia stratiotes*
Water Resources Management Initiative, Puerto
    Rico 30, 39–43
wet scrubbers 65
wetlands, constructed 123–32
wheat production 48, 50
white-rot fungi, enzymes 216
wildlife habitats 131
Witnocol SN70 172
Wolffia spp.
    W. arrhiza 85, 87
    see also Lemnaceae
Wolffiella spp. *see Lemnaceae*
wood, feedstock for pulp and paper industry
    211
World Association of Research and Technological
    Organizations (WAITRO) 16
World Bank 19, 96
World Commission on Environment and
    Development 3

xanthan gum 147
Xanthobacter 183
xenobiotics 91, 184, 188
xylanases 216, 233
xylene 197
    see also industrial solvents

zero/low waste technologies 299
Zimbabwe 245, 248, 299–305
zinc 140, 185
Zoogloea ramigera 148
zooplankton production 94
Zymonomas 249

Printed and bound by CPI Group (UK) Ltd, Croydon, CR0 4YY

23/10/2024

01778249-0001